The Immigrant-Food Nexus

Food, Health, and the Environment

Series Editor: Robert Gottlieb, Henry R. Luce Professor of Urban and Environmental Policy, Occidental College

For a complete list of books published in this series, please see the back of the book.

The Immigrant-Food Nexus

Borders, Labor, and Identity in North America

Edited by Julian Agyeman and Sydney Giacalone

The MIT Press
Cambridge, Massachusetts
London, England

© 2020 Massachusetts Institute of Technology

This work is subject to a Creative Commons CC-BY-NC-ND license. Subject to such license, all rights are reserved.

The open access edition of this book was made possible by generous funding from Arcadia – a charitable fund of Lisbet Rausing and Peter Baldwin.

This book was set in Stone Serif by Westchester Publishing Service. Printed and bound in the United States of America.

Library of Congress Cataloging-in-Publication Data

Names: Agyeman, Julian, editor. | Giacalone, Sydney, editor.
Title: The immigrant-food nexus : borders, labor, and identity in North America / edited by Julian Agyeman and Sydney Giacalone.
Description: Cambridge, Massachusetts : The MIT Press, [2020] | Series: Food, health, and the environment | Includes bibliographical references and index.
Identifiers: LCCN 2019022932 | ISBN 9780262538411 (paperback)
Subjects: LCSH: Immigrants—Social conditions—United States. | Immigrants—Social conditions—Canada. | Ethnic food—Social aspects—United States. | Ethnic food—Social aspects--Canada. | Food habits—United States. | Food habits—Canada. | United States—Emigration and immigration—Social aspects. | Canada—Emigration and immigration—Social aspects. | United States—Emigration and immigration—Government policy. | Canada—Emigration and immigration—Government policy.
Classification: LCC JV6475 .I464 2020 | DDC 338.1/973086912—dc23
LC record available at https://lccn.loc.gov/2019022932

10 9 8 7 6 5 4 3 2 1

This book is dedicated to all who have traveled from the land of their ancestors— forced and / or chosen—following the worn immigration and trade routes carved out by global foodways. We seek to recognize and honor your journeys, your dreams, your skills, your labor, and your foods that in complex and painfully unequal ways have nevertheless created rich cultural fusions that are reflected in all aspects of our food system.

We offer this book in solidarity with you, and recognize each of the countries of origin of the participants, researchers, and authors who have contributed to these pages:

Angola	Jamaica
Argentina	Jordan
Bhutan	Korea
China	Laos
Colombia	Mexico
Democratic Republic of the Congo	Pakistan
Ecuador	Philippines
El Salvador	Poland
England	Puerto Rico
Ethiopia	Somalia
Guatemala	South Korea
Grenada	Syria
Guyana	Taiwan
India	Thailand
Iran	Turkey
Iraq	Vietnam
Italy	Zimbabwe

—From the authors of *The Immigrant-Food Nexus* and all the situational strangers

Contents

Series Foreword ix
Acknowledgments xi

    Introduction: The Immigrant-Food Nexus 1
    Julian Agyeman and Sydney Giacalone

I Borders: Individuals, Communities, and Nations 17

    1 Criminalization and Militarization: Civic World Making in Arizona's Agricultural Borderlands 21
    Kimberley Curtis

    2 Slaughterhouse Politics: Struggling for the Future in the Age of Trump 41
    Christopher Neubert

    3 Contested Ethnic Foodscapes: Survival, Appropriation, and Resistance in Gentrifying Immigrant Neighborhoods 59
    Pascale Joassart-Marcelli and Fernando J. Bosco

    4 Immigrants as Transformers: The Case for Immigrant Food Enterprises and Community Revitalization 81
    Maryam Khojasteh

    5 Food from Home and Food from Here: Disassembling Locality in Local Food Systems with Refugees and Immigrants in Anchorage, Alaska 99
    Sarah D. Huang

II Labor: Fields and Bodies 115

6 Labor and the Problem of Herbicide Resistance: How Immigration Policies in the United States and Canada Impact Technological Development in Grain Crops 119
Katherine Dentzman and Samuel C. H. Mindes

7 Labor and Legibility: Mexican Immigrant Farmers and Resource Access at the US Department of Agriculture 141
Laura-Anne Minkoff-Zern and Sea Sloat

8 Enterprising Women of Mexican American Farming Families in Southern Appalachia 161
Mary Elizabeth Schmid

9 Gender, Food, and Labor: Feeding Dairy Workers and Bankrolling the Dairy Industry in Upstate New York 181
Fabiola Ortiz Valdez

III Identity Narratives and Identity Politics 201

10 The Canadian Dream: Multicultural Agrarian Narratives in Ontario 205
Jillian Linton

11 Planning for Whom? Toward Culturally Inclusive Food Systems in Metro Vancouver 225
Victoria Ostenso, Colin Dring, and Hannah Wittman

12 "Here, We Are All Equal": Narratives of Food and Immigration from the *Nuevo* American South 245
Catarina Passidomo and Sara Wood

13 Boiled Chicken and Pizza: The Making of Transnational Hmong American Foodways 261
Alison Hope Alkon and Kat Vang

14 Recipes for Immigrant Lives: Crossing, Cooking, Cultivating, and Culture at a Shared-Use Commercial Kitchen 281
Situational Strangers

Concluding Thoughts 299
Julian Agyeman and Sydney Giacalone

Contributors 307
Index 311

Series Foreword

The Immigrant-Food Nexus: Borders, Labor, and Identity in North America is the eighteenth book in the Food, Health, and the Environment series. The series explores the global and local dimensions of food systems and the issues of access; social, environmental, and food justice; and community well-being. Books in the series focus on how and where food is grown, manufactured, distributed, sold, and consumed. They address questions of power and control, social movements and organizing strategies, and the health, environmental, social, and economic factors embedded in food-system choices and outcomes. As this book demonstrates, the focus is not only on food security and well-being but also on economic, political, and cultural factors and regional, state, national, and international policy decisions. Food, Health, and the Environment books therefore provide a window into the public debates, alternative and existing discourses, and multidisciplinary perspectives that have made food systems and their connections to health and the environment critically important subjects of study and for social and policy change.

Robert Gottlieb, Occidental College
Series Editor (gottlieb@oxy.edu)

Acknowledgments

This book arose out of a Tufts University Department of Urban and Environmental Policy and Planning class, Food Justice: Critical Approaches in Policy and Planning, in fall 2016. Julian and Sydney, the latter a student at that time, wrote an op-ed for the *Boston Globe* on the likely effects of the then new Trump administration's proposed immigration policies on US agriculture and, more specifically, on immigrant foodways. We quickly realized that there was a book in this, and our friends at MIT Press, especially Senior Acquisitions Editor Beth Clevenger, clearly thought so, too.

The book would not have been possible without the generous help of Dean Alan Solomont of Tufts Tisch College, who awarded us a Tisch Faculty Fellows grant, which employed Sydney as co-editor and researcher for this volume during 2017–2018. It would also not have been possible without the dedication of all the authors we selected from our initial call for papers or the anonymous reviewers who helped us hone and focus our book proposal.

We would also like to thank Anthony Zannino, editorial assistant, for keeping things rolling and on time; Sherry Gerstein, production editor at Westchester Publishing Services; and Susan Clark, MIT Press catalog manager.

Julian would like to thank the excellent students he has had over his 20 years at Tufts University, such as Sydney Giacalone, whose intellectual curiosity, empathy, and meticulous attention to detail made this project possible. This is his seventh book with MIT Press and he would like to thank all the people in the long chain from acquisitions to marketing, people visible and invisible, who make the experience so stimulating. Last but not least he would like to thank his immediate family—Lissette, his wife, Nairobi, their daughter, Oso, their dog—and the wider immigrant Dominican family of which he is proud to be a part.

Sydney would like to thank her family, friends, and mentors who have supported her academic journey into food studies. From her teachers in the Virginia public school system to her professors and mentors at Tufts University, the educators in her life have shaped how she sees the world, forms questions, and builds responses. She thanks Nancy Williams, Steve Gissendanner, Jeff Prillaman, Teresa Wade Tyler, Cathy Stanton, Alex Blanchette, Anne Moore, and Kristina Aikens for their knowledge, guidance, and unwavering commitment to their students. She thanks Joby Giacalone, Jennifer Radar Giacalone, Gabe Giacalone, and her home of Charlottesville, Virginia, for raising her never to stop asking questions. Finally, she thanks Julian Agyeman, whose commitment as a mentor and collaborator to his students never ceases, and to whom she owes the enormous pleasure and humbling responsibility of co-creating this volume.

Introduction: The Immigrant-Food Nexus

Julian Agyeman and Sydney Giacalone

> This story is a single "foodways" thread woven within the collective human story, a thread that speaks of recipes from all cultures, carried in memories, on folded and stained pieces of paper, in pockets and bags, like identity papers, only meaningful to the beholders, only fully real once cooked and eaten.
> —The authors of chapter 14, "Recipes for Immigrant Lives"

These words remind us of the performativity, materiality, and intimacy of food carried across time and space. They remind us of the multiplicity of cultural, religious, and social meanings embedded within the cuisines we create and consume. These foodways are anything but static. Migrants carry complex and life-affirming foodways with them as both memories and dreams, creating an umbilical link between where one is from and where one is now. Food provides a grand stage for the performance of translocal identities, border transformations, belongings, and becomings in a new land.

In the wake of the 2016 US presidential election, media pundits and academics scrambled to provide insights into the likely effects of the new administration's controversial immigration policy goals: stricter enforcement and mass deportation of 11 million undocumented immigrants living and working in the United States. One area of questioning emerged prominently: how would the proposed immigration policies impact our food system?

As two scholars who critically study food systems, we felt this was a question that was painfully absent throughout a campaign season so focused on the hysterical xenophobia of the immigration debate rather than the tangible policy ramifications of this topic. Gradually, however, select media outlets began publishing articles about the agricultural industry's reliance on undocumented immigrants for more than half its labor supply to subsidize prices and profits by using this underpaid and unprotected workforce. With increased deportations and stricter immigration policies throughout the previous decade, newly proposed enforcement policies had the potential to create an even more critical labor gap, forcing farmers to switch crops, decrease production, or increase spending on legal labor. Sources published alarming predictions of the potential costs to be passed on to both US and Canadian consumers. In focusing on price and profit impacts, however, many of these analyses seemed to miss the larger questions surrounding the precariousness of our entire food labor system and immigration status quo.

In February 2017, we proposed the idea for this collection after identifying the increasingly urgent need to recognize immigrant food politics and foodways as essential within these conversations. The impetus for our work was simple: we wanted to show that we cannot talk about immigration without talking about food.

As we and our authors were writing, events over the next two years showed the far-reaching impacts of anti-immigrant and xenophobic sentiment and policies. These events included the 2017 Alt-Right and neo-Nazi terrorism in Charlottesville, Virginia; the attempted repeal of Deferred Action for Childhood Arrivals (DACA); the 2018 "zero tolerance" policy of criminally prosecuting those who cross the US border illegally, resulting in family separation and indefinite detention; and President Donald Trump's 2019 declaration of a national state of emergency over the "national security crisis" on the United States' southern border. These highly publicized events represent the tip of the iceberg; daily experiences of rising anti-immigrant sentiment and xenophobia are also widely documented, as are increased Immigration and Customs Enforcement (ICE) presence, raids, and deportations across the nation.

On June 16, 2018, the US Supreme Court upheld the third version of the president's Muslim-targeted immigration ban. As headlines flooded the news channels and social media, the moderator of a sustainable agriculture list service we subscribe to sent out a call to action to all its members. This message identified the criminalization of immigrants as a crisis and provided a list of training sessions for counteractions happening in major US cities. Moments later, a member replied to this entire online community stating that this was not "the proper forum to be calling for such an action," using derogatory language toward immigrant farm laborers while arguing that the topic of immigration is a topic "peripheral to sustainable agriculture at best." Over the next several days, a debate erupted between the list service moderators and other members sharing opinions and resources on the place of immigration within food conversations—a place many members argued was crucial.

Weeks from the completion of this manuscript, we observed this virtual conflict and subsequent discussion as a fascinatingly timely microcosm of this book's purpose. Whereas we started this volume with the assertion that one cannot talk about immigration without talking about food, we were now witnessing the obverse: can one talk about food without talking about immigration?

Immigrants in Food Scholarship Today

As a conceptual framework for this volume, we have termed the intersection of food systems, immigration policy, and immigrant foodways the "immigrant-food nexus." We feel the nexus is a compelling construct, offering both a theoretical and analytical framework to both pull together and tease apart the questions food and immigration scholars have been asking, albeit in a fragmented way.

However, to contextualize our theoretical intervention, and before we explore the immigrant-food nexus in detail, we must begin with where food and immigration scholarship currently stands. To start, scholars and activists have established that immigrants' role within the US food system is a key topic within the field of food studies and food system change. Inquiries into immigrants within the food system have fallen primarily in the "food justice" subfield of food system studies, a subfield that saw an uptick in research and attention following the rise in attention to food as a political and social topic in the new millennium. The growing environmentally focused food sustainability "movement" (referred to from this point forward as the alternative food movement) was quickly critiqued by academics and activists in the emerging food justice community. This community sought to bring recognition to the ways that low-income communities and communities of color are both disproportionately harmed by industrial food systems and underrepresented in the alternative food movement. To Michael Pollan's famous assertion that to eat healthfully and sustainably in the twenty-first century one should not eat "anything your great-grandmother wouldn't recognize as food," the food justice advocate asks, "Whose great-grandmother?" (Alkon and Agyeman 2011, 3). Positionality matters; some great-grandmothers were given scraps from the table, while others had their food demonized. The food justice movement begins with "an analysis that recognizes the food system itself as a racial project (Omi and Winant 1994) and problematizes the influence of race and class on the production, distribution, and consumption of food" (5).

As a result of this more critical attention toward food, scholars have established that attention to immigrants has occurred more slowly and sporadically than attention to other subtopics. The persistent marginalization of farmworkers, whether on large industrial farms or small, "local" family-owned farms, is harder to "sell" than the term "organic" (Gray 2013; Holmes 2013). Many critical food scholars critique how whiteness has guided alternative food movement concepts of "local," "health," and "access" in ways that negate black, brown, indigenous, and immigrant producers, workers, and consumers within both the conventional and alternative food economies (Alkon and Agyeman 2011; Guthman 2011, 2014; Minkoff-Zern 2018; Leslie and White 2018; Flora et al. 2012; Cadieux and Slocum 2015). Much research on urban food access has been called out for taking a "black and white" approach to the history of resource spatialization based on race, glossing over the nuances of different communities of color within these stories as well as the variety of food sources (such as home gardens and ethnic corner stores) utilized by present-day ethnic groups (Alkon and Agyeman 2011). Food Policy Councils, a growing governance approach to incorporating food policy into US and Canadian city policies and politics, have had mixed success in achieving their goals of diverse member and leader representation and explicit centering of racial equity (Boden and Hoover 2018; Horst 2017).

In addition to documenting the crucial shortcomings of the alternative food movement's attention to immigrants, many scholars have established a growing critical body of work on immigrants within the food system. Scholars have shown how food is mobilized as a tool within the larger projects of settler colonialism and structural racism. Debates about where to locate farmworker housing, for example, reveal underlying tensions to maintain a space's whiteness (Nelson 2007). Gentrification invites cultural appropriation of "authentic" foods, which are co-opted as "exotic" ways to consume others' cultures (Hirose and Kei-Ho Pih 2011; Turgeon and Pastinelli 2002; Zukin, Lindeman, and Hurson 2017; Joassart-Marcelli and Bosco 2017). The media criminalizes and invisibilizes immigrants through inaccurate, dehumanizing narratives of migrant farmworkers and their families (Holmes 2013; Horton 2016; Mitchell 1996; Sbicca 2015; Allen 2008).

Food and immigration scholarship shows us that it is vital to trace the history of the relationship between immigration policy and the daily lived experience of immigrants in order to understand how we have come to where we are today. In the twentieth century, the US agricultural industry began recruiting more heavily from immigrant populations, resulting in a shift from a predominantly black farm labor force to a Latinx one (Mapes 2009; Gray 2013; Mitchell 1996). To formalize and increase access to this labor source, the United States operated the Bracero Program beginning in 1946, issuing over four million guest-worker visas to bring Mexican laborers to US farms (Mapes 2009; Mitchell 1996; Gamboa 2000). Undocumented immigration continued at comparable levels during this time. The program ended in the 1964 alongside a wave of stricter immigration policies and the narrative of immigrant illegality, though the agricultural industry's use of migrant laborers remained strong. In 1986, the Immigration Reform and Control Act (IRCA) criminalized the employment of undocumented workers. Studies have found that the IRCA and the North American Free Trade Agreement (NAFTA) have in fact resulted in an *increase* in Mexican migration by making undocumented workers attractive to employers while at the same time causing economic consequences in Latin America that decreased rural Latinxs' capital and increased incentives for them to find work on US farms (Boucher and Taylor 2007; Zahniser et al. 2018 Aydemir and Borjas 2006). Other studies have investigated the logic of anti-immigrant sentiment that immigrants are filling jobs that would otherwise be filled by US "blue-collar workers," finding on the contrary that immigrants have historically benefited (and are currently benefiting) the US economy (Lewis 2007; Lamphere, Stepick, and Grenier 1994; Dudley 2018).

It is useful to take a comparative look at immigration and food in Canada to provide a wider North American context. Since the 1970s, Canada has been an officially "multicultural" nation, and Canadian immigration policy more generally is seen as multiculturally progressive. In reality, however, Canada's agriculture sector relies on a structure of marginalized immigrant labor and a political language of "othering" immigrants in the same way that the US agricultural sector does. The Seasonal Agricultural Worker Program (SAWP) enables farmers to bring in temporary foreign laborers. It is considered a highly effective program that brings in the same workers cyclically for eight-month periods (Aydemir and Borja 2006). The Canadian government's 2017 "Report of the Standing Committee on Agriculture and Agri-food" devotes one page to the topic of labor reforms, only to focus on labor shortages as a problem for farmers, with no mention of the current marginalization of immigrant laborers through low wages, poor housing, hazardous working conditions, food insecurity, and inability to unionize (Weiler, McLaughlin, and Cole 2017).

While there have been fluctuations in open or closed border stances in the past two decades, US immigration policies and enforcement tactics during the post-9/11 era have increasingly taken on nativist, populist narratives of a "war" against "illegal aliens" on the border and within the nation (Nevins 2010; Horton 2016). This narrative acts as a racial project that defines who "belongs" as a legitimate subject deserving of national protections and who is regarded as an "other." Over the past decade, and particularly since the start of the Trump administration, an increase in deportations and stricter immigration policies has decreased undocumented Mexican migration into the United States and steadily shrunk the United States' agricultural labor force (Zahniser et al. 2018; Martin 2019). Facing financial losses and a drop in production, many immigrant employers and right-leaning congressional representatives have demanded legislation such as H-2A guest-worker reforms to reinvigorate the supply of cheap immigrant labor (Martin 2019). Even with this shortage, the predominance of undocumented immigrants within farm labor has held constant. Today, estimates of the number of undocumented farmworkers range from 1.5 to 2 million, accounting for 50% to 70% of total US farmworkers. H-2A position requests jumped in 2017 and 2018 to their highest levels in recent years, though studies are still anticipating diminishing agricultural productivity if H-2A program reforms do not occur (Martin 2019; Zahniser et al. 2018).

Scholars of food and immigrant studies have documented the extreme marginalization of immigrant farmworkers in the United States historically and today. Among US workers, farmworkers have the fewest labor protections—a structure continually lobbied for by the farm industry and its allies to effectively subsidize food prices and industry profits at the cost of worker marginalization. Farmworkers must endure long hours, are paid well below a living wage and often below the minimum wage, have poor living conditions, and have no vacation, overtime pay, or health insurance in most states (Brown and Getz 2011; Holmes 2013; Gray 2013; Berkey

2017; Mitchell 1996). Scholars have effectively shown that the agricultural industry thus functions through a delicate balance of invisibility created to allow worker exploitation while at the same time protecting their (undocumented) immigration status to keep them within the industry and not integrated into the majority of US public services (Martin 2009). One result of this paradox is that those who grow our food are one of the most food-insecure populations in the nation.

Marginalization is not the only focus of food and immigration scholarship, however. Scholars have developed a rich body of work on immigrant foodways and the sociocultural significance of food. Many scholars have documented how recipes and tastes move across borders and boundaries to provide new ways of performing old and new senses of home, celebration of tradition, and connection with kin and community within new and even hostile spaces (Koc and Welsh 2002; Ray 2004). The notion of "food citizenship" (Baker 2004; Cohen 2011; Hondagneu-Sotelo 2014) considers how in times when the state's definitions of citizenship have failed its inhabitants, food can become an alternative way to perform community care, responsibility, and belonging, as in our epigraph to this introduction, which speaks of recipes as "identity papers" asserting identity through food performance. Food is central to how immigrants (whether first generation or centuries past) interact with the concept of national and self-identity (Gabaccia 1998; Diner 2001; Garcia, DuPuis, and Mitchell 2017). Food also demonstrates the hybridity of these lines of difference and movement, invoking the complexities of the transnational subject (Zavella 1985; Cohen 2011; Holmes 2013; Nagel 2009; Baker 2004; Brandt 2002; Peña et al. 2017) and the translocal cuisine (Komarnisky 2009; Gibb and Wittman 2013; Filson and Adekunle 2017). The term "foodscape" encompasses how food practices are engagements with place making and identity performance, such as in community gardens or ethnic markets, where immigrants perform their transnational identities through the soil and economies of their new homes (Joassart-Marcelli and Bosco 2017; Roe, Herlin, and Speak 2016; Baker 2004; Hondagneu-Sotelo 2014; Pine 2016).

The Immigrant-Food Nexus Framework

Given this broad overview of scholarship on immigration, immigrants, and food systems, we as co-editors asked ourselves what was still missing in the study of this intersection. Since the 2016 election, it has become more pressing than ever to assert the importance of talking about food within immigration and immigration within food. Yet only the most astute observers in the media and in political discourse seemed to incorporate the intersectionalities and interconnectedness of these topics. In our review of the scholarly literature, it was easy to separate most inquiries into two categories: the "macro" approaches, which focused on policy and national narratives, and the "micro" approaches, which focused on daily lived experiences, materiality, personal identity, and cultural practices. These distinctions between foci on the policy narrative versus lived experience mirrored the very tactics being used to marginalize, distance, and dehumanize immigrant and other communities of color in the escalating populism and nationalist policies on the border and within the nation (Gökarıksel, Neubert, and Smith 2019).

What became clear is that multiscalar, intersectional approaches to the connections between food and immigration were desperately needed. Moreover, a framework to bring together scholarship addressing the multitude of facets of this intersection was missing. The immigrant-food nexus encompasses the constantly shifting intersection of food systems, immigration policy, and immigrant foodways. We see the nexus as a multiscalar concept, extending from the macro scale of national policy to the micro scale of the intimate daily performances of culture, community, and individual bodies through food. "Nexus," meaning "connection," encompasses the intimate messiness that becomes apparent when teasing apart the connection between food and immigration.

In taking a critical approach toward questions of food and immigration, we ask: *How can the immigrant-food nexus be understood in our current political climate of rising nationalism, and how does an analysis that transcends traditional micro scales or macro scales from the nation, to the community, to the body provide a new way to think about these issues?*

Through critical, multidimensional research, food and immigration scholars today find themselves at a generative place to bring fact-based, humanized, and multiscalar narratives of the immigrant-food nexus to light. The story of national ICE policy debates, for example, cannot be told separately from the story of an immigrant woman serving tortillas to New York farmworkers while being increasingly surrounded by ICE agents, because these are not separate stories. The concepts we define as "macro" also have real, embodied consequences. The concepts we define as "micro" also

Introduction

have large-scale, important meanings. Our authors recognize this: their work bridges the scales of the nation, community, and individual bodies to "render visible the political tensions about race, agriculture, immigration, and the future of the nation that simmer in everyday life" (Neubert, chapter 2, this volume).

Organization of This Book

This volume consists of three parts, each centered on one guiding theme: "Borders: Individuals, Communities, and Nations," "Labor: Fields and Bodies," and "Identity Narratives and Identity Politics." Individual authors write from a variety of disciplines,[1] and many are members of the immigrant communities they study. When introducing each section, we include a note highlighting how the chapters combine to provide a new perspective on the immigrant-food nexus.

In part I, "Borders: Individuals, Communities, and Nations," the authors explore cultural, physical, and geopolitical borders around immigration and belonging from the scale of the person to that of the nation. Beginning the section and contextualizing the chapters to follow, in chapter 1 Kimberley Curtis provides a case study of the agricultural borderlands of Yuma, Arizona, one of the border areas receiving the greatest focus by the Trump administration, and the area where 90% of the winter greens in the United States are grown. Viewing the border as a site of contention around the livelihoods of farmworker communities, Curtis asks: what kinds of civic worlds are being made as border militarization and immigrant criminalization intensify? Addressing the mobilization of these national politics far from the border, Christopher Neubert in chapter 2 analyzes an ongoing controversy surrounding the construction of a pork processing plant that will use immigrant labor in north-central Iowa. In this strikingly timely case study, this chapter demonstrates how nationalist "fascist body politics" increasingly deployed by far-right elements in the United States come to shape and be shaped by discourses on agriculture and immigration in rural America and leave indelible marks on rural communities.

Shifting from rural boundaries to urban ones, Pascale Joassart-Marcelli and Fernando Bosco in chapter 3 investigate the tensions surrounding immigrant markets in two adjacent urban neighborhoods of San Diego characterized by diverse immigrant populations and gentrification pressures. This chapter shows that while these spaces promote intercultural encounters that may forge connections, they may also exacerbate differences, reflecting broader sociospatial processes associated with class and race within the meaning of "authentic" foods and "foodie" culture. In chapter 4, Maryam Khojasteh continues this focus on spaces of urban immigrant food commerce through case studies of two Middle Eastern grocers in a Buffalo, New York, "food desert" and a transforming market in Philadelphia. This chapter highlights the active role that immigrants play in (re)shaping community food systems while it also reveals the boundaries put up by the lack of support for immigrant food businesses in much of current city food planning. Ending this section, Sarah Huang demonstrates in chapter 5 how foodscapes constructed by immigrants and refugees in Anchorage, Alaska, complicate notions of locality, not only as a marker of space but also as a fluid marker of memories, ideas, and places through time and space. This chapter posits that transnational food identities may reveal nuanced opportunities and barriers through which the local food movement currently excludes nonwhite and nonmigrant identities.

Part II, "Labor: Fields and Bodies," addresses the topic of labor as it connects embodied labor, agribusiness, and immigrant lived experience. To begin the section, in chapter 6, Katherine Dentzman and Samuel Mindes use a comparative US-Canadian approach to study how Mexican immigrant labor shortages are being felt as never before by US grain farmers because of increasing herbicide resistance. With farmers no longer able to rely on the technological labor of herbicides to weed their crops, the situation is forcing new conversations in the industry and engagement with the social side of immigration policy. In chapter 7, Laura-Anne Minkoff-Zern and Sea Sloat show how Latinx immigrant farmers' cultivation practices are at odds with USDA program requirements for agrarian standardization, continuing a legacy of unequal access to agrarian opportunities for nonwhite immigrant farmers. In chapter 8, Mary Beth Schmid then counterconstructs stereotypes of Latinxs' roles in agriculture with examples of binational Mexican American women in southern Appalachian fruit and vegetable farming enterprises who use cooperative, kin-based exchange relations to collectively mitigate consolidation-inducing policies and mediate globalized agrifood political economies. Finally, in chapter 9, Fabiola Ortiz Valdez concludes this part of the book with her ethnography of undocumented male laborers and immigrant women living on New York State dairy farms.

Exploring the intersection of gender, labor, and foodways as forms of resistance, Valdez shows how these women make it possible for workers to maintain a connection to their life back home, providing spaces of sanctuary from law enforcement while also securing the farmers' ability to retain their labor force.

In part III, "Identity Narratives and Identity Politics," the authors focus their analyses of current immigration politics and immigrant experiences on the politics of identity and the power of narrative. In chapter 10, Jillian Linton presents a media discourse analysis of agrarian narratives within the concept of "local food" in Toronto, Canada. Though Canada's agrarian narrative seems on the surface more inclusive than its counterpart in the United States, Linton finds that media coverage of local food continues to position white farming families as neutral, immigrants as "other," and settlement as an uncontroversial, nonviolent action. Shifting to institutional identity, in chapter 11, Victoria Ostenso, Colin Dring, and Hannah Wittman investigate the successes and failures of four food policy councils in Vancouver, Canada, to engage racial justice practices within their work. This chapter considers the complicated tensions occurring within many food-focused organizations in both Canada and the United States, asking what it takes for the alternative food movement to truly prioritize black, brown, indigenous, and immigrant participants within its foundational logic and actions.

In chapter 12, Catarina Passidomo and Sara Wood present oral history excerpts of Central American, Latin American, and Caribbean immigrant cooks and entrepreneurs collected by the Southern Foodways Alliance. Through these stories, the chapter documents the ways in which immigrants use foodways to navigate and make meaning within their economic and social lives in the contemporary, multicultural "New American South." Continuing this focus on identity formation through food, in chapter 13, Alison Hope Alkon and Kat Vang demonstrate the limits of much "dietary acculturation" literature on immigrant foodways. Through an ethnography of Hmong Americans navigating the meanings of food within their daily lives, this chapter advances the concept of translocal food, as it plays a vital role in immigrant identity formation. Finally, chapter 14 follows five immigrants whose lives have become interwoven like ingredients, creating complex individual and collective foodways intersecting at a shared-use commercial kitchen in Connecticut. This chapter illustrates the intricate ways in which each of their stories is a product of both international macro-scale social, political, and economic events and the interplay of micro-scale choices and relationships formed across the kitchen. The authors of this chapter have chosen to remove their names in solidarity with the many research participants in this volume who are forced to remain nameless for their security.

Note

1. In their usage of the term Latino versus Latinx, chapter authors in this volume have made individual decisions based on their participants' own preferences and identity articulations.

References

Alkon, Alison Hope, and Julian Agyeman. 2011. "Introduction: The Food Movement as Polyculture." In *Cultivating Food Justice: Race, Class, and Sustainability*, edited by Alison Hope Alkon and Julian Agyeman, 1–20. Cambridge, MA: MIT Press.

Allen, P. 2008. "Mining for Justice in the Food System: Perceptions, Practices, and Possibilities." *Agriculture and Human Values* 25(2): 157–161.

Aydemir, A., and G. J. Borjas. 2006. "A Comparative Analysis of the Labor Market Impact of International Migration: Canada, Mexico, and the United States." NBER Working Paper 12327, June. Cambridge, MA: National Bureau of Economic Research.

Baker, Lauren E. 2004. "Tending Cultural Landscapes and Food Citizenship in Toronto's Community Gardens." *Geographical Review* 94(3): 305–325.

Berkey, R. E. 2017. *Environmental Justice and Farm Labor*. London: Routledge, Taylor and Francis Group.

Boden, S., and B. M. Hoover. 2018. "Food Policy Councils in the Mid-Atlantic: Working toward Justice." *Journal of Agriculture, Food Systems, and Community Development* 8(1): 39–51.

Boucher, S., and J. Edward Taylor. 2007. "Policy Shocks and the Supply of Mexican Labor to U.S. Farms." *Choices Magazine* 21(1): 37–42.

Brandt, Deborah. 2002. *Tangled Routes: Women, Work, and Globalization on the Tomato Trail*. Lanham, PA: Rowman and Littlefield.

Brown, Sandy, and Christy Getz. 2011. "Farmworker Food Insecurity and the Production of Hunger in California." In *Cultivating Food Justice: Race, Class, and Sustainability*, edited by Alison Hope Alkon and Julian Agyeman, 121–146. Cambridge, MA: MIT Press.

Cadieux, K. V., and R. Slocum. 2015. "What Does It Mean to Do Food Justice?" *Journal of Political Ecology* 22(1): 1–26.

Cohen, Deborah. 2011. *Braceros: Migrant Citizens and Transnational Subjects in the Postwar United States and Mexico*. Chapel Hill: University of North Carolina Press.

Diner, Hasia R. 2001. *Hungering for America: Italian, Irish, and Jewish Foodways in the Age of Migration*. Cambridge, MA: Harvard University Press.

Dudley, Mary Jo. 2018. "Why Care about Undocumented Immigrants? For One Thing, They've Become Vital to Key Sectors of the US Economy." *The Conversation*. http://theconversation.com/why-care-about-undocumented-immigrants-for-one-thing-theyve-become-vital-to-key-sectors-of-the-us-economy-98790.

Filson, G., and B. Adekunle. 2017. *Eat Local, Taste Global*. Waterloo, ON: Wilfrid Laurier University Press.

Flora, Jan L., Mary Emery, DiegoThompson, Claudia M. Prado-Meza, and Cornelia B. Flora. 2012. "New Immigrants in Local Food Systems: Two Iowa Cases." *International Journal of Sociology of Agriculture and Food* 19(1): 119–134.

Gabaccia, Donna R. 1998. *We Are What We Eat: Ethnic Food and the Making of Americans*. Cambridge, MA: Harvard University Press.

Gamboa, Erasmo. 2000. *Mexican Labor & World War II: Braceros in the Pacific Northwest, 1942–1947*. Seattle: University of Washington Press.

Garcia, Matt, E. Elaine DuPuis, and Don Mitchell, eds. 2017. *Food across Borders*. New Brunswick, NJ: Rutgers University Press.

Gibb, Natalie, and Hannah Wittman. 2013. "Parallel Alternatives: Chinese-Canadian Farmers and the Metro Vancouver Local Food Movement." *Local Environment* 18(1): 1–19.

Gökarıksel, Banu, Christopher Neubert, and Sara Smith. 2019. "Demographic Fever Dreams: Fragile Masculinity and Population Politics in the Rise of the Global Right." *Signs: Journal of Women in Culture and Society* 44(3): 561–587.

Gray, Margaret. 2013. *Labor and the Locavore: The Making of a Comprehensive Food Ethic*. Berkeley: University of California Press.

Guthman, J. 2011. "'If They Only Knew': The Unbearable Whiteness of Alternative Food." In *Cultivating Food Justice: Race, Class, and Sustainability*, edited by Alison Hope Alkon and Julian Agyeman, 263–281. Cambridge, MA: MIT Press.

Guthman, J. 2014. *Agrarian Dreams: The Paradox of Organic Farming in California*. Berkeley: University of California Press.

Hirose, Akihiko, and Kay Kei-Ho Pih. 2011. "'No Asians Working Here': Racialized Otherness and Authenticity in Gastronomical Orientalism." *Ethnic and Racial Studies* 34(9): 1482–1501.

Holmes, Seth. 2013. *Fresh Fruit, Broken Bodies: Migrant Farmworkers in the United States*. Berkeley: University of California Press.

Hondagneu-Sotelo, Pierrette. 2014. *Paradise Transplanted: Migration and the Making of California Gardens*. Berkeley: University of California Press.

Horst, M. 2017. "Food Justice and Municipal Government in the USA." *Planning Theory and Practice* 18(1): 51–70.

Horton, Sarah B. 2016. "From 'Deportability' to 'Denounce-ability': New Forms of Labor Subordination in an Era of Governing Immigration through Crime." *PoLAR: Political and Legal Anthropology Review* 39(2): 312–326.

Joassart-Marcelli, Pascale, and Fernando J. Bosco. 2017. *Food and Place: A Critical Exploration*. Lanham, MD: Rowman and Littlefield.

Koc, M., and J. Welsh. 2002. *Food, Foodways and Immigrant Experience*. Toronto: Centre for Studies in Food Security, Ryerson University.

Komarnisky, Sara V. 2009. "Suitcases Full of Mole: Traveling Food and the Connections between Mexico and Alaska." *Alaska Journal of Anthropology* 7(1): 41–56.

Lamphere, Louise, Alex Stepick, and Guillermo Grenier, eds. 1994. *Newcomers in the Workplace: Immigrants and the Restructuring of the U.S. Economy*. Philadelphia: Temple University Press.

Leslie, I., and M. White. 2018. "Race and Food: Agricultural Resistance in U.S. History." In *Handbook of the Sociology of Racial and Ethnic Relations*, edited by P. Batur and J. R. Feagin, 347–364. Cham, Switzerland: Springer International.

Lewis, Ethan. 2007. "The Impact of Immigration on American Workers and Businesses." *Choices Magazine* 21(1): 49–55.

Mapes, Kathleen. 2009. *Sweet Tyranny: Migrant Labor, Industrial Agriculture, and Imperial Politics*. Champaign: University of Illinois Press.

Martin, Philip. 2009. *Importing Poverty? Immigration and the Changing Face of Rural America*. New Haven, CT: Yale University Press.

Martin, Philip. 2019. "Trump, Migration, and Agriculture." *Border Crossing* 9(1): 19–27.

Minkoff-Zern, L. 2018. *The New American Farmer: Immigration, Race, and the Struggle for Sustainability*. Cambridge, MA: MIT Press.

Mitchell, Don. 1996. *The Lie of the Land: Migrant Workers and the California Landscape*. Minneapolis: University of Minnesota Press.

Nagel, Caroline R. 2009. "Rethinking Geographies of Assimilation." *Professional Geographer* 61(3): 400–407.Nelson, Lise. 2007. "Farmworker Housing and Spaces of Belonging in Woodburn, Oregon." *Geographical Review* 97(4): 520–541.

Nevins, Joseph. 2010. *Operation Gatekeeper and Beyond: The War on "Illegals" and the Remaking of the US–Mexico Boundary*. New York: Routledge.

Omi, M., and H. Winant. 1994. *Racial Formation in the United States: From the 1960s to the 1990s*. New York: Routledge.

Peña Devon, Luz Calvo, Pancho McFarland, and Gabriel R. Valle 2017. *Mexican-Origin Foods, Foodways, and Social Movements: A Decolonial Reader*. Little Rock: University of Arkansas Press.

Pine, Adam M. 2016. "Social Reproduction and Urban Competitiveness: How Dominican Bodegueros Use the Care Economy." *City and Society* 27(3): 272–294.

Ray, Kreshnendu. 2004. *The Migrant's Table: Meals and Memories*. Philadelphia: Temple University Press.

Roe, Maggie, Ingrid Sarlöv Herlin, and Suzanne Speak. 2016. "Identity, Food and Landscape Character in the Urban Context." *Landscape Research* 41(7): 757–772.

Sbicca, J. 2015. "Food Labor, Economic Inequality, and the Imperfect Politics of Process in the Alternative Food Movement." *Agriculture and Human Values* 32(4): 1–13.

Standing Committee on Agriculture and Agri-food. 2017. *A Food Policy for Canada*. Committee Report No. 10—AGRI (42–1). Ottawa: House of Commons of Canada.

Turgeon, Laurier, and Madeleine Pastinelli. 2002. "'Eat the World': Postcolonial Encounters in Quebec City's Ethnic Restaurants." *Journal of American Folklore* 115(456): 247–268.

Weiler, Anelyse M., Janet McLaughlin, and Donald C. Cole. 2017. "Helping Migrant Workers Must Be Part of New Food Policy." *Toronto Star*, December 22, 2017.

Zahniser, Steven, J. Edward Taylor, Thomas Hertz, and Diane Charlton. 2018. *Farm Labor Markets in the United States and Mexico Pose Challenges for US Agriculture*. United States Department of Agriculture Economic Research Bulletin 1476-2018-8188, November.

Zavella, Patricia. 1985. "'Abnormal Intimacy': The Varying Work Networks of Chicana Cannery Workers." *Feminist Studies* 11(3): 541–557.

Zukin, Sharon, Scarlett Lindeman, and Laurie Hurson. 2017. "The Omnivore's Neighborhood? Online Restaurant Reviews, Race, and Gentrification." *Journal of Consumer Culture* 17 (3): 459–479.

I Borders: Individuals, Communities, and Nations

Borders, constantly made and remade, signify belonging. Whether a physical wall, a line in the sand, or a glance at a passerby between the farmer's market stalls, seen and unseen borders mark those who belong and those who are "other." Within the current trend (and within the historical ebb and flow) of rising populism and nationalism in United States politics, there is an obsession with borders as precise ways of defining who is able to claim US identity. Likewise, borders that structure who has the authority to define a community are (as they always have been) in constant flux, fighting against or fitting within these larger national politics of who belongs.

Part I of the book excavates the formation and operation of these borders as they reveal themselves within the food system and particularly within the immigrant-food nexus. As political scientists, anthropologists, urban planners, and geographers, our authors see the construction of borders and boundaries as being of paramount importance within food and immigration studies today. How are cultural, physical, and geopolitical borders from the scale of the person, to the community, to the nation formed? Critically, how do these scales come to influence one another? How do national media narratives on an immigrant's "place" within the United States reveal themselves within small-town discourses on a proposed pork plant? How do lines drawn around authenticity of local food businesses speak to deeper frictions between community sovereignty and capitalized consumption of "culture"? When racist language of criminality shifts from slow structural violence to immediate physical threat toward immigrants in the United States, what forms of resistance or spaces of hospitality are possible? The authors of Part I of the book begin to answer these difficult, essential questions.

1 Criminalization and Militarization: Civic World Making in Arizona's Agricultural Borderlands

Kimberley Curtis

Introduction

The agricultural borderlands of Yuma, Arizona, produce 90% of the winter greens consumed in the United States, through the labor of 40,000–50,000 field workers, the majority of whom cross the border daily from Mexico. Rising between midnight and 1 a.m., they dress, prepare lunch, look in on sleeping children, and then make their way to the port of entry. There they wait along with tens of thousands of other field workers for two to three hours, show their papers, and arrive in the fields between 6:30 and 7:00 a.m., when the wage clock begins.

It takes workers between five and a half and six hours just to get to the fields. The return home is another two to three hours. This means seven and a half to nine hours of waiting, of suspended and colonized life, for thousands and thousands of people—mothers, aunts, grandfathers, grandmothers, fathers, and teenagers. There is no real term in our labor lexicon to refer to this time. It doesn't count as wage theft (the denial of wages or benefits rightly owed) or as compensable travel time (which, under the Fair Labor Standards Act, only applies once the workday has begun). It is just the reality of agricultural labor in the borderlands.

In this ethnographic study, I examine the impact of national border security and immigration policies on farmworker communities in the Yuma borderlands. My concerns are with capacities fundamental to community life: mobility and the capacity to maintain familial and social ties, provision, and participate as members in shaping the civic world. There is a stark clarity in the way global dynamics and national policies come to the ground in borderland communities that promises illumination of the troubled nexus of immigration and agriculture.

On Walls and Workers

> Border walls work. Yuma Sector proves it.
> —Elaine Duke, acting secretary of homeland security

The relationship between immigration and large-scale agriculture in the United States has long been fraught. President Donald Trump's 2017 visit to the Yuma borderlands illustrates dimensions of the trouble. On August 22, 2017, just days before announcing the pardon of former Maricopa County sheriff Joe Arpaio, the president visited the Marine Corps air station in Yuma. He spoke with Customs and Border Protection agents, greeted marines, and looked at a Predator drone (newly retired from active combat in Iraq and Afghanistan) in the steeply up-armored Yuma Sector. All the while, the tightly scripted performance completely eclipsed the region's dependence on immigrant labor.

The following morning, the mainstream media reported the administration's story line: before the Secure Fence Act of 2006, with just 5.2 miles of fencing along the sector's 126 miles, the border "was besieged." Today, the narrative continued, apprehensions of illegal border crossers are one-tenth their 2006 levels, thanks to 63 miles of fencing, the tripling in the number of Border Patrol agents, massive increases in roads, electronic mobile surveillance, military hardware, and the construction of second and third layers of walls in urban areas (Carranza 2017; Duke 2017). According to then acting secretary of homeland security Elaine Duke, the Yuma Sector proves that the full border wall proposed by Trump will "turn the tide against the flood of illegal aliens and secure our homeland" (Duke 2017).

The assessment of whether walling works is never as simple as apprehension counts, as scholars have persuasively demonstrated by documenting the dramatic rise in the dangerous criminal economies of drug and human smuggling that *enforcement itself generates* (Andreas 2009). The 6,000 grisly deaths documented by human rights organizations in the first 15 years after border enforcement operations began pushing migrants into ever more dangerous terrain is another case in point (Jimenez 2009). Moreover, the question of where walling might end is chillingly raised by Todd Miller's account of "the 21st century border" in his book on homeland security and migration induced by climate change. In Central America's "northern triangle"—ground zero for climate change in the Americas, drought is heaping unprecedented suffering on already suffering people, pushing farmers into the northward migrant stream (Miller 2017, 71–105). As Homeland Security's quadrennial reports show, the US security apparatus is acutely aware of these climate-induced dynamics. To head off the flow of climate refugees, the United States is funding border enforcement hardware for Mexico and training Mexican immigration agents, police, and the military in border policing, and US Customs and Border Protection agents are physically working in detention centers along the Mexico-Guatemala border. Where does walling end, and who or what does it serve?

Political theorist Wendy Brown convincingly argues that the function of the recent surge of wall building across the world is to stage political sovereignty at a time when globalization has significantly attenuated it. Walls are "theater pieces for national populations specifically unsettled by global forces threatening sovereignty and identity" (Brown 2010, 9). They generate "an imaginary of stable and homogeneous (and sometimes white supremacist) nationhood" (9).

Trump's visit to Yuma and his subsequent pardon of Joe Arpaio, the nation's most visibly racist anti-immigrant sheriff, are performances in this costly nativist theater of state sovereignty. Simultaneously, they participate in a drama long constitutive of an agricultural regime that depends on the "persistent devaluation" of agricultural labor (Brown and Getz 2011). Shifting forms of invisibility and racialized visibility of agricultural communities keep labor costs low and workers cowed, ensuring agrarian accumulation and cheap food. Thus, on the one hand, an acute civic invisibility characterizes their lives; their needs, contributions, and voices are rarely part of public discussion. On the other hand, politicians, bureaucrats, and the media create periods of racialized hypervisibility in the form of "Latino threat narratives" (Chavez 2013) that have laid the groundwork for spasms of deportations (in the 1930s and 1950s) of the majority immigrant, majority Mexican agricultural labor force.

Today we are in the midst of another such spasm—with its own malevolent twist. A kaleidoscope of immigration and security policies have, with growing intensity over the last 30 years, illegalized and criminalized the US agricultural labor force (along with other immigrants), which is 75% foreign born, 68% Mexican, and 50% unauthorized (Martin 2017). The stock character of the "alien invader" overrunning our borders saturates the media (Chavez 2013), and US immigration enforcement increasingly uses the figure of the "criminal alien" to justify expansion of a virulent immigration regime (Cházaro 2016). Thus, as globalization economically displaces and psychologically unsettles populations inside the nation, creating politically exploitable vulnerabilities, the hypervisibility of the racialized criminal invader and the invisibility of farmworker communities dangerously reinvigorate the old drama of "persistent devaluation" of agricultural labor.

Methods

Scholars have argued that borderlands should be studied as sites of struggle within the frame of *fabrica mundi*—world making. In *Border as Method* (2013), Sandro Mezzadra and Brett Nielson suggest we pursue the processes by which the objects of knowledge—borders in both the empirical and cognitive-emotional senses—are constituted. To take the border as method is to adopt an epistemological viewpoint that fosters critical understanding of dispossession, exploitation, and domination *and* attends to the multiple ways that political subjectivity and community are being constructed through struggle. These scholars are (and I am) interested in "the ontological moment"—what kinds of civic worlds are being made as border militarization and immigrant criminalization intensify?

For this study, I conducted 16 open-ended, in-depth interviews with leaders from educational, faith-based, and nonprofit organizations on the US side of the border (see the appendix). All participants are at least loosely connected to Yuma Interfaith, a local affiliate of the Industrial Areas Foundation (IAF), the nation's oldest grassroots organizing network. An organization of organizations, IAF intentionally bridges social divides, bringing people together across races, classes, and ethnicities to work on shared concerns. Having become familiar with IAF as a participant in another local affiliate, I recruited participants through Yuma Interfaith's lead organizer. Informal interviews with growers, field workers, farmworker advocates, and Border Patrol agents, conducted on extended research trips with students to the Yuma borderlands, also inform this study.

In what follows, I contextualize the history of immigration and border security policy significant for agricultural workers and their communities. Then, after an introduction to the study site (see figure 1.1), I discuss my findings regarding the civic worlds emerging in the Yuma borderlands.

Border Security and Immigration Policies

In the following sections, I trace broad historical periods, tending to patterns of invisibility and racialized hypervisibility that have persistently devalued agricultural workers. I draw attention to a "disconcerting history" in which border security and immigration policies have been unrooted in reality (Massey and Pren 2012), shaping an increasingly fascist fictional counterworld (Snyder 2017). Throughout, I illuminate the changing legal logic within which farmworker communities struggle to make a home.

Pre-1965—Circular Migration

Until 1965, US law considered Mexicans to be migrants, expected to follow the labor trail but not to settle. Thus, a circular migration pattern was established, in which agricultural workers came for the season and returned to Mexico when the work ended. The border itself was virtually unpoliced by the state until 1924, when the US Border Patrol was created. While there were periodic mass deportations of these racialized and expendable workers, they were not subject to federal immigration laws like the Chinese Exclusion Act, and this contributed to the early dependence of border state agriculture on seasonal migration from Mexico.

This circular pattern continued in altered form with the Bracero Program. Between 1942 and 1964, 4.6 million Mexicans received guest-worker visas for temporary employment in the border state fields. Equal numbers crossed illegally, a group often preferred by growers because they had absolutely no legal protections. State responses to the presence of these unauthorized workers, based on concerns about lawlessness, veered erratically from waves of legalizations to the deportation of 1.1 million Mexican farmworkers in 1954.

1965–1993—Birth of the Framework of Illegality

The Immigration and Nationality Act of 1965 marked an important shift in US immigration policy. For the first time, Mexicans were considered immigrants, and Mexico received a fixed immigration quota. Yet the quota was far below the number of Mexican field workers that large-scale agriculture had come to depend on. With the United States having simply expelled "the Braceros" the year before, there was no legal way for Mexican field workers to work. Therefore, with border enforcement lax, these long-established migratory flows simply continued, with nearly all workers now without authorization. The number of potentially deportable, largely invisible workers rose to unprecedented levels. The framework of illegality was born.

Although unauthorized immigration leveled out by 1977, the year workers reestablished their Bracero-era levels, a dangerous politics of racialized hypervisibility had been brewing. A "threat narrative" of a "border under siege" emerged in the 1970s with growing ferocity, exploited by politicians and bureaucrats at a time of deepening income inequality and insecurity (Massey and Pren 2012). Studies of national magazine covers and major newspapers from 1965 to 1995 show immigrants depicted as "a tidal wave" poised to "inundate" the United States and "drown" its culture or as an "invasion" against which "outgunned" Border Patrol agents try in vain to "hold the line" (Massey and Pren 2012, 6). A fictional counterworld was in the making.

Enforcement operations and immigration policies appeared at an accelerating rate and with increasing scope. As Massey and Pren show, an "enforcement loop" takes hold, in which enterprising politicians stoke public fear, lawmakers pass increasingly draconian legislation, and border enforcement launches operations. This results in a rise in boots on the ground and other interdiction capabilities, which enables more apprehensions, which provokes more fear, and so on. From 1977 to 1995, the number of Border Patrol agents increased by 2.5 times, the number of linewatch hours doubled, and the Border Patrol budget went up by a whopping factor of 6.5 *despite the lack of any real increase in illegal immigration* (Massey and Pren 2012, 8–14). Thus, Massey and Pren conclude, "a largely invisible circulation of innocuous workers" was transformed into "a highly visible violation of American sovereignty by hostile aliens" that propelled increasingly draconian enforcement operations (Massey and Pren 2012, 8). This treacherous political dynamic, in which policy is unhinged from facts, lays the groundwork for increasingly fascist policies.

Also during this period, the Immigration Reform and Control Act of 1986 (IRCA) created a path to citizenship and made the employment of undocumented workers a crime. Although employer sanctions had almost no teeth, growers preferred to avoid the risk, which fueled the rise of the labor contractor system, in which contractors (usually former field workers) procure and manage labor crews and are the first line of legal culpability for workplace violations. The IRCA also stimulated the underground economy in forged documents, as millions of illegalized workers needed at least the appearance of legal papers. The law propelled already illegalized workers into further illegal acts.

1993–Present: The Criminal Alien

The terrorist attacks that began in the 1990s prompted a series of massive border enforcement operations. Starting in 1993 with Operation Hold the Line, these operations aimed to "seal" urban areas along the border through huge boosts in military equipment, boots on the ground, and budgets. This militarization marks the start of what Aviva Chomsky (2014) calls "an obsession with the border," so evident today.

These militarized operations pushed migrants into longer, more dangerous crossings in remote areas, which in turn increased the need for and the cost of smugglers, making human trafficking more attractive to drug rings (Andreas 2009). They also were responsible for a more than doubling of unauthorized Mexicans in the 1990s and the early part of the following decade. By dramatically raising the costs for migrant workers to return home (including potential death), they encouraged millions to settle, radically reducing circular migration (Massey and Pren 2012).

A long list of reductions in the rights of these hypervisible unauthorized immigrants and increasing criminalization followed. The 2001 PATRIOT Act allowed indefinite detention for noncitizens, while in 2005 Operation Streamline changed unauthorized entry and reentry into the United States from a civil offense to a criminal one, the latter punishable by two years in federal prison. Using the new 1996 powers of "expedited removal" of any noncitizen who crossed the border without documents, Operation Streamline has processed hundreds of thousands of these "criminal aliens" have been processed en masse in special courts, bound together in chains. Critics charge that the proceedings violate their due process right to adequate counsel (Rickerd, n.d.).

Immigrant-only detention centers—a full half of which are private—house those awaiting court proceedings. A growing for-profit economy around criminalized immigrants has raised human rights concerns that also extend to government facilities. The 9th Circuit Court of Appeals recently rejected the Border Patrol's argument that being required to provide mats and blankets to detainees in Arizona, some of whom are held for 72 hours, constitutes "a hardship" for the agency (Fischer 2017a).

Since the War on Terror, deportations of unauthorized immigrants have skyrocketed, rising from a pre-1995 level of 50,000 annually, where it had been for decades, to a peak of 409,000 in 2012 (US Immigration and Customs Enforcement 2016). Although no terrorists have entered the United States through the southern border, no terrorists have been Mexican, and all terrorists entering had legal visas, Mexicans have been disproportionately targeted for deportation by these antiterrorism campaigns, *comprising a shocking 72% of those removed in 2009* (Massey and Pren 2012, 16). In this paroxysm of nativism, policy is unhinged from factual reality.

Under the legalistic guise of opposition to (a manufactured) criminality, racial discrimination against immigrants in the post–civil rights era continues (Chomsky 2014, 14–20). President Trump's January 2017 executive order that makes not only immigrants *convicted* of a crime but also those *charged* with a crime priorities for deportation is a fascist escalation of this logic. It is within this escalating fascism that we should situate the president's sending of thousands of active-duty military and National Guard troops to the border and his 2019 declaration of a state of emergency *at a time when apprehensions at the border were at a 40-year low* (Robinson 2018; US Customs and Border Protection 2019).

Yuma Agricultural Borderlands

Since the mid-twentieth century, extensive agricultural complexes straddling the US-Mexico border have appeared, taking advantage of the steep economic gradient between the two countries. Changes along the border have been swift, with older communities vastly transformed by new members and new towns and cities emerging. Both old and new communities are transnational in character, linked intimately by migratory flows.

Since 1938, the Imperial Dam has diverted 90% of the Colorado River's flow to the desert borderlands. Yuma growers cultivate 230,000 acres. The largest crop is lettuce. Melons, alfalfa, cotton, lemons, seeds, and other labor-intensive crops are also grown. Labor costs as a percentage of total production expenses are high—24% compared to the US average of 10% (Frisvold 2015).

Farmworkers on both sides of the border have close sociocultural and familial ties. Indeed, some 30,000 US residents who work in the Yuma fields live in Mexico because, although wages are 10 times higher in the United

Criminalization and Militarization

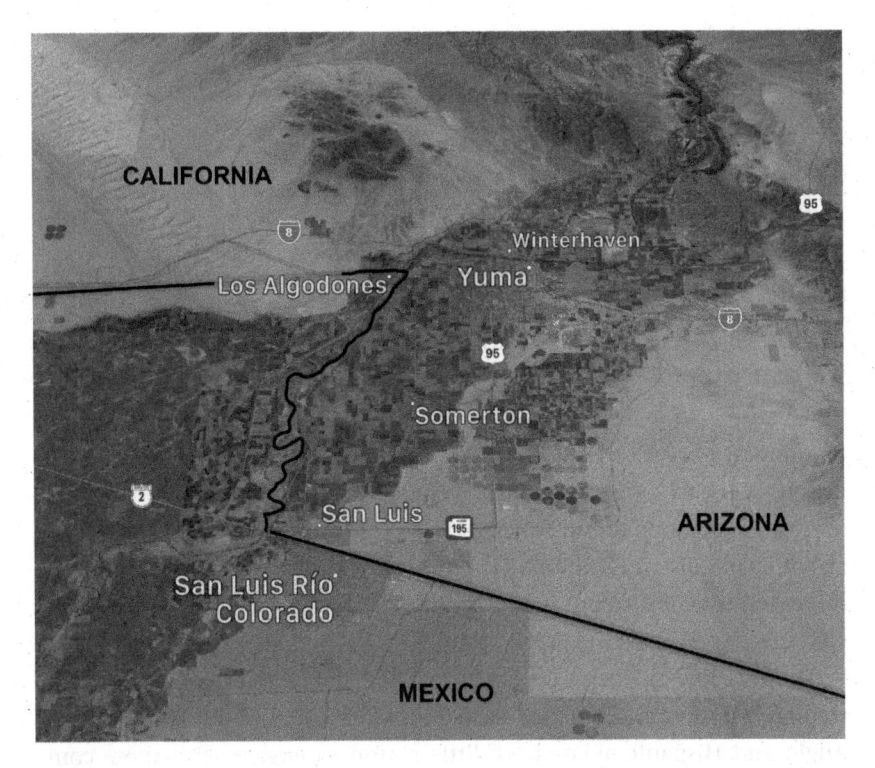

Figure 1.1
Study site.
Source: Apple Maps.

States, housing costs are prohibitive. Moreover, policies barring family members with even minor offenses from living in the United States mean that living in Mexico, despite the life-draining hours spent getting to the fields, is the only way families can stay together (according to a study participant). Along with 2,200 H-2A guest workers, these groups are the backs and arms of a $3.2 billion agricultural industry.

Communities on the US side have developed complex patterns of racial and ethnic enclaving, crosscut by differences in legal status, nativity, and class. The border towns of San Luis and Somerton are farmworker towns (97% Hispanic). Yuma is the county seat, and while 59% Hispanic, it is the heart of Anglo culture and power. White working-class and middle-class retirees flock to Yuma's warmth in the winter months, increasing the population by 50%. Retired military personnel are a large percentage of these migrants. They come to a county where the business of security runs deep:

Table 1.1
Select Demographic Data—Study Site

	Median Household Income ($)	Per Capita Income ($)	Foreign Born (%)	Off Season Unemployment (%)	Poverty (%)
Yuma	46,151	26,000	21.5	21.4*	16.9
Somerton	37,252	13,977	39.9	***	29.2
San Luis	33,767	11,435	49.1	48.0*	27.5

Source: US Census Bureau, *Quick Facts*.
* YCharts, 2018–2019.

two military bases (one of them among the largest in the world), a state prison for felons, a private detention center for immigrants, and 859 border agents. Unemployment in San Luis in the off-season is 48%. Poverty rates in both San Luis and Somerton are close to 30%. In Yuma, by comparison, off season unemployment is 21.4% and the poverty rate is 16.9% (US Census Bureau, n.d.; YCharts, 2018–2019). See table 1.1.

The relationship between Yuma and the farmworker communities to the south is asymmetrical. Unless tied into farm work, residents of Yuma—Anglo and Hispanic alike—have little reason to engage with these communities. Even the Yuma-based growers have insulated themselves from them, relying since the late 1980s on farm labor contractors to organize and oversee farmworker crews. Social distancing from these farmworker communities by second- and third-generation Hispanics in Yuma and even by some Yuma pastors is not uncommon (according to a study participant).

Two annual festivals crystallize this landscape of power, devaluation, and need. "Yuma Lettuce Days," put on since 1998 by the Yuma Tourist Bureau to promote agriculture, had, until 2016 *no recognition of farmworkers*. By contrast, "El Día del Campesino," held since 1997 at 3 a.m. directly on the border and organized by community advocates, celebrates farmworkers and provides essential health and educational services for some 5,000 workers before they head to the fields.

I now turn to a discussion of my interviews with leaders from educational, faith-based, and nonprofit organizations to understand the impacts of border security and immigration policies on these farmworker communities. I find two contrasting civic worlds, treating each in turn.

Fabrica Mundi I—Nativist Security Regime

Study participants report that control over cross-border mobility has increased. Those in their seventies remember that crossing the border used to be easy, enabling communities to sustain relations. One recalled that what the family used to worry about were the avocados her grandmother was smuggling in from Mexico. Another recalled losing his green card and talking to the border agent, who let him pass through. Today it is all about control and surveillance. The man who lost his green card explained, "Everything is worse. Oh, yeah, everything is worse. Behind the wall, they still have fields. They are watching for some traffickers. They watch the people in the fields too, soldiers watching."

Changes in identity cards have made it impossible to use borrowed documents and more difficult to get forged documents to cross into the United States for work. Now, as one participant said, it has to be an inside job—someone within Customs and Border Enforcement does the forgery.

Internal enforcement operations designed to empty the Yuma Sector of unauthorized immigrants have been quite successful. Study participants say that there are few community members on the US side living without documents, most people having moved on if they could, uprooting family and fracturing the community. Others have been arrested, held in detention, and deported. As one organizer in the study explains, "The reason [so few people live without documents in the region] is we are less than 100 miles from the border. There's Border Patrol everywhere. Absolutely everywhere.... You have to be really invisible without documents along the border."

Nonetheless, participants estimate that some 10% of people in the area have become "spatially incarcerated" (Gupta and Ferguson 1992), meaning that because they have no documentation, and because there are Border Patrol checkpoints manned 24/7 and roving patrols carrying out enforcement operations 100 miles into US territory, they literally cannot move about. An immigration services provider in the study shared the story of a 29-year-old woman. Her daughter has a life-threatening medical condition and is on the Make a Wish List, but the woman is unable to accompany her to the hospital in Phoenix. Getting jobs is difficult, and her ex-husband has been threatening her. Police operations designed to clear the land of "criminal aliens" create particular kinds of prisons, leaving people exceedingly vulnerable, creating the conditions for further violation and violence, and fracturing social ties.

Many of the most vulnerable are parents, especially women, who crossed without authorization years ago, when it was easy, and stayed. "Now," as another immigration services provider explained, "the child is 21 and a citizen and wants to petition for her mom to get residency. She can file, but her mom will have to be in Mexico for ten years before she can return." Such a law is aimed directly at families, profoundly disregarding social relationships fundamental to flourishing personhood. Most choose to stay. The immigration services provider explained, "When families hear this, they are so demoralized, and the parent, she is not going to leave. I mean she's been here 15–20 years. And so the family just digs in lower and lower into society and they are very anxious."

In her 2000 ethnographic study of El Salvadoran immigrants, Susan Coutin describes the world inhabited by those denied membership as "spaces of nonexistence." The undocumented exist in a "nondomain," a territory that, like its residents, is and is not there (Coutin 2000, 29). This can produce a radical sense of loneliness and a corrosive feeling that one does not fully exist. As one immigrant advocate in my study put it, "They are invisible people. Do you understand me? They are invisible people. They suffer much humiliation."

Advocates report a new ruthlessness toward those arrested because their papers are not in order. It used to be that the immigration judge would grant them a green card if they had been here a number of years, could establish good character, and show that they had equity in the system such as children born in the United States who rely on them. Today, these things mean little on the border. Instead, detention and deportation are almost certain, along with family separation and loss of social ties. Studies show that 91% of those in the Yuma Hold Room in ICE detention were deported, compared to the national rate of 56% (Transactional Record Access Clearinghouse 2015).

For domestic farmworkers living on the US side, this new ruthlessness is increasing fear and humiliation. Although difficult to gauge, one Yuma grower estimated that 30% of farmworkers use borrowed or forged documents (informal interview), despite the 2008 Arizona law requiring all employers to use E-Verify to confirm their legal status. For these workers, traveling is nightmarish. Farmworker advocates report that at internal checkpoints through which crew buses must pass to get to fields to the north and west, workers (and other travelers) are being hassled more often. A 2015 American Civil Liberties Union (ACLU) report offers corroboration. Based on Department of Homeland Security documents obtained through a Freedom of Information Act lawsuit, the ACLU says that abuse at checkpoints in the Yuma and Tucson sectors is at "epidemic levels." Among their findings are that agents are threatening motorists with assault rifles, electroshock weapons, and knives; destroying and confiscating personal property; and interfering with efforts by community members to videorecord the abuses (Fischer 2017b).

Even permanent legal residents feel threatened. As one immigrant advocate put it, "Before, with a green card, you walked around pretty sure of yourself." Today, rumors circulate that the permanent residency program will be canceled or that residency can be terminated if residents have ever committed a crime. Immigrant advocates report that it is taking longer to renew green cards and that more and more US residents are refusing to leave their houses, drive, or sometimes even go to work when their green card has expired and they've reapplied but have not yet received renewal—a situation that previously would not have caused apprehension. Advocates themselves are not sure that they can be reassuring. An immigration services provider stated, "We do not know what to expect. One day a program is here, the next it is not (TPP, DACA). People are vulnerable. So I too share in the fear that people in the community are experiencing. I fear something very drastic will happen and I won't be in a position to help our community."

Everyone in these farmworker communities knows multiple people caught in some gradation of legal nonexistence, people whose basis for membership in family, community, or nation—blood ties, labor, presence, humanity—has been denied or threatened. This kind of devaluation in farmworker communities is not new. What is new is the terror that is spreading, insinuating itself into the lived experience of all members of these communities, documented or not, creating increasing immobility and a hunkering down into deep inconspicuousness by a widening portion of the community.

As material borders become ever sharper and more extensive, and the consequences of transgressing them more certain and severe, cognitive-emotional worlds shrink, lives become constricted, and people become more alone. The fabric of the civic world depends on trust, abiding relationships, family ties, and the ability to participate, move, dream, grow, care for others, have needs met, take risks, and encounter others. In these farmworker communities, that fabric grows more tenuous as a nativist security regime increasingly bent on clearing the land of immigrants tightens its hold.

In a different way, the fabric of the civic world for everyone else in these borderlands is diminished, for insofar as they normalize the invisibility of their fellows, blocking out the radical reduction of existence taking place around them, the growing terror, they diminish their own ability to grasp the reality of changes in the political order. Informal interviews with growers evince little knowledge of or interest in the labor crews who work their fields. Their practice of not seeing what is in plain sight is a precondition for the violations, chronicled here, of an increasingly fascistic world.

Around such a world, a security economy grows. The Border Patrol actively recruits at community events in these farmworker towns, setting up alongside immigrant rights groups. Immigration advocates in the study report that young people are enamored of the high starting salary and by job descriptions that emphasize being in a position of authority, calling the shots. One stated, "These young people are vulnerable; they buy that." Indeed, half the Border Patrol agents on the southern border are Hispanic. Tragically, they join a security economy that serves the agricultural regime's "persistent devaluation" of the very communities from which most of them come.

The nativist security regime described here contrasts sharply with another civic world emerging in response to militarization and criminalization. I now turn to this counterhegemonic world.

Fabrica Mundi II—Hospitable Community

> We have a lot of poverty, a lot of people barely surviving, but having said that we are an extremely resilient community.
> —Study participant

In the wake of mid-twentieth-century totalitarian states that used techniques of terror to destroy the space between people that gathers and relates them, Hannah Arendt theorized the public world as spaces of appearance (Arendt 1958). Such spaces—civil associations of all kinds—pull those frozen in fear, isolated, and with an attenuated sense of existence or those divided by deep social cleavages *into relationships* where they can shape the civic world.

Increasing control, devaluation, and terror in the borderlands is prompting community leaders in these farmworker communities to create such spaces. In their own relationships, they form a densely networked web; they know each other's stories, collaborate on numerous projects, share an intimate knowledge of the communities they serve, and frequently meet face to face.

These relationships enable critical work: tending to the vulnerable. One participant summarized this work metaphorically by saying, "I take care of lawns...the best way of dealing with a hostile community is, strengthen those who are vulnerable and leave the hostile groups alone. I take care of lawns; I don't kill weeds."

For example, leaders go to the spatially incarcerated. Sometimes traveling 40–50 miles, they bring necessary papers, services, and care. These insistent acts affirm the very existence of the most vulnerable community members, and as their stories are shared within activist groups, they become part of the fabric of the larger world, shaping understandings and informing action. Thus, the cramped material and cognitive borders that so diminish the existence of the most vulnerable are attenuated.

Leaders also tend another vulnerable group. Families have made it clear to educators on the border that they do not want their children to become farmworkers. In response, educational leaders from kindergarten to college have developed a coordinated strategy of high expectations, practical support, and public affirmation to open doors to these children. College administrators and community advocates attend award nights and parent–teacher organization (PTO) meetings and talk at community events. Speaking at an outdoor celebration of Mexican Independence Day, a college representative asks the community, "Are you independent? Are you economically independent? Are you socially independent? I work at the college, and I am waiting for you. I am waiting for your children." Tired farmworker parents respond. PTO meetings in the Gadsden school district (K–8) on the border are standing room only, with 350 attending at a time.

The result has been astounding. Gadsden is the top school district *in the country* for the number of middle school students who qualify for Johns Hopkins University's Center for Talented Youth Program. Year after year, more than 150 kids take the ACT and collectively earn $600,000–$700,000 in scholarships for summer residential academic programs with Johns Hopkins. And waiting for those kids at 7 a.m., when they get to school to take the ACT, are the school honor guard and the marching band.

These educational leaders are creating spaces of appearance where solidarity and possibility emerge. They are undoing the material and cognitive borders that hold farmworker families in a reproductive cycle in which children follow their parents into the fields.

Leaders are also rejecting a rigidly bounded nationalist community. Central American refugees have been crossing the Yuma border in large numbers and turning themselves in, seeking asylum. Yuma Interfaith became involved when parishioners noticed ICE agents dropping off refugees with children and little means of support at the Wal-Mart parking lot.

A pastor explains the community's response: "That led us to officially seek contact [with ICE] to provide a better system than dropping people off at Wal-Mart. It was our way of letting them know we wanted to be in partnership; that these were people who were seeking their way into the US. And let us be hospitable."

Churches held clothing drives; found mattresses, food, and health care; built showers in their churches; and mobilized volunteers. Facing a hostile anti-immigrant public, they kept their work secret for four years. Even within their own communities, parishioners question the work, asking their pastors questions like: "Are they legal? Why should my dollars be used for this purpose? I want our work to benefit the USA. We have enough people that already we can't take care of. How could we possibly take care of more?"

Perhaps the real work is building capacities for hospitable answers. Here again, the pastor quoted earlier explains: "We have to remind them of the ways of Jesus, that he too was an immigrant. And remind them of who puts food on their table. And sometimes they need to hear the stories of everyday fear of being killed, of abuse, slavery. They have to be reminded that life should be an abundant life and you can have that. And to seek that in America should be something anybody can seek."

Capacity building of this kind goes on in another way as well. Yuma Interfaith brings church members—farmworker families, conservative retirees, and Trump supporters—together. Meeting in their churches, they learn to talk with one another about immigration and to find common concerns for civic action. This is hard, slow work, where people who normally don't *exist* for one another except as media caricatures listen to each other's stories and begin to stitch together a different, more realistic sense of their world. Here is how one study participant described it: "What's interesting is that the retirees, many of them are from the great generation of WWII, and they understand what it is to give. You know, you just do it because it's what you are supposed to do; it's the American way. The other population that thinks the same way is immigrant families. So they get each other."

A counterhegemonic civic world is emerging in the borderlands. As community members tend the most vulnerable, support children, welcome the stranger, and work across deep social cleavages, they resist the civic invisibility and racialized hypervisibility that devalue farmworker communities.

Conclusion

The impulse to wall off, scapegoat, devalue, and control immigrants in this country is old, though perhaps never more acute than in this political moment. This study of the Yuma Sector, the country's extreme laboratory for the development of a nativist security regime, suggests that agricultural borderlands are potent sites for understanding the impact of increasingly fascist policies of border militarization and immigrant criminalization on the fabric of our world.

Civic worlds are always in the making and, in its extremity, this local border world speaks to the national one in prophetic tones, foretelling possible futures and illuminating dynamics whose endpoints are as yet unknown. On the one hand is a civic world increasingly constructed through terror and police operations designed to clear the land of long-term immigrants and to stanch the flow of desperate new migrants. For those subject to this new ruthlessness, an ever more reduced social existence, enhanced vulnerability to abuse, social isolation, and civic invisibility follow. Social distancing and acute forms of looking away on the part of US citizens normalize this civic invisibility and, in so doing, disable their ability to grasp the reality of ominous changes taking place in the political order.

On the other hand, border militarization and immigrant criminalization is generating a contrasting thick relational civic world. Characterized by hospitality, reaching across social chasms, tending the vulnerable, and creating spaces where people can be seen and heard, these embattled communities are seedbeds for the regeneration of democratic practices, energies, values, and visions. And it is perhaps in such locations that the social webs critical to farmworker demands for justice are being woven.

Agricultural borderlands, these front lines of practices of terror and oblivion and of unsung and underanalyzed resistance to them, are rich areas for further study. Students of food systems, racial justice, grassroots democracy, immigration, and fascism who seek to understand the dynamics of the present and the contours of possible futures will have much to offer by conducting deep ethnographic work in these frontline communities.

Appendix

Organizations Involved in Study

Faith Based:
Pastoral Campesina—Catholic ministry serving spiritual, social, and educational needs of farmworkers.

Yuma Interfaith—Network of organizations devoted to building power across diverse groups to support public goods.

Education:
Arizona Western College—Junior college, San Luis, Arizona.

Gadsden School District—K–8 district in Arizona on the US-Mexico border.

Migrant Education Program—Federal program.

Nonprofit Service:
Campesinos Sin Fronteras—Integrated service organization for farmworkers.

Chicanos Por la Causa—Community development corporation devoted to family reunification and immigration services.

Comité de Bien Estar—Builds assets and strengthens community, San Luis, Arizona.

Community Legal Services—Provides legal assistance to low-income residents, San Luis, Arizona.

References

Andreas, Peter. 2009. *Border Games: Policing the U.S.-Mexico Divide.* Ithaca, NY: Cornell University Press.

Arendt, Hannah. 1958. *The Human Condition.* New York: Harcourt Brace Jovanovich.

Brown, Sandy, and Christy Getz. 2011. "Farmworker Food Insecurity and the Production of Hunger in California." In *Cultivating Food Justice: Race, Class, and Sustainability*, edited by Alison Hope Alkon and Julian Agyeman, 121–146. Cambridge, MA: MIT Press.

Brown, Wendy. 2010. *Walled States: Waning Sovereignty*. New York: Zone Books.

Carranza, Rafael. 2017. "Trump Renews Push for Wall Funding during His Visit to the Arizona Border." *The Republic*, azcentral.com, August 22, 2017.

Chavez, Leo. 2013. *The Latino Threat: Constructing Immigrants, Citizens and the Nation*. Palo Alto, CA: Stanford University Press.

Cházaro, Angelica. 2016. "Challenging the 'Criminal Alien' Paradigm." *UCLA Law Review* 63(3): 594–664.

Chomsky, Aviva. 2014. *Undocumented: How Immigration Became Illegal*. Boston: Beacon Press.

Coutin, Susan. 2000. *Legalizing Moves: Salvadorans Struggle for US Residency*. Ann Arbor: University of Michigan Press.

Duke, Elaine. 2017. "Homeland Security Secretary: Border Walls Work. Yuma Sector Proves It." *USA Today*, August 22, 2017. https://www.usatoday.com/story/opinion/2017/08/22/homeland-security-secretary-border-walls-work-yuma-sector-proves-it-elaine-duke-column/586853001/.

Fischer, Howard. 2017a. "Border Patrol Must Provide Mats for Detainees to Rest." *Arizona Capitol Times*, December 26, 2017.

Fischer, Howard. 2017b. "Judge Backs ACLU in Lawsuit over Border Checkpoints." *Arizona Daily Sun*, January 28, 2017.

Frisvold, George. 2015. *The Economic Contribution of Agriculture to Yuma County*. Department of Agriculture and Resource Economics, University of Arizona. http://yaac.org/wp-content/uploads/2015/05/Frisvold-12-14.pdf.

Gupta, Akhil, and James Ferguson. 1992. "Beyond 'Culture': Space, Identity and the Politics of Difference." *Cultural Anthropology* 7(1): 6–24.

Jimenez, Maria. 2009. *Humanitarian Crisis: Migrant Deaths at the U.S.-Mexico Border*. ACLU of San Diego and Imperial Counties and Mexico's National Commission of Human Rights. https://www.aclu.org/files/pdfs/immigrants/humanitariancrisisreport.pdf.

Martin, Philip. 2017. *Immigration and Farm Labor: From Unauthorized to H2A for Some?* Migration Policy Institute Issue Brief, August.

Massey, Douglas S., and Karen A. Pren. 2012. "Unintended Consequences of US Immigration Policy: Explaining the Post-1965 Surge from Latin America." *Population and Development Review* 38(1): 1–29.

Mezzadra, Sandro, and Brett Nielson. 2013. *Border as Method, or, the Multiplication of Labor*. Durham, NC: Duke University Press.

Miller, Todd. 2017. *Storming the Wall: Climate Change, Migration, and Homeland Security*. San Francisco: City Lights Open Media Series.

Rickerd, Chris. n.d. "Operation Streamline Issue Brief." American Civil Liberties Union. Accessed July 21, 2019. https://www.aclu.org/other/operation-streamline-issue-brief.

Robinson, Lori. 2018. "The Stats on Border Apprehensions." FactCheck.org, April 6, 2018. https://www.factcheck.org/2018/04/the-stats-on-border-apprehensions/.

Snyder, Timothy. 2017. *On Tyranny: Twenty Lessons from the Twentieth Century*. New York: Tim Duggan Books.

Transactional Record Access Clearinghouse, Syracuse University. 2015. "Detainees Leaving ICE Detention from the Yuma Hold Room." http://trac.syr.edu/immigration/detention/201509/YUMHOLD/exit/.

US Census Bureau. n.d. *Quick Facts*. Washington, DC: US Census Bureau. Accessed July 21, 2019.

US Customs and Border Protection. 2019. *Southwest Border Migration FY 2019*. https://www.cbp.gov/newsroom/stats/sw-border-migration.

US Immigration and Customs Enforcement. 2016. *Fiscal Year 2016 ICE Enforcement and Removal Operations Report*. https://www.ice.gov/sites/default/files/documents/Report/2016/removal-stats-2016.pdf.

YCharts. 2018–2019. https://ycharts.com/indicators/san_luis_az_unemployment_rate.

YCharts. 2018–2019. https://ycharts.com/indicators/yuma_az_unemployment_rate.

2 Slaughterhouse Politics: Struggling for the Future in the Age of Trump

Christopher Neubert

Introduction

Most evenings in north-central Iowa, between segments on the local news, the familiar face of Laurie Johns fills the screen. A former news broadcaster herself, she introduces the "Iowa Minute," a short piece of advertorial produced by the Iowa Farm Bureau. In 2017, many viewers heard the story of Eagle Grove, Iowa (population 3,500), and how a new livestock processing facility would help revitalize the region. Like many small towns in the rural Midwest, Eagle Grove has struggled with a stagnant economy since the farm crisis of the 1980s transformed agriculture in the United States (Foley 2015). The announcement that one of the largest pork producers in the country would soon be building a large processing facility was thus welcomed by many in the area. The Eagle Grove residents featured in the "Iowa Minute" reflected a sense of optimism as they described the potential "growth opportunities" that would herald "a really exciting time." One retailer told viewers that "it's never been a better time to be a resident of Eagle Grove" and invited all Iowans to come to Eagle Grove, while the mayor described the relationship between locals and livestock as "one big happy family."

The optimism expressed here obscures a larger story: the story of how one town's fortunes became bound to the promise of a massive slaughterhouse. Eagle Grove was not the first choice for Prestage Farms, the large agribusiness building the plant. When plans for the plant were first revealed in early 2016, Prestage had selected Mason City (population 28,000), about 60 miles northeast of Eagle Grove, as the site for what they described as a modern and technologically innovative pork processing facility. Initially, the proposal attracted support across the city and state. Business leaders projected that the plant would bring in 1,000 jobs and have an (unspecified) $375 million impact. Governor Terry Branstad, now the US ambassador to China, threw the support of the Iowa Economic Development Authority behind the project, and the Mason City Council put together a benefits package to reduce the plant's tax obligations. All the pieces were in place for swift approval of the plant.

However, a developing opposition moved quickly to organize protests, phone calls, letters, and emails to the city council. They packed council meetings and accused state and local officials of fast-tracking a project that would permanently reshape Mason City and the entire landscape of northern Iowa. Their stated primary concern was that the slaughterhouse could increase the number of concentrated animal feeding operations (CAFOs or, colloquially, "factory farms") near Mason City, potentially contaminating waterways already impaired by manure and fertilizer runoff (Pitt 2015). Activists opposing the plant focused on a May 2016 vote on tax abatements for Prestage Farms, and following debate that lasted late into the night, the council deadlocked, rejecting the package on a tie vote.

The fallout was swift and brutal. Prestage Farms dismissed environmental concerns, accused "kooks" of stirring up racial tensions, and claimed that "racism is alive and well in Mason City" (Brownfield Ag News 2016). Prestage Farms's implication here was clear: those protesting the construction of the pork processing plant were motivated not by concerns about factory farming but instead by racist fears of Latino immigrant laborers moving into the city. By summer's end, Prestage announced their plans to relocate to Eagle Grove, but the political and emotional damage caused by the controversy in Mason City endured. Council members publicly accused each other of betraying the public trust, while letters to the *Mason City Globe Gazette* remained hostile, with each side now accusing the other of anti-immigrant bigotry. For 34-year-old councilman Alex Kuhn, who led the vote against the project, the pressure was overwhelming. On June 5, just ten days after Prestage announced they were moving the plant to Eagle Grove, Kuhn killed himself on a country road outside Mason City. Reports following his suicide and statements from his family revealed that he not only suffered from depression, caused in part by the stress of the Prestage vote, but was also intimidated after his vote by the mayor and other members of the city council (Skipper 2016).

The story of the Prestage plant, the controversy surrounding the Mason City vote, and the fallout and subsequent relocation to Eagle Grove reveal the complex political tensions simmering in parts of the rural United States where agriculture and immigration collide. Importantly, these debates in north-central Iowa occurred just after the 2016 presidential caucus in Iowa, as populist movements on both the left and right were shaking established political norms. On the right in particular, a nativist nationalism emerged as a key motivation behind the rise of Donald Trump. That November, this region of Iowa voted overwhelmingly in favor of the eventual president. However, as scholars studying the relationship between intimacy and geopolitics have long demonstrated, presidential votes, electoral politics, and public policy tell a partial story of ordinary political life (Staeheli et al. 2012).

This chapter will use intimate geopolitics as a framework to discuss how political geographies are realized not through actions taken by distant politicians engaging in statecraft through legislation or nation building but through the messy, difficult everyday relationships that shape and are shaped by discourses of the nation-state. In what follows, I will offer a brief overview of recent interventions in feminist geography on immigration and geopolitics and then return to the Prestage plant controversy and offer a discourse analysis of the discussions as they occurred in 2016. My goal is to demonstrate how moments like this render visible the political tensions about race, agriculture, immigration, and the future of the nation that simmer in everyday life. Such an analysis offers a critical intervention into research investigating the intersections of food and immigration politics as well as broader questions surrounding the current rise in US nationalism, drawing attention to the need among scholars, activists, and advocates for a deeper understanding of how race and capitalism define political life.

Intimacy, Race, and Geopolitics

Feminist interventions in geopolitics interrogate how traditional geopolitical analyses maintain the appearance of a naturalized scalar hierarchy where discourses from above dictate the lives of national subjects (Massaro and Williams 2013; Pain 2015). Such interventions demand attention to how the everyday lives of people are not simply "blank surfaces" waiting for geopolitical discourses crafted by distant statesmen but are actually shaping what it means to be a nation-state through messy, complicated, and personal encounters (Dowler and Sharp 2001, 169; Mountz and Hyndman 2006; Pratt and Rosner 2012). This focus on intimacy raises new questions regarding how states are constantly being produced through aesthetics, emotions, and the movement of bodies (Mountz 2004; Smith 2012; Fluri 2014) and demonstrates that the intimate is as public and political as the global. In fact, the intimate and the global interpenetrate, and it is through this constant interaction that bodies come to be understood as territory and a target site of bordering practices (Smith, Swanson, and Gökarıksel 2016).

In this section, I take such observations about the importance of intimacy in the formation of geopolitical discourse and examine recent geographic scholarship on immigration and agriculture with a particular focus on the rural United States. I want to pose two questions that will guide my exploration of the Mason City/Eagle Grove processing plant in the sections that follow. First, how are geopolitical discourses produced through intimate contestations about the future? And second, how are bodies figured in these tensions as territorial agents?

Gökarıksel and Smith suggest that the nationalist discourses that animated the 2016 US presidential campaign were fueled by a "fascist body politics" that seeks to preserve white male supremacy in the face of perceived demographic threats (Gökarıksel and Smith 2016, 79). This body politics sees "aggression, hardened borders, and violence [as] central to the defense of the nation" and locates threats in the brown bodies present in spaces scripted as white (80). These tropes are certainly not new, and they have been used by politicians for decades to capitalize on anxieties and fears of a changing landscape, but the attacks of September 11, 2001, intensified efforts in the twenty-first century to securitize mobile brown bodies and made the "production of fear" a popular political tactic (Mountz 2010; Hyndman 2012; Silva 2016). Crises are now imagined everywhere, requiring a spectacular response that often means expelling or otherwise walling out the brown bodies that are presumed to be the cause of these (manufactured) crises (Hyndman 2012, 246).

While the US Census Bureau roughly defines a small town as an "urban cluster" with a population between 2,500 and 50,000 (and rural areas as anything below this threshold), Leitner (2012) found in her study of a small town in rural Minnesota that such neat definitions are rarely useful on the ground, where "small town" and "rural" are in fact categories best defined through whiteness connected to a sense of feeling American. Her findings align with similar research demonstrating that alignment between whiteness and American belonging informs white residents' reactions toward immigrants of color, such that even in the absence of lived experience with immigrants or other people of color, white residents reproduce historically racist discourses (Winant 1994; Kobayashi and Peake 2000; Cramer 2016). When encounters do inevitably occur, they are already shaped by generational attempts to securitize the boundaries of white privilege (Leitner 2012, 841; cf. Ahmed 2004). In small towns and rural spaces, which have been mythically constructed as "white spaces" (Agyeman and Spooner 1997; Lawson, Jarosz, and Bonds 2008; Finney 2014), white residents express new feelings of isolation and vulnerability when they encounter bodies that they feel betray "core white American values"—a Protestant work ethic, speaking English, home ownership, public hygiene, family values, and even clothing and bodily movement (Leitner 2012, 837). While these small towns certainly become spaces of mutual transformation (Nelson and Hiemstra 2008; Price 2012), often such encounters do not change white Americans' opinions of groups as a whole even when they speak positively about individual members of such groups (Leitner 2012, 841).

Such observations complicate the "contact hypothesis" advanced by Allport (1954), which suggests that increased contact among different groups reduces prejudice. This hypothesis has enjoyed renewed popularity in a divisive political climate after demonstrating accuracy under controlled settings (Pettigrew and Tropp 2006). However, these settings are often unable to recreate conditions in everyday life, where encounters are less clearly positive or negative (McKeown and Dixon 2017). Furthermore, as the geographers discussed here have demonstrated, racialized social hierarchies persist in these communities through normalizing spatial practices. In the examples that follow, I discuss specifically how this spatial organization has defined encounters that occur when immigrants, primarily from Latin America, move to predominantly white small towns in the midwestern United States.

The Politics of the Slaughterhouse

When the kosher Agriprocessors packing plant began recruiting Guatemalan immigrants 20 years ago, the town of Postville, Iowa, was very similar to nearby Eagle Grove now: predominantly white, with a population of around

3,000. In less than a decade, the Guatemalan population grew from 50 residents to more than 800, and Postville came to symbolize in popular media discourses the dramatic economic, social, and cultural transformations brought on by the vertical integration of livestock agriculture at the end of the twentieth century (Olivos and Sandoval 2015). Meat processing in the United States had shifted from urban, unionized labor toward a reliance on a "global reserve army of labor" composed of farm laborers from Latin America who were expelled from the labor markets in their home countries because of the rise of global agribusiness (Huffman and Miranowski 1997). Olivos and Sandoval found that in communities like Postville, immigrants were scripted into a homogenized "Latino identity" and that, despite the necessity of their labor, they were often considered disposable by their white bosses and neighbors (Olivos and Sandoval 2015, 198).

In 2008, Agriprocessors learned just how indispensable their immigrant labor force was. Early in May, Immigration and Customs Enforcement (ICE) descended on Postville and arrested 389 workers, detaining them in a nearby cattle exhibition center. This would later be found to be just more than half the undocumented workforce, and, following the raid, others either fled or went into hiding (Camayd-Freixas 2009). Postville itself appeared nearly abandoned, and Agriprocessors was unable to staff its plants and eventually filed for bankruptcy before closing altogether. Shalom Rubashkin, who oversaw Agriprocessors's operations in Postville on behalf of his family, was charged with several crimes—including over 9,000 child labor violations—before being convicted on financial fraud charges. He was sentenced to 27 years in prison, but his term was commuted by President Trump after serving just eight—the second use of the president's clemency powers following his pardon of former Arizona sheriff Joe Arpaio (Hawkins 2017).

Unlike the Arpaio pardon, however, the commutation of Rubashkin's sentence was encouraged and welcomed by a bipartisan group of lawyers and legislators who had argued that the sentence was too harsh. Still, it is telling that the earliest commutations issued by President Trump were both for men with close ties to immigration abuses but that the sentence of Rubashkin, who oversaw a slaughterhouse rather than a law enforcement agency, was commuted with comparatively little controversy. To those advocating for commutation, Rubashkin's sentence seemed disproportionate precisely because of how normalized and naturalized his activities were, even as workers in Agriprocessors had long been compelled to work extended shifts in unsafe conditions (Camayd-Freixas 2009; Olivos and Sandoval 2015).

These conditions are not unique to Agriprocessors or Postville, Iowa, of course. The "work of killing" in our modern, industrialized agricultural economy is made invisible to consumers through the production of confinement, separation, and distance—and depends on the enforcement of racialized sociospatial hierarchies to secure that border between life and death (Pachirat 2011, 9). However, Latino immigrant laborers arrive in these new towns not only to perform a job but also to thrive and live full lives, even under extremely difficult conditions. It becomes difficult to make them invisible, even as their "social integration" remains constrained by discourses equating whiteness with belonging (Vega 2012, 206), as the space of cities and towns is organized to keep them spatially marginalized, and as a discourse of "illegality" comes to represent all immigrants as criminal (Nelson and Hiemstra 2008, 327; see chapter 1 of this volume for Kimberley Curtis's analysis of the criminalization of immigrant farm laborers).

This discourse matters because in the absence of any real encounter caused by spatial segregation, white American ideas about racial others are formed through such stories and narratives, which eventually circulate in everyday life. Santa Ana writes that news stories about Latino immigrants follow narrative patterns that either cast immigrants as figures in a Western genre, with Border Patrol agents standing in as the defenders of an "allegorical nation" (Santa Ana 2016, 104), or as tragic figures in a state of "permanent liminality" (108). These mythic formations resonate because they are familiar and reassuring, and they circulate as a sort of "public dream" for white Americans about how immigration functions in their communities (102). Perhaps most importantly, these discourses function to withhold subjectivity from nonwhite residents in small-town communities and shore up the boundaries of whiteness by keeping Latino immigrants scripted into stories either about conflict between demographic groups or about individual hopeless tragedies.

In the following sections, I return to the Mason City/Eagle Grove slaughterhouse with this framework in mind to better understand if and how these discourses about immigration appeared in the ensuing controversy and to examine what they can tell us about the changes occurring in the predominantly white small towns that dot the midwestern US countryside. The Postville slaughterhouse raid, as the largest ICE action at the time—and certainly the most significant immigration enforcement activity ever in Iowa history—must be understood as casting a long shadow over these debates, though obviously much has changed in the 10 years since Postville. How have the views of Iowans regarding slaughterhouses and the immigrant labor that will inevitably staff them changed? How are discourses about race and agriculture interwoven with the fears and anxieties people feel over their town's (and, indeed, their nation's) future? Finally, what might this case be able to tell us about the broader implications of intimate everyday politics?

Borderlands beyond the Border

The research conducted for this chapter consists of a discourse analysis of various media, including original footage of the Mason City Council debates; radio, print, and television news, op-eds, and letters to the editor circulating in Mason City and Eagle Grove; and paid content produced by supporters and opponents of the plant. This material was supplemented by additional informal interviews and my own observations while conducting research in the area from May 2015 to August 2017. As the previous section demonstrated, racism and the desire to securitize the boundaries of whiteness are always present in these "borderlands" far beyond the physical US border. Thus, I want to examine how immigration and immigrants were discussed during these debates about the future of agriculture in Iowa, dwelling on those moments where the lives of the potential workers were or were not made visible. For the most part, nonwhite voices were excluded from these debates, despite the significant presence of Latino communities throughout the cities and towns of north-central Iowa. Thus, these discussions about immigrants and immigration, when they occurred, reflect the attitudes of white residents in these regions that would go on to vote overwhelmingly for Donald Trump in the 2016 election. While Trump and his populist rhetoric did not necessarily influence the politics surrounding this slaughterhouse, these events offer an insight into how such a nationalist discourse dependent on fanciful misrepresentations of demographic shifts (Gökarıksel, Neubert, and Smith 2019) generates fear among white residents in these rural, agricultural communities on a very intimate level.

Rumors that a major agribusiness was seeking to gain a foothold in Mason City had been circulating for months before the official March 20, 2016, announcement that Prestage Farms was planning to build a multimillion-dollar hog processing plant. Attended by Iowa governor Terry Branstad, Ron Prestage, and several dozen city leaders and hog industry supporters, the announcement had the air of a celebration. Details had been carefully crafted prior to the announcement, with the city and state both offering tax incentives. Three required public hearings had been scheduled for the Mason City incentive package: on April 5, April 21, and May 3, when the final vote would occur. Ron Prestage spoke with pride about how his facilities operated in the "most ethical and moral way possible," while Branstad praised the "farm family" values of Prestage and the "state-of-the-art facility" he was preparing to build. Other than subtle hints in the vague promises of up to 2,000 jobs that Prestage "wouldn't discriminate," and that the plant would "follow all the laws," immigration did not come up at this announcement. The event concluded with loud applause, big smiles, and the governor telling Prestage that he had done "a good job." There seemed to be no concern that the path would not be clear for Prestage and his company to begin construction in seven short weeks.

The first obvious sign of trouble for Prestage came just nine days later, at the March 29 meeting of the Cerro Gordo County Board of Supervisors. After the meeting had concluded, two of the three supervisors held an informal discussion with local activists, which was recorded and made publicly available by *North Iowa Today*. Throughout the hour-long recording, both the supervisors and the activists expressed concerns that the proposed plant would negatively impact the rest of the county. The supervisors first focused on the potential for pollution caused by a dramatically increased hog population. Since Prestage Farms is a vertically integrated hog producer—meaning that they own the hogs and control production from birth through death to distribution—the supervisors worried that Prestage would build new hog confinements throughout the area. The fear of pollution produced by these CAFOs was acute for the supervisors, since three nearby counties had recently been sued by the downstream city of Des Moines, Iowa, over the significant costs of removing hog fecal waste from their drinking water. They also discussed their frustration at the apparent lack of any partnership between Mason City and the county, with one supervisor saying the city was pursuing a "lone ranger deal" without any transparency.

For the most part, the conversation focused on agriculture and fears about pollution, until about 40 minutes into the 60-minute meeting, when one supervisor told a story about a friend from nearby Storm Lake, Iowa, where Tyson Foods operates several processing facilities. The supervisor relayed his friend's view that since the plants were built, the town had "deteriorated" to the point where "stink and crime" were prevalent and the low-wage workers were reduced to "stealing all the time." Later in the conversation, Storm Lake reappeared as an example of a place where budgets for jails, courts, human services, and schools had been stretched thin since their plants were built. One supervisor suggested that the increased tax revenue would not be enough for the school district to pay for special education and bilingual teachers. As the meeting concluded, one activist declared, "We know what's going on," and claimed that Cerro Gordo County would be getting all "the scraps" from the proposed plant, while places like "Communist China" reap all the rewards. The rest of the group agreed, and in one final remark yet another activist reported that when his comments appear in the *Mason City Globe Gazette* he will be "beaten into the ground as a racist." The group laughed.

While no one in this group specifically cited Latinos or Latino immigration, the invocation of Storm Lake here is, I argue, a key discursive code acting as shorthand for a community that has seen a spike in population because of immigrant labor working in a slaughterhouse. And just as Leitner (2012) found, conversations about immigrant populations in these small towns are rarely based on any actual encounters but instead on hearsay and rumor that is accepted as widely known truth. What is interesting in this case is that in a conversation that these public officials may or may not have known was being recorded (the recording, of course, is in the public record), they still use coded language to refer to immigrants. This is a theme that was repeated across media during the Mason City debates.

Days after the supervisors' meeting, on April 2, the *Globe Gazette* published an article discussing the changes in Storm Lake and Marshalltown (another nearby Iowa city with a recent surge in immigration) and acknowledged that both school districts were now majority nonwhite districts that struggled to provide "English language learner" classes for nonnative speakers (Colias 2016). Still, the reporter never spoke to any immigrant families, whose lives and stories continued to be inaccessible to the *Globe Gazette*'s readership. The next day, the *Globe Gazette* published a letter to the editor whose author wrote that he had left Storm Lake 15 years earlier, when he "no longer felt safe there." Claiming that he had spoken with "many people in the area," the author further reported that an increase in crime and expensive bilingual teachers have had "detrimental" consequences for Storm Lake residents.

On April 5, the day of the first public city council hearing on the plant proposal, nearly 200 people crowded into the council chambers. At least 50 were given the opportunity to speak during the five-hour meeting, with most opposing the plant. Many expressed concerns about the possibility of air and water pollution from the hogs, but others raised concerns about possible stress to the school district and "cultural clashes," even though the district superintendent had previously expressed her support for the proposal. The council voted unanimously to advance the proposal to the second round of voting.

As the second public hearing approached, letters to the *Globe Gazette* became increasingly contentious, sarcastic, and even mocking. One letter writer offered that "those opposed to a hog slaughtering plant are all vegans" and suggested that a reporter "ask a trucker" about what it is like to work for a large hog plant. In fact, a trucker did write in, and claimed that residents of Mason City should prepare for truck washes that would flush hog waste into the local sewage system and produce constant foul odors. The letters published varied between opposition and support, with supporters cheering the potential for growth and jobs, and opponents raising concerns about pollution, odor, and the stress to social services an underpaid labor force might cause. These last concerns continued to use coded language, innuendo, and rumor to talk about Latino immigrants, and one writer claimed that her friends in Storm Lake and Marshalltown told her "social ills and stress on the schools" changed the "entire dynamic" of those communities. She encouraged her readers to "talk to your friends in communities with slaughterhouses.... I'm not hearing good things." Regular news reports, however, continued to contradict these statements, with officials in both Storm Lake and Marshalltown observing that they had seen no unreasonable increase in crime.

On April 21, after seven more hours of public testimony in another completely full city council chamber, the council voted again to advance the project, but this time the vote was 5–1. Alex Kuhn, then a second-term councilman, cast the lone vote against the project, citing concerns that the incentive package was too high a price to pay considering the likely burden to taxpayers.

A day later, KIMT, the local CBS affiliate and the only television station based in Mason City, aired a lengthy report about the proposed plant and spent more than half the six-minute piece focusing on the potential consequences of an increase in immigrant labor. Through interviews with city officials in Storm Lake and Marshalltown, the reporter makes clear that there is a consensus that crime has not been a significant problem in these communities. In fact, the police official in Storm Lake acknowledged that some forms of violent crime had decreased. The rest of the piece focused on concerns about teaching English to immigrant students in local schools.

Throughout the piece, the reporter made several rhetorical choices to avoid naming the specific populations he was talking about. In six minutes, he never spoke the words "Latino," "immigrant," "immigration," or even "Spanish," and only once acknowledged that some students are "Hispanic." Instead, he vaguely uses some form of the words "ethnic" (five times), "diversity" (twice), "culture" (twice), "minority" (twice), or "new neighbors/arrivals" (twice). Perhaps of most concern, the report claimed, without any evidence, is that many of these workers came from "cultures where the police aren't trusted" and that "those same cultures have trouble with authority members in school." One of his interview subjects, a schoolteacher in Marshalltown, added the additional claim that some students had "backgrounds where school isn't that important." The piece concluded with the Storm Lake official offering that, despite some tension and conflict, immigration has made Storm Lake a place where residents have "experienced what the world's really like." In his final remarks, the reporter suggested that despite the piece KIMT had just aired, most concerns expressed at the previous night's city council meeting were about odor and pollution.

The language employed by the KIMT reporter—referring to nonspecific "ethnic groups" and "cultures"—is common among many of the letters, articles, and statements made by those who were uncomfortable with the idea of new people moving into Mason City. This language performs a specific purpose: to maintain the invisibility of immigrants of color by creating a distance through language, thus avoiding "precise descriptions of repugnant things, inventing instead less dangerous names and phrases for them" (Pachirat 2011, 9). Acknowledging the presence of such language in the Mason City debates demands an acknowledgment that, for some, the idea of immigrant bodies crossing the borders of their town was repugnant. Similarly, Prestage's promotional documents constantly refer to their planned facility as "modern," "clean," "high tech," and "state-of-the-art," all to conceal that the plant remains, essentially, a place of death and dismemberment meted out by workers who are exploited and underpaid. Moreover, there is an assumption in these publications that "the locals" and "the citizens" refer specifically to white Iowa residents, despite the fact that about 10% of the population of Mason City identified as "Hispanic" in the 2010 census. The voices of Iowa's immigrants—in Mason City, Storm Lake, Marshalltown, and elsewhere—were completely removed in these discourses about their communities' futures.

City Hall was packed again on May 3, 2016, for the third and final vote on the incentive package. Again, both supporters and opponents of the plant delivered comments late into the night. On the 10:00 p.m. newscast, KIMT reported that a decision had yet to be reached, and it was not until around midnight that the final vote was called and the shocking result—a 3–3 tie—was reached. Since city rules require that tax incentive packages receive a clear majority of votes, the tie meant the package was dead. Opponents were jubilant, while Ron Prestage and his supporters vowed that the fight was not over.

Days later, Prestage took his message to the airwaves. Clearly irritated with the outcome of the vote, he argued that "kooks" and "racists" had hijacked the democratic process in Mason City, spreading "misinformation and unjustified fears" (Brownfield Ag News 2016). While the environmental activists who marched, organized their neighbors, and showed up at city council meetings strongly denied and rejected such claims, the analysis presented here demonstrates that there is some truth to the claim that a "thinly-veiled racism," in Prestage's words, motivated the actions of the overwhelmingly white opposition.

Conclusions

Throughout my examination of the discourses employed during the debates in Mason City, three observations became clear. First, these debates about the future of the community completely excluded nonwhite voices. Second, attempts to securitize whiteness were often made through subtle language that worked to make the actual immigrant workers invisible. Finally, fear and anxiety regarding the possibility that the boundaries of whiteness might collapse become intimately tied to the success or failure of the agricultural economy in these communities. Certainly, there are contradictions in these conclusions. White residents are deeply concerned about the future success of agriculture and worry that immigration will somehow displace them, yet immigration remains central to the production of food that is at the heart of cultural, economic, and social life in these rural regions. Ultimately, these contestations about the future of agriculture, and what constitutes success, are struggles over multiple meanings of success: increasing profits by maximizing food production and hiring precarious labor, securitizing an idealized space and way of life, or raising crops and livestock in an environmentally sustainable way. The analysis in this chapter demonstrates that any discussion of the future of food and agriculture must also consider the intertwined social and spatial practices that have structured the everyday lives and landscapes of those people most intimately associated with food production.

It is important to remember that these communities, while small and occasionally isolated, are still complex global spaces where people encounter each other, engage in tension or struggle, and participate in mutually transformative and beneficial processes. The point of this chapter is not to indict the activists who fought the Prestage plant in Mason City but rather to acknowledge how attempts to securitize whiteness in communities scripted as white for generations will persist in discussions about the future of food and agriculture. It is imperative that activism and scholarship working toward a just food system address this reality. In the Mason City case, while the organized opposition did not openly align with anti-immigration sentiments—at times condemning such statements—neither did they actively seek to challenge the underlying assumptions linking race to capitalist food production, which limited the discussion of this processing plant before its construction was even announced.

While some would undoubtedly argue that the short timeframe through which the plant was being pushed necessitated a quick response, I would counter that the messy entanglements of race, food justice, and agricultural politics mean we absolutely *should* demand more from such activism, especially when most of the activists are white and the bodies of racialized others are removed from the conversation even as their labor is central to this economy. Further research must also proactively prioritize nonwhite voices from these landscapes, certainly a shortcoming even in this chapter, and this will require collaboration and engagement that incorporates participatory methodologies beyond the scope of what I have presented here. Given that rural US communities are so often scripted as white, regular attempts must be made to disrupt these narratives, and scholars and activists must ask who is being included in the agricultural future that these political debates are often contesting and what it means when protecting the environment also does the work of securitizing whiteness. There can be no sustainable food system brought forth through environmental justice that does not also confront the racial hierarchies embedded in the economy (Pulido 2017). The fallout from the Mason City proposal demonstrates the consequences of such an attempt.

As the Prestage plant moves forward in Eagle Grove, many of these contradictions have yet to be resolved. With rapid approval of various tax incentives, construction on the slaughterhouse began in Eagle Grove just months after Mason City rejected Prestage. Despite the mayor's claim on the "Iowa Minute" that they are all part of "one big happy family," tensions are rising. Two local sports broadcasters were fired from their positions at the end of 2017 for openly mocking the Spanish-sounding names on the Eagle Grove High School basketball team, and on my most recent visit I witnessed a truck driving through downtown with a large confederate flag prominently hung from the tailgate. If Prestage or the community has any plans to address these tensions, they have yet to discuss them publicly. Considering how rarely immigrants were actually acknowledged as such during the Mason City debate, this is perhaps not surprising. Unfortunately, it appears that the lesson Eagle Grove's elected officials learned from the Mason City proposal was to discuss the plant as little as possible and move even faster through the approval process. Nevertheless, as Prestage opens in 2019, and as the Trump administration continues to curtail immigration with support from a base much like the white residents of Eagle Grove, these inevitable encounters will transform this community and ultimately offer key insights into how visions of the future of agriculture are contributing to political efforts to securitize space for whiteness.

References

Agyeman, Julian, and Rachel Spooner. 1997. "Ethnicity and the Rural Environment." In *Contested Countryside Cultures*, edited by Paul Cloke and Jo Little, 197–217. London: Routledge.

Ahmed, Sara. 2004. "Collective Feelings: Or, the Impressions Left by Others." *Theory, Culture and Society* 21(2): 25–42.

Allport, Gordon W. 1954. *The Nature of Prejudice*. Oxford: Addison-Wesley.

Brownfield Ag News. 2016. "Ron Prestage Discusses Future Plans for Iowa Pork Plant." https://brownfieldagnews.com/rural-issues/ron-prestage-discusses-future-plans-iowa-pork-plant/.

Camayd-Freixas, Erik. 2009. "Interpreting after the Largest ICE Raid in US History: A Personal Account." *Latino Studies* 7(1): 123–139.

Colias, Meredith. 2016. "As Mason City Schools Eye Meatpacking Plant, Lessons from Storm Lake, Marshalltown." *Mason City Globe Gazette*, April 2, 2016.

Cramer, Katherine J. 2016. *The Politics of Resentment: Rural Consciousness in Wisconsin and the Rise of Scott Walker*. Chicago: University of Chicago Press.

Dowler, Lorraine, and Joanne Sharp. 2001. "A Feminist Geopolitics?" *Space and Polity* 5(3): 165–176.

Finney, Carolyn. 2014. *Black Faces, White Spaces: Reimagining the Relationship of African Americans to the Great Outdoors*. Chapel Hill: University of North Carolina Press.

Fluri, Jennifer L. 2014. "States of (In)Security: Corporeal Geographies and the Elsewhere War." *Environment and Planning D: Society and Space* 32(5): 795–814.

Foley, Michael Stewart. 2015. "'Everyone Was Pounding on Us': Front Porch Politics and the American Farm Crisis of the 1970s and 1980s." *Journal of Historical Sociology* 28(1): 104–124.

Gökarıksel, Banu, Christopher Neubert, and Sara Smith. 2019. "Demographic Fever Dreams: Fragile Masculinity and Population Politics in the Rise of the Global Right." *Signs: Journal of Women in Culture and Society* 44(3): 561–587.

Gökarıksel, Banu, and Sara Smith. 2016. "'Making America Great Again'? The Fascist Body Politics of Donald Trump." In "Banal Nationalism 20 Years On," *special issue, Political Geography* 54 (September): 79–81.

Hawkins, Derek. 2017. "How Trump Came to Commute an Ex-Meatpacking Executive's 27-Year Prison Sentence." *Washington Post*, December 21, 2017, sec. Morning Mix.

Huffman, Wallace, and John Miranowski. 1997. "Immigration, Meatpacking, and Trade: Implications for Iowa." Iowa State University, Department of Economics, Staff Paper Series 285, January.

Hyndman, Jennifer. 2012. "The Geopolitics of Migration and Mobility." *Geopolitics* 17(2): 243–255.

Kobayashi, Audrey, and Linda Peake. 2000. "Racism Out of Place: Thoughts on Whiteness and an Antiracist Geography in the New Millennium." *Annals of the Association of American Geographers* 90(2): 392–403.

Lawson, Victoria, Lucy Jarosz, and Anne Bonds. 2008. "Building Economies from the Bottom Up: (Mis)Representations of Poverty in the Rural American Northwest." *Social and Cultural Geography* 9(7): 737–753.

Leitner, Helga. 2012. "Spaces of Encounters: Immigration, Race, Class, and the Politics of Belonging in Small-Town America." *Annals of the Association of American Geographers* 102(4): 828–846.

Massaro, Vanessa A., and Jill Williams. 2013. "Feminist Geopolitics: Redefining the Geopolitical, Complicating (In)Security." *Geography Compass* 7(8): 567–577.

McKeown, Shelley, and John Dixon. 2017. "The 'Contact Hypothesis': Critical Reflections and Future Directions." *Social and Personality Psychology Compass* 11(1): 1–13.

Mountz, Alison. 2004. "Embodying the Nation-State: Canada's Response to Human Smuggling." *Political Geography* 23(3): 323–345.

Mountz, Alison. 2010. *Seeking Asylum: Human Smuggling and Bureaucracy at the Border*. Minneapolis: University of Minnesota Press.

Mountz, Alison, and Jennifer Hyndman. 2006. "Feminist Approaches to the Global Intimate." *Women's Studies Quarterly* 34(1–2): 446–463.

Nelson, Lise, and Nancy Hiemstra. 2008. "Latino Immigrants and the Renegotiation of Place and Belonging in Small Town America." *Social and Cultural Geography* 9(3): 319–342.

Olivos, Edward M., and Gerardo F. Sandoval. 2015. "Latina/o Identities, the Racialization of Work, and the Global Reserve Army of Labor: Becoming Latino in Postville, Iowa." *Ethnicities* 15(2): 190–210.

Pachirat, Timothy. 2011. *Every Twelve Seconds: Industrialized Slaughter and the Politics of Sight*. New Haven, CT: Yale University Press.

Pain, Rachel. 2015. "Intimate War." *Political Geography* 44(Supplement C): 64–73.

Pettigrew, Thomas F., and Linda R. Tropp. 2006. "A Meta-analytic Test of Intergroup Contact Theory." *Journal of Personality and Social Psychology* 90(5): 751–783.

Pitt, David. 2015. "Polluted Iowa Waterways Rise 15 Percent in 2 Years." *Des Moines Register*, May 14, 2015.

Pratt, Geraldine, and Victoria Rosner, eds. 2012. *The Global and the Intimate: Feminism in Our Time*. New York: Columbia University Press.

Price, Patricia L. 2012. "Race and Ethnicity: Latino/a Immigrants and Emerging Geographies of Race and Place in the USA." *Progress in Human Geography* 36(6): 800–809.

Pulido, Laura. 2017. "Geographies of Race and Ethnicity II: Environmental Racism, Racial Capitalism and State-Sanctioned Violence." *Progress in Human Geography* 41(4): 524–533.

Santa Ana, Otto. 2016. "The Cowboy and the Goddess: Television News Mythmaking about Immigrants." *Discourse and Society* 27(1): 95–117.

Silva, Kumarini. 2016. *Brown Threat: Identification in the Security State*. Minneapolis: University of Minnesota Press.

Skipper, John. 2016. "Unbeknownst to Most, Alex Kuhn Battled Depression; the Prestage Debate in Mason City Didn't Help." *Mason City Globe Gazette*, December 18, 2016.

Smith, Sara. 2012. "Intimate Geopolitics: Religion, Marriage, and Reproductive Bodies in Leh, Ladakh." *Annals of the Association of American Geographers* 102(6): 1511–1528.

Smith, Sara, Nathan W. Swanson, and Banu Gökarıksel. 2016. "Territory, Bodies and Borders." *Area* 48(3): 258–261.

Staeheli, L. A., P. Ehrkamp, H. Leitner, and C. R. Nagel. 2012. "Dreaming the Ordinary: Daily Life and the Complex Geographies of Citizenship." *Progress in Human Geography* 36 (5): 628–644.

Vega, Sujey. 2012. "The Politics of Everyday Life: Mexican Hoosiers and Ethnic Belonging at the Crossroads of America." *City and Society* 24(2): 196–217.

Winant, Howard. 1994. *Racial Conditions: Politics, Theory, Comparisons*. Minneapolis: University of Minnesota Press.

3 Contested Ethnic Foodscapes: Survival, Appropriation, and Resistance in Gentrifying Immigrant Neighborhoods

Pascale Joassart-Marcelli and Fernando J. Bosco

Introduction

In many urban neighborhoods, "ethnic markets" are the primary—if not the only—source of food. Most are small stores operated by owners who belong to groups socially constructed as "ethnic"—a fluid and situational category related to immigration, race, and class and involving both self-identification and classification by others. These enterprises not only make food accessible to local residents and thereby reduce food insecurity but also create economic opportunities for immigrants, contribute to a sense of place and community, and help revitalize neighborhoods (Joassart-Marcelli, Rossiter, and Bosco 2017).

Nevertheless, ethnic markets are often vilified as selling overpriced, unhealthy, and low-quality food—a perception that contributes to the stigmatization of ethnic neighborhoods and the devaluation of immigrant food practices (Joassart-Marcelli, Rossiter, and Bosco 2017). Paradoxically, urban ethnic markets and eateries have also recently become a terrain for foodies to distinguish themselves by their cosmopolitan, democratic, and adventurous attitudes (Johnston and Baumann 2010; McClintock, Novie, and Gebhardt 2017).

As we began writing this chapter, these contradictions were laid bare by intense conflicts surrounding a proposed *fruteria* (a fruit and juice shop) in the Barrio Logan neighborhood of San Diego—one of our study sites. When a young white woman—known on social media as the "barefoot bohemian"—proclaimed her intention to "create an urban sanctuary" and "bring healthy options to the barrio," she stirred up a storm in the community, both online and offline, with people accusing her of appropriating

Mexican culture, contributing to gentrification, and behaving as a "white savior" (Zaragoza 2017).

These contradictions reflect historical ambivalence regarding the meaning of ethnic food and the place of immigrants in cities. They also reveal new developments in the political, economic, and cultural geographies of cities. Global migration, neoliberal urban politics, and economic restructuring, including rising inequality and the expansion of cultural economies based on consumption and lifestyles, have deeply transformed contemporary cities and encouraged new forms of gentrification (Amin and Thrift 2007; Theodore, Peck, and Brenner 2011). Today, food and lifestyles in general have become means to brand places and generate economic value but also to differentiate and exclude (Joassart-Marcelli and Bosco 2018a).

In this chapter, we examine the place-based tensions surrounding food and ethnicity in the context of gentrification. We define gentrification as the transformation of low-income urban areas into upper-middle-class residential and commercial use that is accompanied by the displacement of residents (Lees, Slater, and Wyly 2008). In particular, we draw attention to the contested meanings of ethnic food and taste in urban neighborhoods that have been historically associated with immigrant communities and are now being transformed into trendy consumption sites by an influx of capital and new residents. We explore the contradictions between racialized descriptions of immigrant neighborhoods as "food swamps" and "food deserts" on the one hand and the simultaneous praise of their authenticity, diversity, and sense of place on the other hand. We ask, who shapes the discursive and material production of ethnic food, and to what ends?

We begin with a critical overview of the concept of an ethnic foodscape, using a historical example from New York's Lower East Side. We then turn our attention to City Heights and Barrio Logan, two San Diego neighborhoods characterized by large and diverse immigrant populations, active ethnic food economies, and varying levels of gentrification pressure. Our goal is to emphasize the fluid, relational, and contested nature of these two ethnic foodscapes.

Ethnic Foodscapes as Fluid and Contested Spaces

The concept of a foodscape is useful to contextualize understandings of food in particular places and draw attention to the social, economic, political, and cultural factors shaping its significance (Joassart-Marcelli and Bosco

2018b). It highlights the material and discursive environments in which food acquires meaning and emphasizes the importance of perspective—the angle from which landscapes are observed, including the unique lens of ethnicity. According to Joassart-Marcelli, Salim, and Vu (2018), ethnic foodscapes consist of physical places (e.g., ethnic enclaves, home kitchens, restaurants, markets, gardens), people (e.g., immigrant cooks, street vendors, shopowners, tourists, experts), objects (e.g., ingredients, spices, seeds, cookware), discourses (e.g., understanding of health, domesticity, belonging, ethnicity), and sensual elements (e.g., tastes, smells, sounds, memories) that are associated with the foodways of ethnic groups in particular places.

Historically, restaurants, street carts, and food stores have been central elements of ethnic foodscapes, providing economic opportunities to the first and subsequent generations of immigrants, helping feed families, and structuring social interactions within neighborhoods (Gabaccia 1998). New York's Lower East Side provides a poignant example of how food, ethnicity, and place are coproduced. At the turn of the twentieth century, hundreds of thousands of Eastern European and Russian Jews moved into its cramped tenements, replacing previous generations of German immigrants. Pushcart vendors, street markets, coffeehouses, kosher bakeries, delis, butcher shops, and restaurants emerged throughout the neighborhood (Lobel 2015). They provided residents with a source of income, contributed to social reproduction, strengthened the cultural fabric of the neighborhood, and shaped the rhythm of the streets. They also attracted the attention of outsiders, including cultural elites drawn by the bohemian atmosphere (Beck 2014). Since then, most Jewish families have relocated elsewhere, making room for new waves of immigrants (see chapter 4 of this volume for Maryam Khojasteh's analysis of this dynamic in the context of Philadelphia). Over time, Latino bodegas, Middle Eastern stores, and Chinese restaurants replaced previous businesses, yet Jewish food continues to frame the Lower East Side's sense of place. However, it now does so in a nostalgic and commoditized form that reflects gentrification trends and the tastes of more affluent consumers attracted by the historic character of the neighborhood (Beck 2014). Remaining Jewish restaurants such as Katz's Delicatessen, Russ & Daughters, and The Original Yonah Schimmel Knishery have been praised by food critics and have reached cult status, attracting locals and tourists in search of an "authentic" New York experience.

The different meanings associated with ethnic food reflect a fundamental tension between what could be described as the *use value* of ethnic foodscapes (their ability to fulfill a physical and social need for residents of ethnic enclaves) and their *symbolic value* (their contribution to social status, distinction, and identity). In contrast to use value, which cannot be easily monetized, symbolic value has become a major source of profit in today's cultural economy (Baudrillard 2000). Along those lines, Bourdieu (1984) argued that the symbolic value of food—or taste—is a socially produced means of class distinction. In rapidly changing neighborhoods, food—which is both biologically necessary and deeply symbolic—has become a tangible medium for newcomers and earlier residents, as well as immigrants and natives, to relate to each other. While ethnic food may promote encounters and forge connections between different groups, it also tends to exacerbate differences, reflecting broader sociospatial processes associated with class, race, ethnicity, and different understandings of "good food."

This chapter seeks to unpack these tensions by investigating current developments surrounding ethnic food businesses in San Diego. Our analysis rests on fieldwork conducted during the past five years in City Heights and Barrio Logan. In 2015, we conducted audits of all food stores in City Heights ($n=82$). In summer 2016, we followed up with interviews of owners ($n=24$) and customers ($n=67$) of ethnic businesses in that neighborhood, working with interviewers speaking English, Spanish, and Vietnamese. While we do not have the same extensive primary data for Barrio Logan, we have an intimate knowledge of the neighborhood, as well as observational, media, and census data, which warrant using it as a comparative study site.

Ethnic Foodscapes of City Heights and Barrio Logan

Despite their unique histories, City Heights and Barrio Logan are representative of many so-called inner-ring urban areas in US cities. They first emerged as middle-class residential areas on the outskirts of the central business district in the late 1800s and prospered until discriminatory planning decisions and political-economic forces changed their fate. Their foodscapes therefore must be understood within these shifting political and economic contexts.

In Barrio Logan, economic decline occurred in the early twentieth century, when the railroad brought factories and warehouses to the area. As lumberyards, canning facilities, and shipyards were added to the landscape, partly because of zoning regulations favorable to such industries, wealthier residents left their Victorian homes and relocated to more desirable areas (Rosen and Fisher 2001). Most of those who remained were Latinos employed in local industries and restricted from owning or renting a house in other areas by racially biased housing covenants (Guevarra 2012). Immigrants from Mexico settled in the neighborhood in large numbers throughout the first half of the century. By 1950, Barrio Logan was one of the largest Mexican American communities in California (Delgado 1998). In the 1960s, the construction of Interstate 5 and the Coronado Bridge galvanized the community in the Chicano movement (Rosen and Fisher 2001) but also contributed to the loss and deterioration of housing, the exit of more affluent residents, the worsening of environmental conditions, and the economic decline of the neighborhood—trends that continued until very recently (Le Texier 2007).

Until the 1960s, City Heights was a typical white middle-class suburb, with an active commercial main street and a majority of single-family houses. However, rapid suburbanization subsidized by housing and transportation policies spurred the relocation of the most affluent residents and businesses, leaving behind lower-income households (Wolch, Pastor, and Dreier 2004). Here, too, decisions by city planners to rezone the area for multifamily residences, and state and federal highway construction, transformed the community (Ford and Griffin 1979), heralding a long era of decline and neglect (Bliesner and Bussell 2013).

Today, both neighborhoods are home to many immigrants and racial minority groups who have taken up residence in the multiunit buildings that replaced single-family homes. Barrio Logan maintains a distinct Chicano and Mexican identity. Although it also has a large Latino population, City Heights is much more ethnically diverse: in the past 30 years, it gradually became one of the largest refugee resettlement areas in California, welcoming successive waves of refugees from Vietnam, Cambodia, Russia, Ukraine, Ethiopia, Somalia, Sudan, Iraq, and, more recently, Burma and Syria. Despite differences in their ethnic compositions, both neighborhoods have very small proportions of non-Latino white residents—9.6% in Barrio Logan and 11.7% in City Heights (US Census Bureau 2016).

Figure 3.1 shows the location of City Heights and Barrio Logan and for each neighborhood highlights the high proportion of foreign-born residents relative to the rest of the region in 2016 (US Census Bureau 2016). Data from the US census in 1980, 1990, 2000, and 2010 consistently reveal significantly

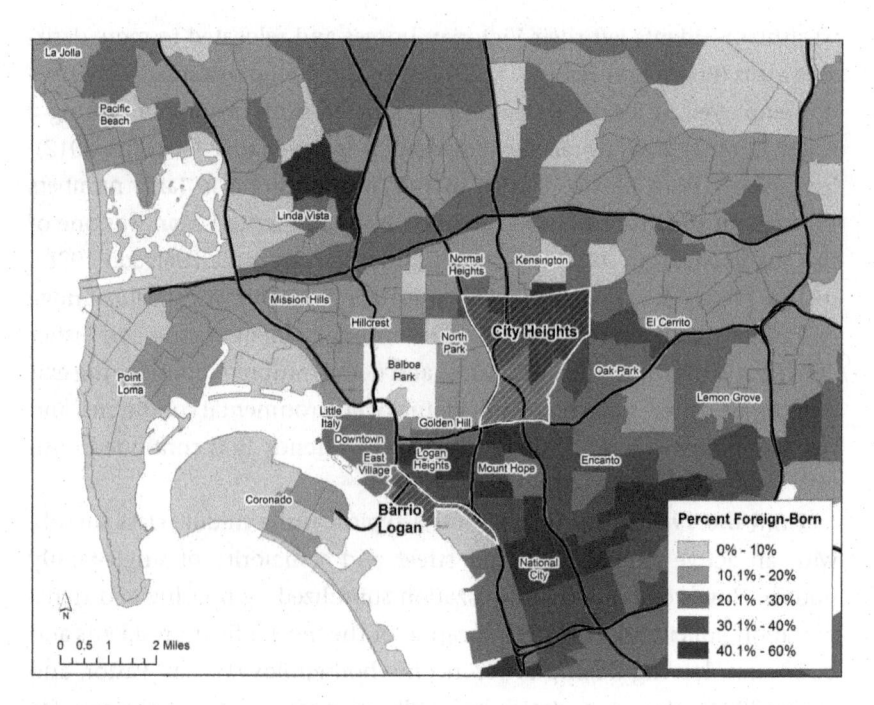

Figure 3.1
Proportion of foreign-born residents by census tract, central San Diego, American Community Survey 2012–2016.

higher proportions of immigrants and nonwhite residents, higher poverty rates, larger household sizes, and lower home ownership rates in these two neighborhoods (and the surrounding areas) than in San Diego as a whole.

By the 1970s, the neighborhoods south and east of downtown, including our two study areas, had come to be viewed by the general population as dangerous and dysfunctional places and by entrepreneurs and financial institutions as too risky for investment, reflecting earlier patterns of redlining (Ford and Griffin 1979). Meanwhile, the city of San Diego began concentrating its redevelopment efforts in the downtown area, banking on tourism and real estate to stimulate growth and ignoring growing problems in surrounding communities (Chapin 2002).

The foodscapes of City Heights and Barrio Logan reflect these unique histories of marginalization and neglect. Both have recently been described by residents, local media, and community-based organizations alike as food deserts, despite the presence of numerous small grocery stores and restaurants (Joassart-Marcelli, Rossiter, and Bosco 2017), reflecting inherent biases in the food desert concept (Shannon 2014; Widener 2018). It is true that, by 1980, very few food retailers remained. The city of San Diego's historic business license records indicate that there were only three licensed food stores in City Heights at that time, including a grocery store, a convenience market, and a gas station, and three restaurants, serving pizza, fast food, and Mexican fare (City of San Diego 2017). In Barrio Logan, there were six licensed food stores and three restaurants. In the subsequent decades, the majority of food businesses that opened were so-called ethnic businesses. By 2015, City Heights was home to 82 food stores and 74 restaurants, most of which (38 stores and 57 restaurants) had a visible ethnic association, including many Mexican, Vietnamese, Chinese, and East African businesses. Barrio Logan's foodscape has historically been dominated by Mexican businesses, which by 2016 comprised 8 of the 10 small convenience or specialty stores and 20 of the 38 restaurants located in the area. Still, neither place had a supermarket.

Very recently, community-led efforts and tax incentives have succeeded in attracting a new supermarket in each neighborhood. Both opened in 2016 and brand themselves as Mexican markets, which pleases a majority of residents but also frustrates others, including people who wished for stores that catered to other ethnic groups and newcomers who would have preferred trendier and presumably healthier markets. Several food stores, cafés, and restaurants also have opened within the past few years, offering "upscale" or "authentic" ethnic food "with a twist."

At the same time, community gardens and farmers markets have become part of organized local efforts to increase food security, especially in City Heights, where refugee organizations have embraced urban agriculture as a mechanism for integration and resettlement (Bosco and Joassart-Marcelli 2017). These recent changes in the food environment are more than a mere reflection of demographic trends; they are leading the way in the transformation of both neighborhoods and, as a result, have become sites of tension.

The Use Value of Ethnic Foodscapes in the Everyday Lives of Immigrants

Although the food environments of City Heights and Barrio Logan are often depicted in similarly negative terms, they play important roles in the everyday life of these communities. In Barrio Logan, *fruterias*, *tortillerias*, *carnicerias*, and other Mexican businesses have dominated the landscape, while food establishments in City Heights replicate the ethnic diversity of its residents, with Asian, East African, and Latino businesses operating side by side. Although there are concerns, particularly from public health experts, regarding the affordability, healthfulness, and quality of food, these small establishments have been instrumental in providing local residents with culturally appropriate food—often at prices competitive with those of supermarkets in surrounding areas (Joassart-Marcelli, Rossiter, and Bosco 2017). In addition, these businesses are often important in creating a sense of place, promoting social interaction, and fostering community within the microgeographies of city blocks.

Figure 3.2 illustrates the location of food stores and restaurants in each neighborhood, distinguishing between stores without any specific ethnic association and those visibly associated with very broad ethnic categories. The maps suggest the presence of active ethnic food economies and challenge traditional depictions of "food deserts" as devoid of accessible food, to the extent that most residents have (at the minimum) geographic access to a variety of food retailers (see also Joassart-Marcelli, Rossiter, and Bosco 2017). Our interviews with customers of ethnic food stores in City Heights show that, although some residents shop outside their neighborhoods, the majority patronize local businesses on a regular basis—3.5 times per week on average.

For immigrants, and new refugees in particular, the relationship between shopkeeper and customer goes well beyond economic transactions. We witnessed numerous informal and friendly conversations taking place in the stores, suggesting that people knew and cared about each other. Natalia, who owns a Mexican store in City Heights, states, "[My business] is a community.... I know a lot of people here because I've been here fourteen years, so they know me and I know them, but yeah it's like family, you know?" Numerous customers also referred to the store in which they had just made a purchase and the surrounding neighborhood as a "community." Some customers also noted the lack of stigma associated with using electronic benefit transfers (EBT) in these ethnic markets, and owners confirmed that participating in the EBT program was essential to their business.

Despite the significance of ethnic businesses in the daily lives of residents, our research indicates that many of these stores struggle to stay open and be profitable, leading to a high rate of turnover and widespread

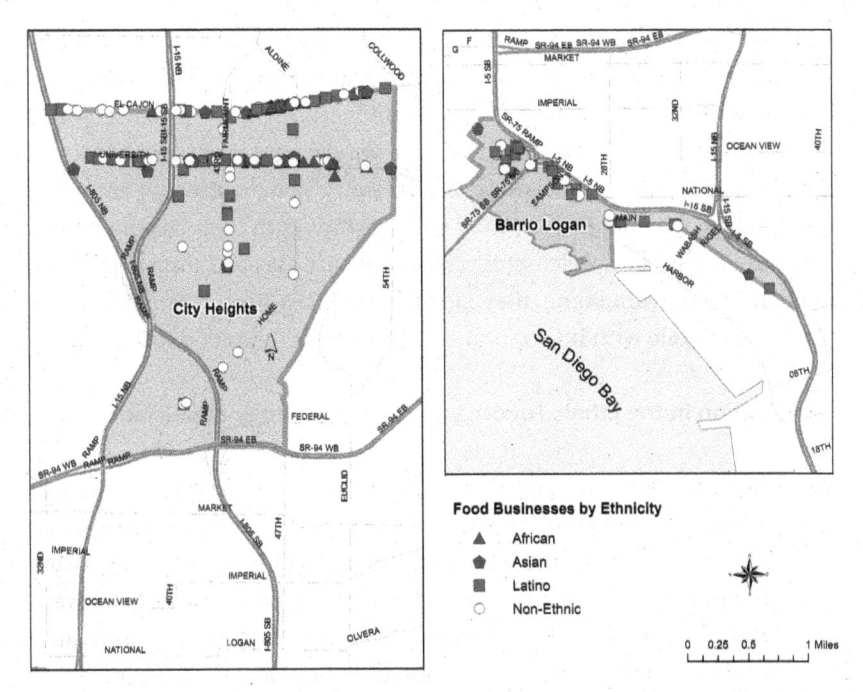

Figure 3.2
The ethnic foodscapes of Barrio Logan and City Heights. A variety of food businesses can be found in both neighborhoods, representing many of the dominant ethnic groups and providing geographic access to food for most residents.

self-exploitation. Indeed, numerous owners, mostly foreign-born men, told us that they were working 70 to 80 hours per week, often with help from a spouse, child, or sibling. Many do not count their hours and do not pay themselves a set income. In very small businesses, personal and business finances are often mixed without clear accounting; owners take money from the cash register to buy dinner or put gas in their car and rely on their own savings to maintain cash flow. Similarly, many owners told us that they never borrowed money from a bank or received any type of financial or technical assistance with their business but instead used their own savings and family contributions to support their professional activities. The family's economic and social integration is so deeply intertwined with the success of the business and its role in the community that the lines between the three spaces of family, business, and community are often blurred; the hard work is often justified by future economic gains as well as the positive role that the store plays in the neighborhood. For instance, the owner of a Mexican store told us how much he appreciates it when "[his customers] say thank you for being here, thank you for having good produce. And I love that, when they say that...because we do work long and hard, put in so many hours here." These interactions sustain the livelihood of shopowners, who depend on good relationships with their customer base, including providing a welcoming, culturally sensitive, and supportive environment. At the same time, they facilitate the daily lives of families, many of whom struggle with integration, discrimination, and food insecurity.

Contestation in the Ethnic Foodscape: Appropriation and Resistance

Recent changes in Barrio Logan and City Heights have led to tensions in the communities and contestation in the foodscapes. The return of white and affluent residents to these neighborhoods is putting upward pressure on rents and property values, contributing to the displacement of older residents (Delgado 2017). At the same time, gentrification is putting into question the place of ethnic markets and the taste of older residents, including immigrants who fashioned the foodscape in past decades. While newcomers are attracted to the diversity, simplicity, and authenticity of older food establishments, they also seek novelty, exoticism, and social distinction—a trend associated with cosmopolitanism (Johnston and Baumann 2010). This results in the transformation of food spaces to cater to the tastes of new residents, often by "improving," "revamping," and "glamorizing" ethnic foods.

The photographs in figure 3.3 illustrate how these tensions play out in the urban landscape, with new food establishments eclipsing older ones. Immigrants and older residents are typically excluded in this "revitalization" process, since they do not own the new businesses, do not have a say in how the food they serve is prepared, usually cannot afford it, and feel culturally and emotionally disconnected.

Although the owners of older stores and restaurants we interviewed saw economic opportunity in gentrification, many feared competition from new businesses and questioned their ability to meet new demands. In particular, owners became defensive when asked questions about healthy food—a heavily classed and raced concept (Hayes-Conroy and Hayes-Conroy 2018; Guthman 2008) associated with gentrifiers and outsiders. Many pointed out that the new residents wanted more variety of local and organic produce,

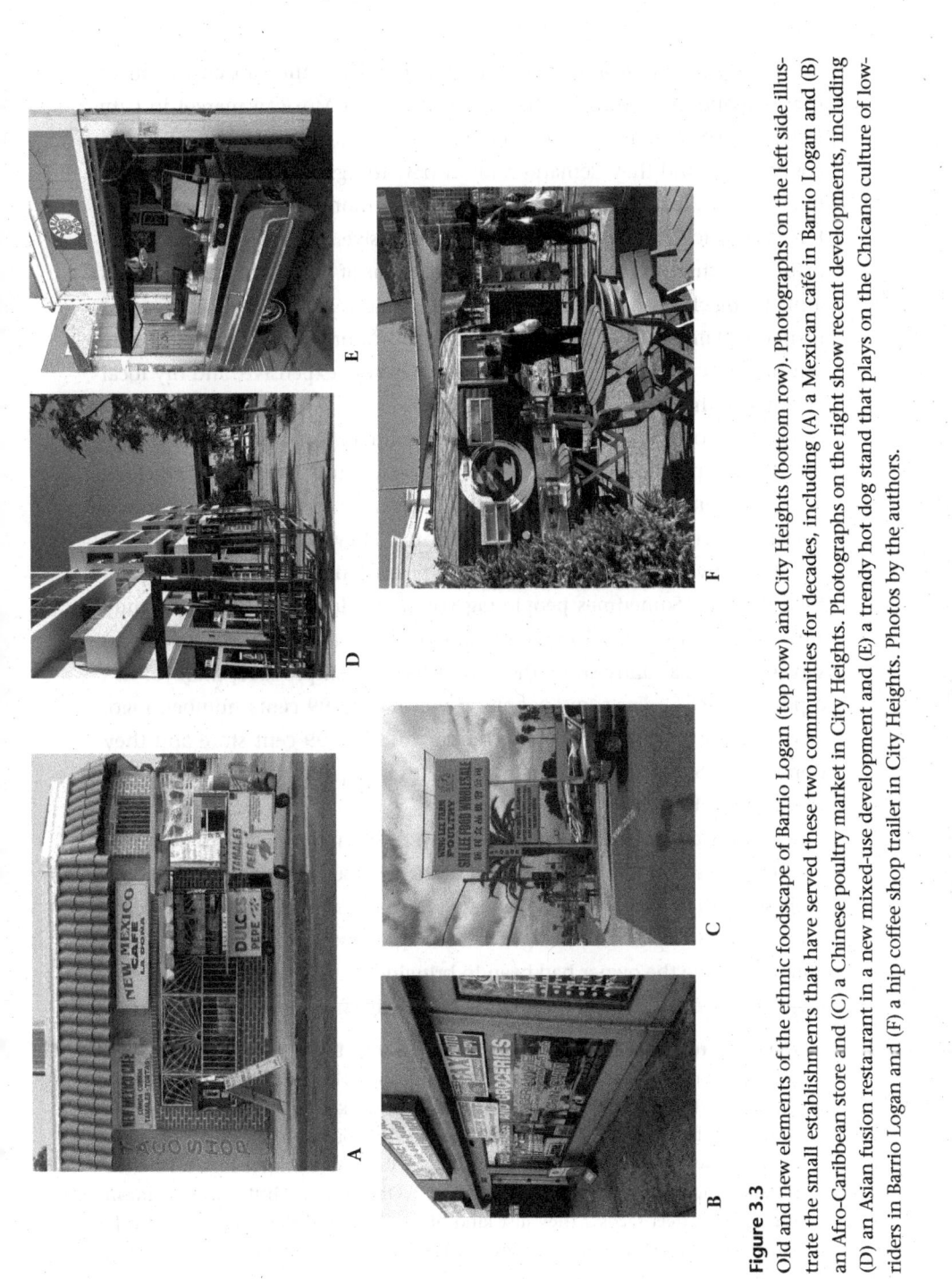

Contested Ethnic Foodscapes

Figure 3.3
Old and new elements of the ethnic foodscape of Barrio Logan (top row) and City Heights (bottom row). Photographs on the left side illustrate the small establishments that have served these two communities for decades, including (A) a Mexican café in Barrio Logan and (B) an Afro-Caribbean store and (C) a Chinese poultry market in City Heights. Photographs on the right show recent developments, including (D) an Asian fusion restaurant in a new mixed-use development and (E) a trendy hot dog stand that plays on the Chicano culture of low-riders in Barrio Logan and (F) a hip coffee shop trailer in City Heights. Photos by the authors.

but the owners indicated that they did not have the capacity to meet that demand. According to the owner of a small Mexican market in City Heights, "The neighborhood is changing a lot. There is a lot of new people coming in...and they demand a lot of new things. I try to provide them, but it's really hard because I don't have the money to purchase them. I have the space, but having inventory is expensive....I would like to open a café, but...there's going to be expenses in terms of putting in new windows and buying chairs....I just don't have the money." Another ethnic business owner explained, "Well now everybody asks for organic....I have to figure out how I'm going to order it, the quality is very expensive, and my local customers, they don't want to spend that."

Others, however, were adamant that their traditional customers were in fact already purchasing healthy foods, such as fruits, vegetables, chicken, and other unprocessed ingredients, to cook at home—although perhaps packaged, served, and/or priced differently. They resented being labeled negatively by new residents who stereotype certain ethnic markets as "junk food stores": "Sometimes people tag you as a 'Mexican store' or a candy store or piñata store....I've seen people walk by and talk between themselves, 'oh, it's a candy store, they don't have [fresh produce] there' or 'it's a piñata place.'...For some people, if you have a 99 cents number, logo, anything on your front, people assume you are a 99 cent store and they don't come. Some of them might say no, it's low quality, which I am trying to stay away from that."

In contrast, owners of new businesses were often quick to claim their pioneering role in bringing healthy and authentic food to the neighborhood, suggesting that these types of foods were not available until their recent arrival. For example, the manager of a new grocery store told us how savvy and innovative the owner had been in bringing "quality" ethnic food to the neighborhood, often by appropriating knowledge from smaller producers:

> Nobody's really trying it [tortillas made fresh in the store]. And when [the owner] opened [the store], he didn't have tortillas. He wasn't making them. He wanted to do it, but he had no idea how....He just kept trying to talk to people to find out how you do it,...he went someplace in Southeast San Diego...bought all a [Mexican] guy's stuff, and then the guy came to work for him for a little bit and taught him how to use everything....Out front our little tacos...I mean it's Tijuana street tacos...they just kind of went out and started looking at it [a Tijuana taco shop], and then...they went and hired some of those people to start the thing for them.

As this quotation illustrates, there is an unspoken hierarchy between Mexican food workers, who might know how to cook, and entrepreneurs, whose valuable business skills can turn that knowledge into a large-scale profitable endeavor that attracts a much wider clientele. The Mexican-inspired décor and products of this particular business conceal the fact that the owners also own a number of other mostly nonethnic grocery stores in the region, including a successful chain focused on fresh, healthy, and organic food. This phenomenon of nonethnic business owners selling "authentic" ethnic food can be interpreted as a form of cultural appropriation (McClintock, Novie, and Gebhardt 2017). To the extent that authenticity is commoditized and valued by affluent outsiders, both as a way to earn profit and as a sign of taste and distinction, it is linked to gentrification (Zukin 2009). In fact, recent media attention to the foodscapes of Barrio Logan and City Heights has attracted a growing number of adventurous foodies eager to sample its authentic food. For example, the food guide *Zagat* (Horn 2016) named Barrio Logan as San Diego's "next hot food neighborhood," and *San Diego Magazine* (Ram 2015) described City Heights as "a central urban nabe lay[ing] claim to authentic international eats, along with live music venues, craft beer, coffee, and outdoor fun."

Residents of City Heights have only recently begun to resist cultural appropriation. In Barrio Logan, however, where there is a longer history of community activism as well as stronger gentrification pressures (Le Texier 2007), people have been more vocal against this trend. In fact, as we noted in our introduction, a recent event in Barrio Logan drew our attention because of how well it illustrated rising community tensions regarding the cultural appropriation of ethnic food and its relationship to gentrification. In October 2017, a white and seemingly privileged woman posted a video on her Kickstarter site to raise funds for La Gracia Modern Fruteria—a "plant-based cocina and vegan coffee bar" that would "help [us] improve San Diego and bring a healthy option to the Barrio." The video (available in Zaragoza 2017) was so rife with stereotypes that it almost looked like a parody; it included glamorous shots of the blonde entrepreneur strutting in front of historic Chicano murals, prepping colorful fruit bowls, exotic smoothies, and "wellness lattes" in her upscale kitchen, and vacationing at luxurious Mexican beach resorts—all with a Spanish guitar soundtrack. It received instantaneous and fierce backlash from the community, particularly on social media, where race and gentrification became central to the conversation. Deeply offended by what was described as "a blatant grab at capitalizing off of Mexican culture" and "yet another signifier of gentrification in Barrio Logan" (quoted in Fokos 2017), thousands of people reacted by posting remarks on various platforms. In the comment section of one of the first articles to report outrage on this issue (Zaragoza 2017), one person wrote:

> Nothing in the hood is broken white lady. We already have fruterias. Just cause the hood is cool, and you like to resort in Mexico doesn't mean you can go all gentrification on the people. Take that shit to Encinitas, Solana Beach or Del Mar. Theres plenty of white ladies just like you who would love your $15 fruit cups.... I dont want my rent to go up, i dont want neighbors complaining about our fiestas, i dont want a vons where my carniceria used to be. They dont sell the queso seco i like. Stop trying to turn our hoods into Kensington [a historic, quaint, affluent, and primarily white urban neighborhood of San Diego].[1]

Another commenter (quoted in Fokos 2017) urged residents, "Don't support...white owned businesses in Barrio....Don't be friendly to them. Don't give them your money. Shut them down by any means necessary." The *Defend Barrio Logan* Facebook page also posted several critical entries on La Gracia and other similar food ventures, which they clearly view as cultural appropriation and an integral aspect of gentrification—one they argue is worth resisting in order to protect the character of the neighborhood and maintain affordable housing for its residents. In the midst of this social media frenzy, activists spray-painted "no gracias" (no thanks) on the front window of the store that had been leased for La Gracia. As a result of this backlash, the official video and promotional materials were taken offline and the project was put on hold.

To be sure, some comments on Zaragoza's article (2017) were supportive of La Gracia, claiming that "[business] is good for everyone!" and "the free market economy will decide if she makes a success of it or not." Some also noted that the entrepreneur "seemed to be in the best of humanitarian intentions" and "showed appreciation towards our culture" and "interest in the community." To a handful of commenters, opposing her business was seen as a form of "reverse racism." These remarks underscore the complexity of pinpointing racism, or even discussing it, in a context where free market and color-blind ideologies dominate. The failure to acknowledge the racial power differentials that exist in the marketing of ethnic food and foodscapes allows gentrifiers to talk about appreciation for authenticity and support for the community without taking responsibility for contributing to the resulting displacement.

The customers we interviewed in City Heights also showed ambivalence toward the changes taking place in their neighborhood. For many customers, especially recent immigrants, ethnic markets have a social or emotional significance as spaces of social reproduction and economic livelihood. That meaning is lost on customers who come in and out to buy a can of soda, a bag of chips, or a pack of cigarettes, without speaking more than a few words or even looking at the cashier. We observed and interviewed a number of such shoppers, most of whom were not immigrants, had recently moved into the neighborhood, and were typically more affluent. They explained to us that they were only shopping at ethnic markets because it was convenient, but many reported not liking "the smell of the place," how "dirty" it was, "the lack of healthy options," and the "high prices." In some instances, these interviewees related these negative characteristics to race, implying that people of certain ethnic backgrounds had different tastes—presumably inferior to theirs, since they tolerated low-quality food. It was striking to us how polarized customers' perceptions were; what was a "friendly neighborhood place" to some was a "dump" to others.

This bifurcation was primarily related to race and class, with affluent and white residents more likely to criticize the foodscape and its ethnic businesses. People's relationship to their neighborhood's foodscape, however, was more complex than this simple dichotomy. Many of the newer residents claimed to have moved to City Heights or Barrio Logan because of its diverse and vibrant food culture. At the same time, some longtime residents seemed to resent the very ethnic markets on which they rely almost daily, for a variety of reasons associated with the lack of diversity, affordability, and quality of food. These complaints need to be understood in a context where the foodscapes of City Heights and Barrio Logan are gentrifying while simultaneously being scrutinized by health advocates and stigmatized as unhealthy and inferior through the enduring food desert metaphor. Despite the observed availability of fresh produce in many shops, shame transpired in both business owners' remarks regarding the simplicity of their businesses and customers' denigrating comments about the paucity of good food options in their neighborhoods. For instance, an East African store manager told us: "I am sorry, this is very small and a little messy....I need to paint and make repairs. The store does not look very good right now, but we East Africans are used to this. You probably would not shop in a place like this. It's not for everybody. I am trying to change it, but it's hard.... I know what I need to do to make it better.... I just don't have the money."

Some consumers echoed this idea that one would not want to shop in certain City Heights stores if they had other choices. It is highly possible that these statements were motivated by study participants' assumptions about our positionality, which they may have perceived as aligned with gentrifiers or health professionals. The various expressions of shame, embarrassment, or resentment need to be deciphered in light of power differentials between the different actors shaping ethnic foodscapes, including immigrants, low-income residents, and newcomers, as well as researchers and policymakers.

Conclusion: Resisting the Gentrification of Ethnic Foodscapes

The evidence we gathered through interviews, audits, surveys, public data, and media content indicate that two of San Diego's most iconic immigrant neighborhoods are undergoing a critical transformation in which food plays a central role. The quest for authenticity, exoticism, and adventure is bringing affluent and mostly white consumers back to these communities and altering their foodscapes, with significant consequences for low-income and minority residents.

Although this influx is assumed by some to have a positive effect on existing businesses and to signal the spread of multiculturalism, our research shows that it contributes to gentrification and displacement of older businesses and residents for whom the foodscape represents an integral part of their livelihoods and social lives. This reflects the tension between the use and symbolic values of ethnic foodscapes. Several urban scholars have highlighted these contradictions with regard to housing and the built environment of low-income urban neighborhoods, showing for example that an old building will have significantly more market value as a symbol of yesterday's urban village where new elite lifestyles can be sold than it would as affordable housing for multiple households (Zukin 2010). We argue that a similar dynamic applies to changing ethnic foodscapes where the symbolic value of ethnic food and its capacity to signify authenticity, cosmopolitanism, and social distinction for middle- and upper-class consumers is more valuable to investors and policymakers than its role in the social reproduction of immigrant and minority communities.

In a context where the symbolic value of ethnic food is increasingly relevant, the question of who controls the way food is represented, prepared, and marketed is central to understanding how foodscapes evolve and who ultimately benefits from them. Under current circumstances in the United States, immigrants who are perceived as "ethnic"—a racial code for nonwhite—are more likely to be poor, marginalized by immigration policy, excluded from mainstream finance, and omitted from urban planning decisions. As a result, their ability to capitalize on the cultural heritage and symbolic value of their own food and neighborhood is limited. Instead, outsiders with more capital and business experience, often acquired in other "up-and-coming" neighborhoods, have the capacity to take advantage of this opportunity by appropriating the food and omitting the histories, struggles, and cultures of immigrants. The erasure of these legacies is felt viscerally by many immigrants, for whom the food has a different meaning that is intimately tied to those very histories.

Although changes in immigration and urban dynamics have exacerbated inequality, urban communities around the country are finding ways to resist this new form of food-related gentrification (Anguelovski 2016). First, there is a growing awareness that gentrification is a multifaceted process that often begins with seemingly innocuous projects such as community gardens, cafés, craft breweries, farmers markets, healthy food stores, biking trails, or art galleries. Although these projects ultimately threaten them, local residents and community organizations have often encouraged and facilitated them because they were viewed as a resource for the neighborhood. Increasingly, however, local actors have become weary of initiatives and developments that usurp control from the community and may lead to displacement. Active antigentrification campaigns in Barrio Logan and other immigrant neighborhoods across US cities reveal a growing suspicion of cultural appropriation and a rejection of the commodification of ethnic food cultures. This is illustrated by the boycott of places like La Gracia Modern Fruteria, described in the previous section. This awareness has prompted entrepreneurs to work with the community in creating and supporting businesses that provide jobs for residents, meet their daily needs for healthy and affordable food, and create space for community gatherings.

A number of community organizations have allocated resources to help existing businesses revamp their storefronts, add refrigeration, expand their selection of whole foods, and build connections with local growers, instead of advocating for corporate retailers and supermarkets (Mari 2016). Some are working toward creating co-ops or other forms of community-owned businesses that would serve residents and reinvest profits locally, inspired by "For Us, By Us" approaches (McCutcheon 2011) and examples such as Mandela Marketplace in Oakland, California (Figueroa and Alkon 2017), and the Ujamaa Food Co-op in Detroit, Michigan (White 2011). However, these initiatives require economic resources as well as collective awareness of and mobilization against gentrification. Although City Heights benefits from nonprofit and philanthropic resources, antigentrification efforts are stronger in Barrio Logan, where there is a longer history of local activism against environmental racism, poor planning decisions, and municipal neglect (Le Texier 2007).

At the same time, it is also important to recognize that food is only one element—albeit an increasingly important one—in the gentrification process, which is both cultural and economic. Indeed, for Smith (1996), gentrification is first and foremost a process of capital accumulation in which culture greases the wheels. Without the influx of capital and the expectation of high returns, it is unlikely that a produce market or ethnic restaurant could transform an entire neighborhood. Accordingly, fighting food-related gentrification of ethnic neighborhoods requires a redistribution of economic resources that empowers residents to stay put through labor and housing policies (e.g., community land trusts, living wage initiatives). This is particularly relevant for immigrants, who often earn poverty wages, have low rates of home ownership and limited access to credit, and are therefore more vulnerable to rising rents and evictions—a concern exacerbated by the lack of well-paying jobs and affordable housing programs for this population. In addition, unlike the typical solutions to food deserts that seek to attract outside investors, programs that help immigrants invest in their foodscapes are instrumental in ensuring that these enterprises benefit the community and meet the needs of residents first. Without conscious efforts to support small ethnic businesses and empower immigrant communities, the ethnic foodscape is likely to become a commodified landscape of consumption that will benefit outside investors and consumers at the expense of current residents.

Note

1. To preserve the anonymity of people who posted comments online but did not explicitly agree to participate in this research project, we do not include citations for those comments except for the main article in response to which they were written.

References

Amin, Ash, and Nigel Thrift. 2007. "Cultural-Economy and Cities." *Progress in Human Geography* 31(2): 143–161.

Anguelovski, Isabelle. 2016. "Healthy Food Stores, Greenlining and Food Gentrification: Contesting New Forms of Privilege, Displacement and Locally Unwanted Land Uses in Racially Mixed Neighborhoods." *International Journal of Urban and Regional Research* 39(6): 1209–1230.

Baudrillard, Jean. 2000. "Beyond Use Value." In *The Consumer Society Reader*, edited by Martyn J. Lee, 19–30. Malden, MA: Blackwell.

Beck, Peter. 2014. "Tasting a Neighborhood: A Food History of Manhattan's Lower East Side. https://sophiecoeprize.files.wordpress.com/2014/07/beck-sophiecoe2014-tastinganeighborhood1.pdf.

Bliesner, James, and Mirle R. Bussell. 2013. *The Informal Economy in City Heights*. San Diego, CA: City Heights Community Development Corporation.

Bosco, Fernando, and Pascale Joassart-Marcelli. 2017. "Gardens in the City: Community, Politics and Place." In *Global Urban Agriculture: Convergence of Theory and Practice between North and South*, edited by Antoinette WinklerPrins, 50–65. Boston: CABI International.

Bourdieu, Pierre. 1984. *Distinction: A Social Critique of the Judgment of Taste*. Cambridge, MA: Harvard University Press.

Chapin, Tim. 2002. "Beyond the Entrepreneurial City: Municipal Capitalism in San Diego." *Journal of Urban Affairs* 24(5): 565–581.

City of San Diego. 2017. *Business License Records*. San Diego: Office of the City Treasurer.

Delgado, Emanuel. 2017. "Unintended Consequences of 'Gentefication' in Barrio Logan." MS thesis, Department of Geography, San Diego State University.

Delgado, Kevin. 1998. "A Turning Point: The Conception and Realization of Chicano Park." *Journal of San Diego History* 44(1). http://www.sandiegohistory.org/journal/1998/january/chicano-3/.

Figueroa, Meleiza, and Alison H. Alkon. 2017. "Cooperative Social Practices, Self-Determination, and the Struggle for Food Justice in Oakland and Chicago." In *The*

New Food Activism: Opposition, Cooperation, and Collective Action, edited by Alison H. Alkon and Julie Guthman, 206–231. Berkeley: University of California Press.

Fokos, Barbarella. 2017. "E-lynched in Barrio Logan." *San Diego Reader*, November 15, 2017. https://www.sandiegoreader.com/news/2017/nov/15/city-lights-e-lynched-barrio-logan/.

Ford, Larry, and Ernst Griffin. 1979. "The Ghettoization of Paradise." *Geographical Review* 69 (2): 140–158.

Gabaccia, Donna R. 1998. *We Are What We Eat: Ethnic Food and the Making of Americans*. Cambridge, MA: Harvard University Press.

Guevarra, Rudy P., Jr. 2012. *Becoming Mexipino: Multiethnic Identities and Communities in San Diego*. New Brunswick, NJ: Rutgers University Press.

Guthman, Julie. 2008. "'If They Only Knew': Color Blindness and Universalism in California Alternative Food Institutions." *Professional Geographer* 60(3): 387–397.

Hayes-Conroy, Jessica and Allison Hayes-Conroy. 2018. Critical Nutrition: Critical and Feminist Perspectives on Bodily Nourishment. In *Food and Place: A Critical Exploration*, edited by Pascale Joassart-Marcelli and Fernando J. Bosco, 129–146. Lanham, MD: Rowman and Littlefield.

Horn, Darlene. 2016. "7 Reasons Why Barrio Logan Is San Diego's Next Hot Food Neighborhood." *Zagat*, February 23, 2016. https://www.zagat.com/b/san-diego/7-reasons-why-barrio-logan-will-be-the-next-san-diego-hotspot.

Joassart-Marcelli, Pascale, and Fernando J. Bosco. 2018a. "Food and Gentrification: How Foodies Are Transforming Urban Neighborhoods." In *Food and Place: A Critical Exploration*, edited by Pascale Joassart-Marcelli and Fernando J. Bosco, 129–146. Lanham, MD: Rowman and Littlefield.

Joassart-Marcelli, Pascale, and Fernando J. Bosco. 2018b. "A Place Perspective on Food: Key Concepts and Theoretical Foundations." In *Food and Place: A Critical Exploration*, edited by Pascale Joassart-Marcelli and Fernando J. Bosco, 13–29. Lanham, MD: Rowman and Littlefield.

Joassart-Marcelli, Pascale, S. Jaime Rossiter, and Fernando J. Bosco. 2017. "Ethnic Markets and Community Food Security in an Urban "Food Desert." *Environment and Planning A* 49(7): 1642–1663.

Joassart-Marcelli, Pascale, Zia Salim, and Vienne Vu. 2018. "Food, Ethnicity, and Place: Producing Identity and Difference." In *Food and Place: A Critical Exploration*, edited by Pascale Joassart-Marcelli and Fernando J. Bosco, 211–233. Lanham, MD: Rowman and Littlefield.

Johnston, Josée, and Shyon Baumann. 2010. *Foodies: Democracy and Distinction in the Gourmet Foodscape*. New York: Routledge.

Lees, Loretta, Tom Slater, and Elvin Wyly. 2008. *Gentrification*. New York: Routledge.

Le Texier, Emmanuelle. 2007. "The Struggle against Gentrification in Barrio Logan." In *Chicano San Diego: Cultural Space and the Struggle for Justice*, edited by Richard Griswold del Castillo, 202–221. Tucson: University of Arizona Press.

Lobel, Cindy. 2015. "Lower East Side." In *Savoring Gotham: A Food Lover's Companion to New York City*, edited by Andrew F. Smith, 351–353. New York: Oxford University Press.

Mari, Elle. 2016. *Improving San Diego's Urban Food Landscape*. University of California San Diego Center for Community Health and Live Well San Diego Community Market Program. https://ucsdcommunityhealth.org/wp-content/uploads/2017/06/LWCMP-June-2017-1.pdf.

McClintock, Nathan, Alex Novie, and Matthew Gebhardt. 2017. "Is It Local…or Authentic and Exotic? Ethnic Food Carts and Gastropolitan Habitus on Portland's Eastside." In *From Loncheras to Lobsta Love: Food Trucks, Cultural Identity, and Social Justice*, edited by Julian Agyeman, Caitlin Matthews, and Hannah Sobel, 285–310. Cambridge, MA: MIT Press.

McCutcheon, Priscilla. 2011. "Community Food Security 'For Us, By Us.'" In *Cultivating Food Justice: Race, Class, and Sustainability*, edited by Alison H. Alkon and Julian Agyeman. Cambridge, MA: MIT Press.

Ram, Archana. 2015. "Neighborhood Guide: City Heights." *San Diego Magazine*, May. http://www.sandiegomagazine.com/San-Diego-Magazine/May-2015/Neighborhood-Guide-City-Heights/.

Rosen, Martin D., and James Fisher. 2001. "Chicano Park and the Chicano Park Murals: Barrio Logan, City of San Diego, California." *Public Historian* 23(4): 91–112.

Shannon, Jerry. 2014. "Food Deserts: Governing Obesity in the Neoliberal City." *Progress in Human Geography* 38(2): 248–266.

Theodore, Nik, Jamie Peck, and Neil Brenner. 2011. "Neoliberal Urbanism: Cities and the Rule of Markets." In *The New Blackwell Companion to the City*, edited by Gary Bridge and Sophie Watson, 15–25. Oxford: Wiley.

US Census Bureau. 2016. *2011–2015 American Community Survey 5-Year Estimates*. US Census Bureau, American Community Survey Office. https://factfinder.census.gov.

White, Monica M. 2011. "Environmental Reviews & Case Studies: D-town Farm: African American Resistance to Food Insecurity and the Transformation of Detroit." *Environmental Practice* 13(4): 406–417.

Widener, Michael J. 2018. "Spatial Access to Food: Retiring the Food Desert Metaphor." *Physiology and Behavior* 193:257–260.

Wolch, Jennifer, Manuel Pastor, and Peter Dreier. 2004. *Up against the Sprawl: Public Policy and the Making of Southern California.* Minneapolis: University of Minnesota Press.

Zaragoza, Alex. 2017. "A Chicano Community Is Outraged over a White Woman's Attempt to Open a Modern Fruteria." *Mitú,* October 27, 2017. https://wearemitu.com/things-that-matter/san-diegos-barrio-logan-community-is-fighting-against-a-modern-fruteria-and-the-gentrification-it-embodies/.

Zukin, Sharon. 2009. "Changing Landscapes of Power: Opulence and the Urge for Authenticity." *International Journal of Urban and Regional Research* 33(2): 543–553.

Zukin, Sharon. 2010. "Gentrification as Market and Place." In *The Gentrification Debates: A Reader,* edited by Japonica Brown-Sa, 37–44. New York: Routledge.

4 Immigrants as Transformers: The Case for Immigrant Food Enterprises and Community Revitalization

Maryam Khojasteh

Introduction

> In the 20th century, the Rust Belt housed industrial powerhouses like the U.S. steel, coal, and auto industries, but today it is entrepreneurial partnerships between immigrants and local communities that are fueling the region's economies.... The WE Global Network recognizes that if we're to remain competitive in the global economy, we must support and maximize the efforts of local initiatives that welcome, retain, integrate, and empower immigrant communities.
>
> —David Lubell, executive director of Welcoming America

David Lubell's words at the launch of "The Welcoming America Global Network"—a regional initiative that promotes the economic contribution of immigrants in the Midwest—represents the core of numerous government programs that have been established in recent years to attract immigrant communities and tap into their potential for economic development (McDaniel 2014). Some initiatives, such as "Welcoming Pittsburgh," aim to provide an immigrant-friendly environment for the newcomers, while others, such as Chicago's "New Americans Plan," focus on promoting immigrant entrepreneurship. Scholarly works, for the most part, echo these initiatives in regard to the impact of immigrants on communities by showing how newcomers have revitalized commercial corridors, occupied vacant housing, and stimulated real estate markets, especially in areas that have experienced disinvestment (Vigdor 2017; Schuch and Wang 2015).

While these studies and initiatives provide a promising case to support immigrant revitalization, the complex and sometimes contested mechanisms and processes through which immigrants reshape their communities are less understood. This is because the contribution of immigrants to community revitalizations is viewed largely through an economic lens, which leaves out the multifaceted ways that immigrants build their communities. The most visible, yet underappreciated, way that immigrants define the boundaries of their environment is by establishing businesses that offer cultural (food) products. The visibility of immigrant food businesses not only (re)shapes the physical environment, but also restructures social and racial processes embedded within the built environment. This chapter explores the contribution of immigrant food entrepreneurs to community development in two neighborhoods, one in Buffalo, New York, and the other in Philadelphia, Pennsylvania. In this chapter, community development encompasses a broader definition to include the ways that immigrants impact the health and well-being of residents by providing increased access to affordable, healthful, and cultural foods. The conditions under which these food enterprises operate (Buffalo) and the interaction between the newcomers and the established residents (Philadelphia) provide an opportunity to unpack some of the complexities that exist in the relationship between immigrants and their environments.

Public health and food-related scholarship has paid a great deal of attention to understanding how the predominant food environment of the host country impacts the health of newcomers. These studies largely focus on the impact of dietary acculturation on health outcomes of immigrants (Zhang et al. 2019). The health and living conditions of immigrants who work on farms, in processing plants, and in the food service industry have also been subjects of many inquiries that illuminate how labor regulations—or the lack thereof—impact some of the most marginalized, yet crucial, actors in the food system (Moyce and Schenker 2018; Mucci et al. 2019). While these studies are important in making visible the cruelty of the food industry, they often present immigrants in the US food system as victims of abuse and discrimination. This narrative focuses on how immigrants are shaped by their new physical environment and does less to argue how newcomers are active agents in reshaping the food systems of their new communities.

Similarly, literature on immigration's effects on receiving communities tends to view immigrants as those in need of assistance and not as agents of change. More often than not, immigrants are understood through a bifurcated cost-benefit framework of analysis (Brettell and Hollifield 2014). This narrative, which runs strong in political debates and public perception of immigrants, centers on the fiscal implication of immigrants and provides a static understanding of the ways immigrants could shape their new communities. In reality, the relationship between immigrants and their new environment is dynamic and interactive. Through pathways such as entrepreneurship, immigrants play an active role in (re)shaping their new environments. The decisions made by immigrant entrepreneurs on where to locate, what products and services to offer, and who to serve have definite impacts on the well-being of communities.

Some local governments celebrate immigrant entrepreneurs as heroic actors who build up communities (Waters 2018). However, they fail to incorporate immigrant businesses in the overall economic development strategies. As a result, the particular needs and challenges faced by immigrant entrepreneurs are off the radar of officials charged with the promotion of small business ownership, such as local chambers of commerce (Center for an Urban Future 2007). To most local governments and scholars, the appreciation for immigrant-run businesses ends with their contribution to generating revenue and increasing the local tax base, and is less concerned with the nature of services and products they provide. All the while, immigrant food entrepreneurs are shaping community food systems around their own needs, culture, and social relations.

The latest wave of immigrants from Asian and Latin American countries has had a significant role in transforming the US food system at both national and community scales. On the national level, these entrepreneurs have introduced new seeds and agricultural practices, formed alternative food supply chains, and shaped the pattern of food production in the United States (Imbruce 2015). On the community level, immigrant entrepreneurs impact the quality and quantity of foods available in a neighborhood by operating food-related enterprises (Emond, Madanat, and Ayala 2012). This is especially important since many immigrant-run businesses are located in places often underserved by the general market, where access to healthy and affordable food is scarce (Kim 2010).

Most public health scholarship that informs public policy and interventions undermines the value of small-scale food stores in healthy food provision, largely because of an overemphasis on supermarkets as the prime outlet for healthy foods (Zhang et al. 2016; Gordon et al. 2011). However, the mixed results of recent supermarket interventions to promote positive diet-related health outcomes have pushed public health scholars to consider the diverse ways that individuals procure food besides shopping at national chain supermarkets (Dubowitz et al. 2015; Cummins, Flint, and Matthews 2014). A handful of studies have examined the potential of immigrant-run food businesses to increase food access and contribute to food security (Emond, Madanat, and Ayala 2012; Short, Guthman, and Raskin 2007). These businesses, despite their small size, often carry a diverse range of products, from meat to fresh produce. A study of Latino grocery stores (*tiendas*) in San Diego, California, compared the availability, quality, and cost of fresh produce at *tiendas* with supermarkets and concluded there is no significant difference between the two food markets in terms of access to fresh produce. In fact, the cost of meeting the USDA's recommended weekly produce serving was three dollars lower in *tiendas* compared with supermarkets (Emond, Madanat, and Ayala 2012). Research shows that particular ethnic diets are beneficial to health, as they are largely based on plant-based dishes (Ooraikul, Sirichote, and Siripongvutikorn 2008). A study of food shopping venues in a diverse community supports this claim by demonstrating an association between shopping at ethnic markets and a lower body mass index among Guyanese participants, arguing that ethnic markets offer access to a diet that protects against obesity (Hosler, Michaels, and Buckenmeyer 2016).

Immigrant food businesses in urban environments increase access to culturally appropriate, affordable, and healthy foods, and rebuild communities while doing so. For example, these businesses provide access to jobs and income, and transform communities by turning once abandoned properties into places for business transactions and social interactions. Immigrant merchants play a critical role in community building by supporting ethnic-based organizations and using their power to influence planning and political decisions that may impact their communities. The success stories of immigrants revitalizing disinvested communities have been embraced and celebrated by local governments and planners (Center for an Urban Future 2007). However, this narrative of immigrant revitalization reflects an incomplete understanding of the dynamic, and sometimes contentious, relationships between immigrants, their new environments, and the established residents.

The remainder of this chapter sheds light on some of the existing challenges that immigrant food entrepreneurs face in creating healthier food environments (Buffalo) and contributing to community and economic revitalization in urban neighborhoods (Philadelphia). The Buffalo case study is a qualitative examination of two Middle Eastern grocers located in a "food desert" who manage to carry healthy and affordable foods in a low-resource environment. The promising examples from Buffalo argue for inclusion of ethnic groceries in healthy food initiatives and financing to leverage these existing community assets in creating healthier urban food environments. The Philadelphia case study uses historical literature, business inventories, field observations, and interviews with storeowners to demonstrate how recent Mexican entrepreneurs successfully revitalized a run-down public food market that used to be owned and managed by Italian immigrants.

My status as a university-affiliated immigrant from the Middle East both helped and hindered my access to these immigrant communities. In Buffalo, I had relatively easy access to Arab business owners. However, my identity did only so much in gaining their trust; the participants were hesitant to speak freely about their perception of the local government. In Philadelphia, my university-affiliated status was more important than my immigrant status, since to many (undocumented) workers it was not clear "who I am working for." I am grateful to my translator, who not only aided with conducting the interview but also helped to gain the trust of the Mexican community.

Buffalo, New York: The Role of Immigrant-Run Grocery Stores in Shaping Urban Food Environments

> Fruits and vegetables are good items for everyone. Everyone wants fresh produce, even American[s]....If they know I have fresh produce, everyone will come.
> –Ali,[1] a new Buffalonian grocer of Iraqi heritage

Realizing the unmet demand for fresh produce in his neighborhood, Ali established a small-scale grocery store in Buffalo, New York, that provides cultural products for customers from Iraq, Turkey, India, and Pakistan and brings access to fresh produce for a broader range of customers who may not have walkable access to healthy food retailers. Ali and his fellow immigrant food entrepreneurs who have settled in Buffalo in recent years are shaping the city's food environment one block at a time.

Buffalo, a Rust Belt city characterized by abandoned and vacant properties, outmigration, and departure of businesses, has seen a slow resurgence in recent years, with the population increasing by 5% from 2005 to 2009 (US Census Bureau 2009). These recent changes can be partially attributed to the arrival of new immigrants and refugees who have diversified their traditional destinations and now settle across the United States. The foreign-born population of Buffalo increased by 40% from 2005 to 2009, a growth rate eight times higher than that of the city overall. These newcomers have changed their neighborhoods in small and incremental ways, by starting businesses and offering newly available products to serve community needs. Despite their role in rebuilding communities, these actors are often invisible in the local government's economic development strategies, which often focus on physical improvement projects to attract large companies and businesses, such as the downtown and waterfront revitalization projects (Shibley, Hovey, and Teaman 2016).

Buffalo's recent economic growth has had little translation into actual development for low-income communities. Today, only a few supermarkets serve the city, and they are often out of the physical reach of low-income residents who don't have access to a private vehicle (Widener, Metcalf, and Bar-Yam 2011). Despite the lack of access to large-scale food outlets, immigrant neighborhoods are home to many small-scale grocery stores that offer fresh produce and culturally appropriate foods. The increasing population of immigrants and refugees begets a growth in demand for cultural products unmet by the general market. As a result, Buffalo has witnessed visible growth in immigrant-run eateries and food businesses throughout the city. As of 2013, the most recent year for which statistics are available, there were 56 ethnic food places, a significant enough number to draw the attention of news outlets and magazines about the growing diversity of Buffalo's food landscape (Kelly 2013).

Using a private business vendor, store audits, and consultation with a community partner, two full-service Middle Eastern grocery stores were chosen from the northern neighborhoods of Buffalo with a foreign-born population share similar to that of the city (7% vs. 9%). The nearest supermarket is located outside the neighborhood, in an adjacent suburb with limited pedestrian and public transportation access. While it is classified by the USDA Food Atlas as a "food desert," the neighborhood is anything but "deserted"; it is home to a vibrant commercial corridor of ethnic eateries and grocery stores.

The stores provide a wide variety of dairy products, dry goods, fresh and frozen meat, poultry, and seafood, beans, grains, herbs, spices, and fruits and vegetables. The diverse array of fresh produce sets these stores apart from other small-scale urban food stores. While the quantity of fresh produce may be less than at larger food markets, its availability is critical to the success of these businesses. Both storeowners showed an acute understanding of the market conditions and unmet demands of the residents for fresh produce. They shared the fact that fresh produce attracts customers to their stores, shaping about 10% of their total sale, and increases the overall sale, as those who come for the fresh produce often end up purchasing multiple items.

The variety of foods mimics the line of products available at supermarkets but with lesser-known brands and names. The store audits show that besides a few specialized food items, the main inventory is staple food products found in many cuisines. Thus, it is not surprising that both stores reported that nonethnic customers represent 5%–10% of their customer base. The immigrant food entrepreneurs, aware of a need to diversify their clientele and services, allocate parts of their stores to catering services. This strategy helps them survive as a small business with low profit margins and to advertise their name to a wider population. The catering divisions also provide face-to-face interaction among community members, which builds social connections and facilitates information sharing among the immigrant communities (Liu, Miller, and Wang 2014).

To evaluate how conventional metrics would assess these stores' provision of "healthy" food, I used a modified version of a previously validated tool known as the Nutritional Environmental Measure Survey (Glanz et al. 2005). Both stores scored relatively high (38 and 24) on a scale of 0 (no healthy foods) to 54 (no unhealthy food), mainly because of the availability of fresh produce and absence of tobacco and alcoholic beverages. However, the limits of this type of metric are significant. For example, the tool does not consider a store's capacity to enable customers to prepare a homemade meal. To Zahra, an Iraqi storeowner with 25 years of business experience, providing convenience for residents where they can meet most of their needs with a one-stop shopping trip is an intentional goal when she selects products for her store. She explains, "Because I am a cook and I am a mom, I know what people will need to prepare meals and I bring them…my customers even ask me to give them recipes."

To provide this breadth of ingredients, these businesses need to navigate multiple challenges to offer culturally appropriate fresh produce at their stores. Ali reported how the absence of ethnic wholesalers in this region impacts his business: "It confuses our customers. They need produce on [a] daily basis, but if they come and they don't find it and they don't know when we will get it again, they may go and never come back again." Since these stores are small, it is not profitable for distributors to provide them with direct delivery. The immigrant storeowners overcome this challenge by making multiple trips to the nearest city with an established network of ethnic wholesalers; both storeowners travel twice a month to Detroit, Michigan, to access suppliers. Making these trips comes with additional costs and challenges for the entrepreneurs, such as arranging truck rentals, incorporating long-distance and time-consuming procurement trips into their schedules, and delegating work responsibilities. Despite the personal efforts that the storeowners make to meet this demand, other factors, such as weather or truck availability, could affect their commitment to offer fresh produce on a fixed schedule.

The immigrant entrepreneurs also need to overcome the challenge of navigating the regulatory environment and establishing relationships with local government agencies. Ali noted that the challenge is to "stick to [the regulations]. The laws are changing every now and then so we have to keep up with changes." To Zahra, the bigger challenge is establishing a relationship with government agencies: "If you are new and nobody knows you, you are gonna get people to trust you… [running the business] used to be harder 'cause they didn't know what we are gonna sell. So we had to explain it to them. We get a lot of stuff in bulk, and we have to package it and label them; it is required by law. They send inspectors to check if we do." Local agencies, such as the County Health Department, may be unfamiliar with ethnic products and their handling, which could delay the licensing process, hence restraining the entrepreneurs' ability to sell and generate revenue in a timely manner.

While not-for-profit organizations, such as the Small Business Development Center, help immigrant entrepreneurs with the start-up process, a staff member from the Office of Licenses shared that local government does not have any formal program or strategy to attenuate their particular challenges. This is especially important since local governments have the capacity to offer financial support to foster healthy food retailing. For example, the FRESH program in New York City offers a combination of tax reductions, sales tax exemptions, and grants for infrastructure changes to operators who meet particular health guidelines. Importantly, the local government could play a central role in connecting local and regional growers with ethnic grocers and distributors/wholesalers. Such a strategy could promote the operation of small-scale grocery stores throughout the city and provide a viable commercial strategy to support the new refugee farmers and growers, currently supported by grant-based nonprofits, with limited capacity to scale up their food production. Finally, the local government could improve their understanding of the needs and challenges of immigrant entrepreneurs by establishing partnerships with ethnic community organizations (see chapter 11 of this volume for Ostenso, Dring, and Wittman's analysis of food policy councils' successes and failures toward this goal in Vancouver). Through such relationships, the local government could better connect entrepreneurs to existing resources, share information about federal and state funding opportunities, and notify entrepreneurs about changes in regulations and laws.

Philadelphia, Pennsylvania: Continuous Wave of Immigrant Food Entrepreneurs and Community Revitalization

South Philadelphia, a reemerging immigrant neighborhood, is a prime example of how interactions between the new wave of immigrants and the children of the earlier European immigrants have created a vibrant, yet contentious, urban food landscape. This study focuses on two food stores, a Mexican-owned fish market and an Italian American butchery, in the South 9th Street Food Market (hereafter the Market) that provide staple food products and represent enterprises owned by the new wave of immigrants (hereafter new immigrant entrepreneurs) and by the descendants of the earlier immigrants (hereafter old immigrant entrepreneurs).

Noticing the largely African American clientele and diverse array of seafood in a small fish market, one may not immediately assume that a Mexican family owns and runs this store. Jose—a Mexican immigrant who arrived in the United States in 1998—is no stranger to managing his own business. He had years of experience producing and selling garden products prior to his migration to the United States. Describing his experience entering the US food industry, he explains, "I had to shift my business because of need... the need to survive when you arrive to this country. In my first [work] opportunity here I was introduced to the fish market." That fish market is still owned by an Italian American entrepreneur located not too far away from Jose's own business today. Working at the fish market not only met Jose's "need to survive" but also gave him an opportunity to learn new skills and gain necessary market information to establish his own business 20 years later.

The relationship between the old Italian food entrepreneurs and the new immigrant entrepreneurs, largely from Mexico, is a dynamic that has shaped and rebuilt the Market. In the late nineteenth and early twentieth centuries, South Philadelphia was the second-largest Italian enclave on the East Coast, becoming a gateway neighborhood for European immigrants. Upon their arrival, Italian immigrants formed a food market along South 9th Street. The Market was a major employment center for the incoming Italians, often single men, who worked at the Market and lived in nearby boardinghouses. The prosperous years of the Market were short lived; it faced a persistent and gradual decline following World War II, with outmigration of Italian immigrants and their children to the surrounding suburbs. However, the post-1965 wave of new immigrants and refugees from Asian and Latin American countries provided a renewed source of population, labor, and entrepreneurial activity for the Market. There were few Italian American businesses left at the Market when the new immigrants started to open businesses. The contribution of the newcomers to the Market and the real estate in the area flipped the predominant narrative from one that was formally identified as "blighted" by the Philadelphia Redevelopment Agency to one centered on community-led revitalization.

These community changes, however, did not happen in a vacuum. The community-led development occurred in the midst of racial controversies. Jose, in fact, cited his experience of abuse and discrimination at work as a motivation for him to seek self-employment: "I was bad treated because of my look and I am thankful for that because it motivated me to do more things and move forward. I had to go to school to educate myself because of Geno my neighbor. He said that if in his store they speak English then I should learn it too. He was saying the truth and instead of being mad I did the opposite and learned the language with a positive attitude."

Exemplifying this racist sentiment, nearby business owner Joe Vento attracted national attention to the Market over the years with his anti-immigration rhetoric. A third-generation Italian American himself, who held his own grandparents as the frame of reference for "doing it the right way," Joe Vento took issue with the undocumented immigrants and their lack of "assimilation," threatening those who hired undocumented workers and supporting multiple anti-immigration legislative actions locally and nationally. Joe's tactics were particularly loud and aggressive, but he was not alone in his stance. Many Italian American merchants threatened by the influx of new immigrants used a variety of tactics (e.g., unified awnings showing the business establishment's years in operation) to remind visitors of the Italian identity of the market and ultimately of its ownership (Vitiello 2014). Geno, son and heir of Joe Vento, did not take the "Speak English or Press 2 for Deportation" sign off his (in)famous steak shop until 2016, against the dying wish of his deceased father.

While these narratives center on tensions over ethnic identity, a closer look into the reciprocal relationship between the old and new immigrant food entrepreneurs provides a detailed picture of the processes that enabled the Market's revitalization. Many Italian American property owners rent their vacant business spaces to the new immigrant entrepreneurs. The already established physical commercial infrastructure—such as storefronts and warehouses—helps new immigrants acquire spaces for commercial purposes. In return, new businesses have revitalized this once run-down commercial corridor, stimulating growth and bringing vitality back to the Market. This vitality is a result of the diverse products now found at the Market. Once entirely a food market, the Market now provides access to other essential products and services, such as clothing, sporting goods, and entertainment, that serve the growing immigrant community. The combination of these businesses with the Italian food vendors—which often offer specialty food items such as cheese, olive oils, and handmade pastas—attracts both regular and loyal customers as well as clientele seeking a "cosmopolitan" experience.

The increased diversity of customers has worked in favor of the Italian American entrepreneurs. Francesco, a fourth-generation Italian American butcher, takes pride in serving a racially and economically diverse clientele: "We have a diverse group...we have a large black following...a lot of welfare...we have a large yuppie following now. We have whites, blacks and not as many Asians as we used to have, they now shop at their own supermarket around the corner. We have a mixed Jamaican clientele...they go for their own traditional stuff. You go from a person on welfare to an executive of a company and everybody gets along here. The store on Saturdays is a big melting pot." Francesco has responded to this increased diversity by expanding his business to tap into the growing demand of his customers. Francesco expanded his great-grandfather's small butchery into a large store with two new divisions (catering and poultry) that are managed by his wife and son. Similarly, many Italian businesses have benefited from the increased diversity, as it has forced them to keep expanding their businesses to stay competitive in the market. For example, a small Italian cheese store across the street that opened in 1939 has now evolved into a multidepartment company with its own importing division supporting more than 300 employees.

Instead of following most Mexican food vendors in providing cultural products, Jose chose the fish market based on his personal experience and the profitability of this sector. He explains, "People ask me do you know what color the business is?...I said business is neither black nor white. While the black people say it is black, the white people will say it's white, [I say] it is green, whoever has the money can buy the product." He points to the opportunity he found in the fish market by adding that "African Americans consume a lot of fish. As Mexicans, because of our geographic zone [at home] we don't have it a lot on our daily menu...but African Americans, Africans and Jamaicans have it. I have some products that come from Jamaica. If they ask for a product I will find it for them."

Successful new Mexican food entrepreneurs have managed to open additional stores in the surrounding neighborhoods. Upward mobility centered around individuals' successes has a special appeal in the American narrative of immigration, but the benefit of these success stories goes beyond an individual or a household to influence the community at large. For example, Jose has already helped two of his relatives establish their own fresh produce businesses by providing them with market information, financial support, and access to facilities such as refrigerators. Jose also contributes by shortening and facilitating the self-employment pathway of other community members and by creating additional businesses at the Market, which function as a continuous source of income and taxes. Italian American entrepreneurs have a similar trajectory in sustaining and expanding their businesses, but with a distinct difference: instead of helping their families and relatives to have their own businesses and spread horizontally across the market, they train, employ, and retain them within the same family business. This often results in a vertical and consolidated growth of Italian businesses, providing them with access to increased prosperity and subsequently more power and leverage over the market.

To Francesco and his fellow Italian American entrepreneurs, the Market functions as a "melting pot," which provides them with greater opportunities to grow their businesses. To these businesses, the increasing popularity of the Market among "yuppies" is not a concern but rather a business opportunity. However, beneath the surface of this "melting pot," there exists a complex range of challenges to the success and growth of the Mexican entrepreneurs. To many Mexican businesses, continuous growth of the Market, combined with the fast pace of gentrification in the area, makes them unable to keep up with the increasing rents. Consequently, the very same people who contributed to the growth of the Market may be forced out of their businesses if they do not compete in this "prosperous" market. To be sure, the Market still provides opportunities for Mexican entrepreneurs, as new Mexican businesses have opened in the Market since this study was initiated. However, the growth of businesses does not seem to go hand in hand with their increased power: of the 12 seats on the board of directors at the South 9th Street Business Association, all are allocated to entrepreneurs with Italian heritage. The embedded racial conflicts in the Market and the unequal distribution of power could ultimately cost the Market its newly regained vitality and its "cosmopolitan" status.

Conclusion

The case studies presented showcase how immigrant food entrepreneurs are increasing access to healthy foods, contributing to community wealth, and rebuilding social and economic dimensions of their receiving communities. These examples demonstrate the multifaceted impact of immigrants on community revitalization and that the process through which the newcomers interact and reshape their new environments is filled with physical challenges (e.g., lack of access to infrastructure) and racial conflicts (e.g., lack of power and control over their environment).

The Buffalo case study shows how small-scale immigrant-run grocery stores are improving food access in urban environments. These ethnic grocery stores are offering healthy produce while navigating multiple systemic barriers, such as lack of access to ethnic distributors and the need to gain the trust and approval of government agencies. Both issues highlight how Buffalo, unlike traditional gateways such as New York City, is relatively new to hosting a diverse group of entrepreneurs. In traditional gateways, immigrants benefit from already established networks and civil societies that play a great role in incorporating immigrants. In Buffalo, such a connection between multiple actors—whether that be government agencies, ethnic community centers, or refugee settlement institutions—needs to be established. Local government—particularly planning, economic development, and public health departments—can play a significant role in initiating such a network by taking a proactive and comprehensive approach toward strengthening the community food system. In the absence of local government support, nonprofit organizations fill this gap. In 2018, United Way of Buffalo and Erie County allocated funding to 13 organizations to strengthen the community's food system, of which two work directly with refugee farmers (Somali Bantu Community Organization) and immigrant food businesses (West Side Bazaar). However, the degree to which these organizations work and speak to one another is unclear (Niagara Frontier 2018). Connecting ethnic grocers to refugee producers and regional farmers, supporting the establishment of a food hub, training health and licensing staff about ethnic food products, and effectively disseminating information are among the many ways that local government could play an active role in promoting the businesses of these community actors.

The Philadelphia case study provides an example of how immigrant food entrepreneurship is playing a central role in market revitalization. In fact, the interaction between the old and new immigrant food entrepreneurs accomplished the same goals sought by many community economic development professionals who focus on both people and place-based strategies, such as training a new generation of entrepreneurs and stimulating development in a previously run-down area.

The city government in Philadelphia acknowledges the contribution of immigrants to the city's growth and strives to provide a welcoming environment to protect the well-being of the immigrant communities, whether through mandating language access to all city services or declaring Philadelphia a sanctuary city. However, the local government tends to focus more on the outcome (i.e., revitalization) rather than the process that could enable or hinder the incorporation of the newcomers. Scholars argue that a promising outcome of immigrant revitalization is conditioned on the social, political, and economic context of their receiving communities, where a hostile and anti-immigration attitude could hamper immigrants' incorporation (Reitz 2002). This has been observed to some degree in the case of the South 9th Street Market. On the surface, the Mexican entrepreneurs have become economically incorporated by joining the Market and establishing businesses. However, the existing racial conflicts have prevented the newcomers from becoming fully integrated within the Market, as they are often excluded from decision-making processes that could ultimately impact their businesses. In this case, the overall pro-immigration approach of the city government could only do so much to protect immigrants, leaving the newcomers to navigate ethnic conflict and tensions with the established residents on their own. A prime example of individual-level effort is the owner of South Philly Barbacoa, a Mexican restaurant in the Market that has gained national recognition after being featured on multiple cooking shows and documentaries. The owner uses her own undocumented status and newly gained platform to advocate for the rights of undocumented workers in the food industry, even if that means being more visible to the Immigration and Customs Enforcement agents who have been actively present in South Philadelphia since 2017.

This chapter demonstrates the opportunities that immigrant food entrepreneurs create in their new environments while giving a reminder that urban food environments are extremely diverse, complex, and contentious places. Until the day that these community actors' successes, needs, and challenges are formally incorporated into cities' plans and policies, local policymakers will continue to create policy for urban food environments that remains ignorant of some of those environments' most transformative actors.

Note

1. All names are changed to protect the identities of the study participants.

References

Brettell, Caroline B., and James F. Hollifield. 2014. *Migration Theory: Talking across Disciplines*. New York: Routledge.

Center for an Urban Future. 2007. *A World of Opportunity*. New York: Center for an Urban Future.

Cummins, Steven, Ellen Flint, and Stephen A. Matthews. 2014. "New Neighborhood Grocery Store Increased Awareness of Food Access but Did Not Alter Dietary Habits or Obesity." *Health Affairs* 33(2): 283–291.

Dubowitz, Tamara, Madhumita Ghosh-Dastidar, Deborah A. Cohen, Robin Beckman, Elizabeth D. Steiner, Gerald P. Hunter, Karen R. Flórez, Christina Huang, Christine A. Vaughan, and Jennifer C. Sloan. 2015. "Diet and Perceptions Change with Supermarket Introduction in a Food Desert, but Not Because of Supermarket Use." *Health Affairs* 34(11): 1858–1868.

Emond, Jennifer A., Hala N. Madanat, and Guadalupe X. Ayala. 2012. "Do Latino and Non-Latino Grocery Stores Differ in the Availability and Affordability of Healthy Food Items in a Low-Income, Metropolitan Region?" *Public Health Nutrition* 15(2): 360–369.

Glanz, Karen, James F. Sallis, Brian E. Saelens, and Lawrence D. Frank. 2005. "Healthy Nutrition Environments: Concepts and Measures." *American Journal of Health Promotion* 19(5): 330–333.

Gordon, Cynthia, Marnie Purciel-Hill, Nirupa R. Ghai, Leslie Kaufman, Regina Graham, and Gretchen Van Wye. 2011. "Measuring Food Deserts in New York City's Low-Income Neighborhoods." *Health and Place* 17(2): 696–700.

Hosler, Akiko S., Isaac H. Michaels, and Erin M. Buckenmeyer. 2016. "Food Shopping Venues, Neighborhood Food Environment, and Body Mass Index among Guyanese, Black, and White Adults in an Urban Community in the US." *Journal of Nutrition Education and Behavior* 48(6): 361–368.

Imbruce, Valerie. 2015. *From Farm to Canal Street: Chinatown's Alternative Food Network in the Global Marketplace*. Ithaca, NY: Cornell University Press.

Kelly, Erin. 2013. "Map for Adventurous Eaters." *Buffalo News*, May 20, 2013. https://www.buffalorising.com/2013/05/a-map-for-adventurous-eaters/.

Kim, Wook-Jin. 2010. *Korean Immigrant Entrepreneurs in Inner-City Minority Neighborhoods: Who Are Those Who Stay and Who Are Those Who Leave?* Chicago: University of Chicago Press.

Liu, Cathy Yang, Jonathan Miller, and Qingfang Wang. 2014. "Ethnic Enterprises and Community Development." *GeoJournal* 79(5): 565–576.

McDaniel, Paul. 2014. "Immigration and Economic Revitalization in America's Cities." Immigration Imapct. http://immigrationimpact.com/2014/06/02/immigration-and-economic-revitalization-in-americas-cities/.

Moyce, Sally C., and Marc Schenker. 2018. "Migrant Workers and Their Occupational Health and Safety." *Annual Review of Public Health* 39:351–365.

Mucci, Nicola, Veronica Traversini, Gabriele Giorgi, Giacomo Garzaro, Javier Fiz-Perez, Marcello Campagna, Venerando Rapisarda, Eleonora Tommasi, Manfredi Montalti, and Giulio Arcangeli. 2019. "Migrant Workers and Physical Health: An Umbrella Review." *Sustainability* 11(1): 232–254.

Niagara Frontier. 2018. "United Way of Buffalo & Erie County, General Mills Foundation Funding 13 Organizations to Strengthen Community's Food System." *Niagara Frontier Publications*, April 12, 2018. https://www.wnypapers.com/news/article/current/2018/04/12/132237/united-way-of-buffalo-erie-county-general-mills-foundation-funding-13-organizations-to-strengthen-communitys-food-system.

Ooraikul, Buncha, Anchalee Sirichote, and Sunisa Siripongvutikorn. 2008. "Southeast Asian Diets and Health Promotion." In *Wild-Type Food in Health Promotion and Disease Prevention*, edited by Fabian DeMeester, 515–533. Totowa, NJ: Humana Press.

Reitz, Jeffrey G. 2002. "Host Societies and the Reception of Immigrants: Research Themes, Emerging Theories and Methodological Issues." *International Migration Review* 36(4): 1005–1019.

Schuch, Johanna Claire, and Qingfang Wang. 2015. "Immigrant Businesses, Place-Making, and Community Development: A Case from an Emerging Immigrant Gateway." *Journal of Cultural Geography* 32(2): 214–241.

Shibley, Robert, Bradshaw Hovey, and Rachel Teaman. 2016. "Buffalo Case Study." In *Remaking Post-industrial Cities: Lessons from North America and Europe*, edited by Donald K. Carter, 25–45. New York: Routledge.

Short, Anne, Julie Guthman, and Samuel Raskin. 2007. "Food Deserts, Oases, or Mirages? Small Markets and Community Food Security in the San Francisco Bay Area." *Journal of Planning Education and Research* 26(3): 352–364.

US Census Bureau. 2009. Total Population: 2009 American Community Survey 5-Year Estimates. Washington, DC: US Census Bureau.

Vigdor, Jacob L. 2017. "Estimating the Impact of Immigration on County-Level Economic Indicators." In *Immigration and Metropolitan Revitalization in the United States*, edited by Domenic Vitiello and Thomas J. Sugrue, 25–39. Philadelphia: University of Pennsylvania Press.

Vitiello, Domenic. 2014. "The Politics of Place in Immigrant and Receiving Communities." In *What's New about the "New" Immigration? Traditions and Transformations in the United States since 1965*, edited by Marilyn Halter, Marilynn S. Johnson, Katheryn P. Viens, and Conrad Edick Wright, 83–110. New York: Palgrave Macmillan.

Waters, Stephanie. 2018. *Immigration Policies*. Philadelphia: City of Philadelphia Office of Immigrant Affairs. https://beta.phila.gov/2018-01-08-immigration-policies/.

Widener, Michael J., Sara S. Metcalf, and Yaneer Bar-Yam. 2011. "Dynamic Urban Food Environments: A Temporal Analysis of Access to Healthy Foods." *American Journal of Preventive Medicine* 41(4): 439–441.

Zhang, Qi, Ruicui Liu, Leigh A. Diggs, Youfa Wang, and Li Ling. 2019. "Does Acculturation Affect the Dietary Intakes and Body Weight Status of Children of Immigrants in the US and other Developed Countries? A Systematic Review." *Ethnicity and Health* 24(1): 73–93.

Zhang, Y. Tara, Barbara A. Laraia, Mahasin S. Mujahid, Samuel D. Blanchard, E. Margaret Warton, Howard H. Moffet, and Andrew J. Karter. 2016. "Is a Reduction in Distance to Nearest Supermarket Associated with BMI Change among Type 2 Diabetes Patients?" *Health and Place* 40:15–20.

5 Food from Home and Food from Here: Disassembling Locality in Local Food Systems with Refugees and Immigrants in Anchorage, Alaska

Sarah D. Huang

Introduction

Sitting at an English-language school in Anchorage, Alaska, a middle-aged Angolan woman called Nattat[1] distinguished the multifaceted meanings of "local food." Local food in Angola is "food from home," whereas local food in Anchorage is "food from here." Her definitions contrasted the mainstream understanding of "local"[2] that I came to query in Anchorage, where the Alaskan government has mobilized to increase in-state food production in order to address the state's food security challenges. For Nattat, her foodscape is constructed through these distinctions of familiarity and unfamiliarity. She practices these distinctions through her desire to try local foods in Alaska and by interchanging traditional Angolan flavors for more commonly found ingredients to make fufu. Fufu, a pastelike corn flour dish widespread in Africa, is her favorite dish from home. In one sense, Nattat's fufu utilizes ingredients that evoke a familiarity with her home in Angola. She associates local food with the connection between fufu and Angola and her home. However, when describing how she makes fufu in the United States, she said, "You can find the corn flour here. It's actually the Spanish version, which is the maize flour that the Spanish people use and that's what I use." When I asked if the taste compares to eating it at home in Angola, she said, "No, not really. But it's edible," identifying unfamiliarity in the ingredients disassociated from home.

Nattat's retelling of making fufu demonstrates her distinction of food from here and food from home. While she can't find all the ingredients to make fufu in Alaska, she reassembles the recipe with food from here in order to try and make a dish that is comforting and familiar in an unfamiliar place. These re-creations of familiarity are reassembled through the global movement of foods and recipes from her home country to Anchorage. Komarnisky (2009) terms this a *foodscape*, where food plays the dual role of being connected to a place and connecting places. This dual role of food is important in understanding how "local foods" create an invisible barrier to participation and comfort in the daily food practices of a growing population of immigrants and refugees residing in Anchorage, Alaska.

Conversations about "local food" within race, class, and nutrition inequalities suggest that food systems require understanding the experiences and categories of belonging that can inadvertently exclude immigrant and refugee communities. Since the rise of the term "local food" in the first decade of this century, in movies like Robert Kenner's *Food, Inc.* (2009), popular books like Michael Pollan's *The Omnivore's Dilemma* (2006), and quick industry usage in marketing, it has been dominated by imagery of food sourcing within state borders or predefined spatial or geographic boundaries. "Local food" has come to be associated with increased traceability, quality of food, and trust in exchanges between producer and consumer (Hinrichs 2015). While these values benefit overall environmental health and community health, the rhetoric of "local" as a specific distance mobilizes notions of food locality that perpetuate a specific white imaginary around food systems (Guthman 2008a). The alternative food discourses that are supported in farmers markets and "local food" systems disregard subjects and communities that utilize practices not yet recognized by dominant narratives within local food movements. Purchasing and cooking practices are further marginalized through race and class inequalities in health and nutritional access (Slocum 2006; Drewnowski and Specter 2004; Pothukuchi and Kaufman 1999) and land resource access in urban food systems (White 2011a, 2011b) and urban food justice movements (Loo 2014; Gottlieb and Joshi 2010). While outside the legal legitimation that occurs in "agricultural racial formations" or the everyday experiences of racial exclusions and "othering" (Minkoff-Zern et al. 2011), I argue that a local food movement similarly constitutes an acceptable type of local food that makes visible subjects of those who participate in purchasing and consuming local foods grown within a specific geographical distance.

Nattat's dual meaning of "local food" shows what is lost when neglecting diverse understandings and practices of local food. The established use of the term creates a binary of acceptable and unacceptable food consumers, segregating those who do not or cannot support local food (DuPuis and Goodman 2005; Valiente-Neighbours 2012). I use ethnographic cases of immigrants' and refugees' experiences to show the limits of local food movements mobilized in the popularity of farmers markets, community-supported agriculture programs, the US Department of Agriculture's defining of local and regional food systems, and the sourcing of local foods by large food retailers in Alaska (Hinrichs 2015). Immigrants and refugees challenge the concept of "local" in the food movement and push sustainable and local food systems[3] to better consider the migration of foods and peoples. Excluding how foods and peoples move becomes especially harmful when discourses of cultural foodways are minimized in sociopolitical decision making regarding what counts as local food and what it means to participate in growing local food.

To complicate the local food concept (Hinrichs 2015, 2003; Ostrom 2006) within the reality of the local-global interplay of cultural identity and the creation of local diets (Ohnuki-Tierney 1994; Watson and Caldwell 2005), scholars offer solutions for a deeper cross-cultural understanding of the unique needs and contributions of immigrant communities. These solutions take into account how food is utilized in "longing for home" and the "painful struggle to accommodate to new ways of being in the world" when placed within the context of Michael Pollan's "food rules" (Pollan 2008; Mares 2012, 335). Nattat's description at the start of this chapter of local food as food from home and food from here encapsulates how these shifts in purchasing and cooking practices to accommodate unfamiliar ingredients requires that localness not be static or monolithic but move beyond spatial boundedness toward relationality, contextuality, and translocalism (Appadurai 1996; Komarnisky 2009). I utilize the definition of translocality, building from the work of Holtzman (2006), Choo (2007), and Conradson and McKay (2007), to describe the relationship between food and people in the process of ongoing emplacement, commitments to families and communities, and emotional and material affiliations. Nattat's story is only one example of how food can express familiarity and unfamiliarity, where "the absence of familiar material culture, and its subtle evocations of home, is surely one of the most profound dislocations of transnational migration" (Law 2001, 277). This ethnographic project details how unfamiliarity exists in different modes for immigrants and refugees navigating the multiple layers of nonbelonging created through "local food" systems.

Food becomes the material object for reflection on the arrangements of power that allow for the inclusion and exclusion of peoples. It acts as a symbol of belonging to a food movement or recognition of how cultures establish a local food system not represented by the dominant food movements (Alkon and Vang 2016; Mintz 1985; Law 2001). This chapter aims to create greater understanding of how immigrant and refugee communities understand their roles in local food production while adapting familiarities of foodscapes from their home countries to Anchorage, Alaska.

Researching Transnationalities in Anchorage, Alaska

This chapter is based on ethnographic research conducted with immigrants representing 19 countries and refugee farmers from Bhutan residing in Anchorage, Alaska, in summer 2015. Given its geographic location, Alaska is not widely recognized for its immigrant and refugee communities, but it has recently gained press coverage in state and national newscasters' and academic scholars' reporting of Anchorage's diversity (Saleeby 2010; Jessen 2011; O'Malley 2015; Allen-Young 2014). Among Alaska's growing immigrant population, Native Hawaiians and Pacific Islanders made up the fastest-growing communities, increasing by 73% between 1990 and 2000. And the Latino/a community saw substantial growth of 45.2% in that same time frame (Bibbs 2006). In addition, refugees began arriving in Alaska in the early 1980s and have been supported by Catholic Social Services' Refugee Assistance and Immigration Service, a state and federally funded resettlement program in Alaska (Saleeby 2010; Tsong 2004).

Since the 1970s, Alaska has been addressing food security through sustainable agricultural systems given Alaska's precarious geographic and political location, caught between government protection of resource extraction, subsistence hunting shortages, and environmental uncertainties resulting from record high temperatures in both summer and winter (Hodges Snyder and Meter 2015; Stevenson et al. 2014). Alaska's food security is addressed through programs and research conducted by community food organizations, government agencies, academics, cooperative extension agents, and the Alaska Food Policy Council (Stevenson et al. 2014). These models for addressing food security would benefit from an understanding of how diverse peoples participate within these parts of Alaska's food system.

To explore these questions about inclusion of a growing immigrant and refugee community in Anchorage's local food movement, I utilized participant observation and semistructured interviews at various sites across Anchorage in summer 2015, including community gardens, farmers markets, English-language centers, and refugee assistance program sites. To recruit participants, I relied on connections with local organizations, including a farm-to-market program and an English-language center. These two sites were essential in connecting with participants and providing sites at which to conduct participant observation in spaces directly related to the area's local food scene. Through semistructured interviews, my goal was to better understand the nuances of lived experiences as an immigrant or refugee in gaining access to food needs and desires, participation in local food practices, and understanding of how local food differs between their home countries and Anchorage. I asked questions to compare experiences between their home country and Anchorage regarding access to foods, food practices, "good foods," and "local foods." I interviewed 20 people, 14 women and 6 men, who have lived in the United States between three months and 38 years. Of these participants, seven people work in food-related jobs at farmers markets, grocery stores, or restaurants.

In this chapter, I first situate Anchorage, Alaska, as an important site for exploring current trends in urban food systems. I then examine how refugee farmers from the Growing Community Gardens (GCG) program and other immigrant residents understand what it means to be part of a "local food" system. I close with the challenges and opportunities through which food scholars, food movement participants, and policymakers can better connect meanings of locality, as food from here and food from home, to create a local food system accessible to immigrants and refugees living in urban spaces.

Constructing Food Landscapes and Connecting Alaska

Alaska's long histories of a locally producing agricultural sector (Davies 2007; Francis 1967; Lewis and Pearson 1990) and subsistence-based Alaska Native cultures (Loring and Gerlach 2009; Lee 2003, 2002; Kancewick and Smith 1991) set the context for Alaska's diverse food system guided by state laws, federal laws, and food development programs. After the US government purchased Alaska from Russia in 1867, Congress passed the Homestead Act of 1898 to set up experimental agricultural stations throughout the state, build agricultural infrastructure, and move a workforce of 202 families from the Lower 48 states to generate Alaska's agricultural industry (Davies 2007; Haycox 2002). With the economic downturn leading into the Great Depression, providing food for the state's population became a priority. Federal policymakers provided moving costs and supplies to families transplanted from the Midwest into preconstructed barns and houses to farm north of Anchorage-Matanuska Valley (Meter and Phillips Goldenberg 2014). This access to farmland and agricultural development from the government started a history of family farms, homesteaders, and land-use practices that later influenced today's food system meant to sustain the state's food sufficiency and supply Anchorage markets.

Food security in Alaska, as defined by Loring and Gerlach (2009, involves a food system that supports biophysical, social, and ecological health, yet continues to be strained by fluctuations in oil prices and global costs of food (Meter and Phillips Goldenberg 2014). In order to address these strains, Meter and Phillips Goldenberg (2014) suggest that Alaska's future food system must be fueled through local resources and a renewed interest and investment in local capacity building or residents' food access may suffer from the scarcity and rising costs of fossil fuels. However, Alaska's food system is currently reliant on energy-intensive imported food and climatic changes that have the potential to strain 95% of the food that Alaskans purchase from out of state (Alaska Cooperative Extension Service 2006; Meter and Phillips Goldenberg 2014). The state's strategy to localize Alaska's food system, defined as in-state production, is to reduce the amount of food imported from the Lower 48 states by developing small-farm, greenhouse, and aquaculture production.

The Alaska Department of Natural Resources (2013) developed the Alaska Grown labeling campaign to promote the production and consumption of food grown within the state. This project stems from the US Department of Agriculture's definition of "local food" as "marketing of food to consumers produced and distributed in a limited geographic area" (USDA 2016). This model does not explicitly assign a specific distance to the term "local" but still aligns with dominant definitions of locality, effectively limiting perceptions of the local to a confined space within Alaska and belonging to Alaska. As this research investigates how Alaska's residents take part in local food system programming where food is grown within the state, it becomes clear that this limited perception of "local food" ostracizes the cultural histories and traditions that Alaska's residents participate in and actively construct.

Navigating the Familiar and Unfamiliar

Fatima fumbled with the large bundle of keys in her hand as she sat across from me in the lobby on the University of Alaska—Anchorage's campus. "In America, for me, I have to shop like either twice a week or once a week 'cause of the work schedule, my three children, and the stores are bigger. I have to jump from store to store to find what I want to cook. [In Jordan], we don't have to worry about it, even if you work, you just walk by the store and pick up what you're gonna cook" (interview, August 7, 2015). Her discomfort is visible in her body language, and the keys still shifting in her hand amplify her discomfort in having to shop for food in a food system unfamiliar to her. Here, she is describing two places. One place is *home*, a foodscape that was built for the types of foods that she recognizes and can easily find. Another place is *here*, a foodscape that is not recognizable to her and results in the invisible discomforts that marginalize her food decision making.

Fatima describes how to make beef and rice stews or couscous and grilled chicken down to the spices in each dish. Her description of these processes builds a menu of foods that she describes as local food in Jordan. Locality, as Fatima distinguishes it, is defined as the dishes frequently served in homes in Jordan, dishes that she knows well enough to be able to provide me with a step-by-step process for preparing them. But when asked to define "local" in Alaska, she stuttered and searchingly said, "I don't think there is anything local. Mmmm local I think... when you say local I think the things that grow up in Anchorage, or Alaska" (interview, August 8, 2015). In her unfamiliarity with Alaska, she assumes that local foods are those that are produced within the state, disassociated from the foods cooked in people's houses or the relationality of food sharing she associates with Jordan's local foods. While her definition of local food aligns with the Alaska Grown definition, she points to an important distinction: that local food is food from a place that evokes a sense of familiarity and home. Fatima embodies these distinctions of local food through her discomfort in talking about grocery shopping in Alaska and her ease in talking about shopping in Jordan. She describes shopping at a halal store on Mountain View Drive in Anchorage: "Even though it's like an African store, [the storeowners] have items that have Arabic writing. If it has English and Arabic writing, I know what it is. But some of them have their own language or [the storeowners] tell you, 'I don't know what this is, how do you guys use this?'" (interview, August 7, 2015). She is able to navigate different shops to find familiar foods to create the local Jordanian dishes in her home in Alaska. But more importantly, her desire to be able to find familiarity through local foods, as food from home, showcases the type of exclusion that creates discomfort in a place or emphasizes an inability to perform certain food purchasing or consumption practices (Slocum 2007). By seeking the familiar through various ingredients found in stores throughout Anchorage, Fatima must navigate barriers of unfamiliarity that exclude her from Anchorage's food system and the types of foods that count as local in Alaska and in Jordan.

These landscapes of invisible barriers to access are expressed in locations of large-box grocery stores like Fred Meyer or Carr's. Carlos, a middle-aged Colombian man, describes how these stores differ in accessibility of foods based on the neighborhood where they are located. He gave the example of the Penland Park Carr's, which although considered a large-box grocery store sells nopanelles and yucca, produce that he says you cannot find in other Carr's stores or other large-box grocery stores:

> Then if you go three miles into Midtown and shop at the Midtown Carr's, like in the very center of the Sears mall, that one will have almost nothing compared to the Penland Park one. It won't even seem like the same store. You'll have like your apples, your oranges, but you're not going to find that much of a variety in the ethnic foods. And it gets even weirder when you go down Northern Lights heading west maybe another 1.4 mile[s] away to Carr's Aurora Village, where that one is really, it's like maybe a yuppie food store. They have lots of produce, then they have a big organic section, but they're definitely thin on sort of the ethnic foods like foods that are specific from certain regions or where people will recognize specific labels. (interview, August 9, 2015)

Understanding where ingredients from home are located in Anchorage can better serve immigrants and refugees in food access (Taylor and Ard 2015). However, this funneling of food access into particular neighborhoods creates borders of inclusion and exclusion in food systems. This form of exclusion, such as the racial makeup of neighborhoods and familiarity with stores and types of food, can obscure avenues of access to "local foods" and familiarity. When these neighborhood particularities are couched within histories of whiteness in the food system (Guthman 2008b), then exclusions by ethnicity and race and food create perceptions of what foods and food resources are recognized as part of Anchorage's local food movement. Fatima shows how familiarity with stores and certain types of stores allows her access to her local foods in Jordan that she cooks at home. But traveling between stores like the halal market or large-box grocery stores showcases the ethnic and racial boundaries of Anchorage's urban landscape. Places like the halal market or a Korean grocery store, frequently visited by GCG farmers, are not given the same recognition in local food movements as the Alaska Grown products found in some large-box grocery stores, nor are they equally accessible on public transportation routes or main roads. The distinctions between these large-box grocery stores of carrying either organic foods or "ethnic foods" further pushes categorizations of places of belonging within the "food rules" of purchasing organic or "local" foods. Aside from these large-box grocery stores in Anchorage, farmers markets became important sites for the marketing of Alaska Grown products and engagements of refugee farmers in Anchorage's local food movement.

While refugee farmers rely on purchasing networks within an immigrant and pan-Asian community in Anchorage to acquire foods from home, these differ from the foods from here, or the vegetables that they sell at three farmers markets in Anchorage. These differences between foods from home and foods from here are based on how the refugee farmers understand what is considered local food based on the dynamics they see at the markets. I frequented three farmers markets in Anchorage based on where GCG farmers sold their produce. These examples showcase the most visible forms of exclusion in the creation of a "local food system."

Puja, a Bhutanese farmer at GCG, believes that "local food" is grown in Alaska, much like the kale he was selling at the market. After the market, Puja hands me a plastic grocery bag filled with leftover kale, suggesting I take it home. The program coordinator at GCG said, "After we decided to grow kale at the garden, because I knew that it would sell well at the farmers markets, they wanted to take it home to eat. They replaced it for a green in a Bhutanese dish, but came back the next day saying, 'Anne, why do you like this? It's not good.'" The gardeners at GCG continue to grow kale because it is one of their best-selling products at the markets. Individual gardeners like Puja and Suba now grow kale in their own gardens to sell at the markets.

Suba said, "Puja and I sell this one (pointing to sesame leaves on the market table) instead of [sesame leaf] seeds because no one eats [sesame leaf seeds]." Puja and Suba's affinity for vegetables that they perceive as nonpreferable by white customers at the farmers markets is supported by a normalness of cultural practices that go unacknowledged within community food movements, such as local food movements (Slocum 2006). Hegemonic discourses of "local" perpetuate normal behaviors, such as those performed at farmers markets, by overlooking how refugee farmers are excluded in the consumption of local foods despite being involved in their production. Puja and Suba navigate the unfamiliarity of local foods by trying to eat kale, but, similar to Nattat and Fatima, resort to finding ways to make familiar foods in the confines of their homes while engaging in a food system unfamiliar to them. Thus, Puja and Suba point to a disconnect of local food systems: while farmers markets bring consumer and producer closer, they only do so for particular types of consumers and producers who can access the local food movement.

Suba started gardening with other Bhutanese refugee farmers in 2012 and has continued to farm with GCG while also growing produce for her own consumption at other community gardens. Her understanding of access to farmers markets is determined by what she sees as common behaviors of vendors at the three farmers markets that I visited with her. The GCG program coordinator, Anne, told me that when Suba started selling her own produce, she was adamant about selling underneath the same tent as GCG because she was not able to purchase a tent on her own and she wanted to look like all the other farmers. Anne said, "I've been trying to get Puja and Suba to sell at [the other market] because they live over there, but Suba doesn't want to because they don't have a tent. She thinks that she doesn't look *official* [Anne's emphasis] unless she has a tent. I told them I wouldn't mind buying them a $30 tent so that they could sell their stuff separately at the market, but they declined" (conversation from August 5, 2015). For Suba, the unspoken rules of farmers market displays act as a barrier to feeling like she could be a legitimate vendor at the farmers markets.

At a Thursday market, Suba nudged me to the front of the table as an older white woman approached the GCG market stand. The woman looked directly at me and asked me about Suba's sesame leaves that were neatly bundled with a rubber band at the stem. I encouraged Suba to tell the woman about how she cooks them at home. Suba nervously looked at me and then began talking to the woman about eating sesame leaves with rice. Suba nodded me over to where she was standing and leaned close to my ear, whispering, "Sarah, I need you to speak with [the customers]. You know English better than me and can sell." She was afraid that if she couldn't speak English well enough, then she wouldn't be able to sell her sesame leaves. Suba's idea of belonging at the farmers market developed from her perceptions of white farmers who speak to customers in a native-English accent and display their produce under tents. In these cases, Suba and Puja's experiences show how the outward appearance of their participation in the local food scene of Anchorage hides the barriers that they, as immigrants and refugees, are facing from truly feeling like agents within this scene. The inclusion of refugees and immigrants as producers of local food falls into the "local trap," defined as an inherently good quality of the "local" (Born and Purcell 2006). Without attention to the weaknesses or complexities of what it means for these immigrants and refugees to produce local food, the local trap makes it easier to see only the benefits of producing local food while ignoring how the participation of immigrants and refugees in mostly white farmers markets puts them on the peripheries of local food movements.

Conclusion

While the immigrants and refugees I met while selling at farmers markets, growing vegetables in community gardens, and working in Anchorage's restaurant industry have carved out a niche for themselves in Anchorage, they remain on the peripheries of dominant definitions of the "local food" movement. For them, "local food" is not just something that is grown in Alaska, or "of Alaska" as Nattat described. Rather, it is transformed as it moves between places called home or here, where immigrants and refugees find opportunities to bring familiar flavors, produce, and practices from their home countries to Alaska. However, these familiar and unfamiliar practices further illuminate the invisible barriers by which the "local food" movement further ostracizes the translocal nature of how these immigrants and refugees create their own local food systems. Despite providing the labor that grows these local foods, immigrants and refugee farmers continue to exist just outside the borders of alternative food practices by providing labor and vegetables to fill a whiter and wealthier demographic's table. These local food spaces do not resonate with their food needs and result in their having to obtain multiple garden plots to grow "food from here" for white customers at the market and "food from home" for their own families. These farmers earn an income from farmers markets but are doubly burdened with expending labor in local food spaces while facing barriers to inclusion in a local food system. These limiting factors of spatial proximity disproportionately favor family farms, which own the majority of agricultural land, and exclude refugee and immigrant farmers expending their energies and land resources to feed a local food movement that does not accept their own practices and vegetables as part of that system.

This chapter has offered insight into the lives of a few immigrants and refugees living in Anchorage, Alaska, and their understandings of the dynamic definition of "local food." These narratives are personal and relatable in that they describe the complexities of how immigrants and refugees make sense of their new places of settlement through food practices and food choices. Thus, they reveal the importance of access to culturally important foods. By rooting my analysis of local foods in these narratives, I show the importance of understanding the multiple layers of local foods that are applied in home countries, in Anchorage, Alaska, in farmers markets, and in homes. I call for greater attention to those who remain at the peripheries of these dominant food movements in the United States and how these movements continue to foster exclusions regarding what it means to be "local." Alongside many of my fellow chapter authors in this volume, my recommendation is not for a new definition of "local" but rather for more attention to how immigrants and refugees are carving out spaces of familiarity by connecting home countries and countries of settlement through the multiple meanings of food that can extend translocal foodscapes into "local food" systems. Through a more attentive perspective of what "local food" means, policymakers, scholars, and food movement participants may be able to foster greater opportunities for these populations that recognize that "local foods" can exist as "food from home" and "food from here."

Notes

1. The names of all persons and organizations have been changed.

2. I utilize "local" in opposition to the mainstream definition of local foods as foods within a specific geographic distance.

3. As illustrated by Pollan (2008), Nestle (2013), Berry (2010), and Lappe (2010).

References

Alaska Cooperative Extension Service. 2006. *The Agricultural Industry in Alaska: A Changing and Growing Industry—Identification of Issues and Challenges.* Anchorage: University of Alaska.

Alaska Department of Natural Resources. 2013. "Alaska Grown Program." http://dnr.alaska.gov/ag/ag_AKGrown.htm.

Alkon, Alison Hope, and Dena Vang. 2016. "The Stockton Farmers Market: Racialization and Sustainable Food Systems." *Food, Culture and Society* 19(2): 389–411. doi: 10.1080/15528014.2016.1178552.

Allen-Young, Corey. 2014. "Study Calls Anchorage Schools America's Most Diverse High Schools." *KTUU*, February 27, 2014.

Appadurai, Arjun. 1996. *Modernity at Large: Cultural Dimensions of Globalization.* Minneapolis: University of Minnesota Press.

Berry, W. 2010. *Bringing It to the Table: On Farming and Food.* Berkeley, CA: Counterpoint Press.

Bibbs, RaeShaun. 2006. *Guide to Alaska's Cultures: 2006–2007 Edition.* Anchorage: Alaska Conservation Foundation.

Born, Branden, and Mark Purcell. 2006. "Avoiding the Local Trap: Scale and Food Systems in Planning Research." *Journal of Planning Education and Research* 26(2): 195–207. doi: 10.1177/0739456X06291389.

Choo, Simon. 2007. "Eating Satay Babi: Sensory Perception of Transnational Movement." *Journal of Intercultural Studies* 25(3): 203–213.

Conradson, David, and Deirdre McKay. 2007. "Translocal Subjectivities: Mobility, Connection, Emotion." *Mobilities* 2(2): 167–174. doi: 10.1080/17450100701381524.

Davies, Darcy Denton. 2007. "Alaska's State-Funded Agricultural Projects and Policy—Have They Been a Success?" Thesis. School of Natural Resources and Agricultural Sciences, University of Alaska Fairbanks.

Drewnowski, A., and S. E. Specter. 2004. "Poverty and Obesity: The Role of Energy Density and Energy Costs." *American Journal of Clinical Nutrition* 79(1):6–16.

DuPuis, E. Melanie, and David Goodman. 2005. "Should We Go 'Home' to Eat? Toward a Reflexive Politics of Localism." *Journal of Rural Studies* 21(3): 359–371. doi: http://dx.doi.org/10.1016/j.jrurstud.2005.05.011.

Francis, Charles. 1967. "Outpost Agriculture: The Case of Alaska." *Geographical Review* 57(4): 496–505.

Gottlieb, Robert, and Anupama Joshi. 2010. *Food Justice.* Cambridge, MA: MIT Press.

Guthman, Julie. 2008a. "Bringing Good Food to Others: Investigating the Subjects of Alternative Food Practice." *Cultural Geographies* 15(4): 431–447.

Guthman, Julie. 2008b. "'If They Only Knew': Color Blindness and Universalism in California Alternative Food Institutions." *Professional Geographer* 60(3): 387–397. doi: 10.1080/00330120802013679.

Haycox, Stephen. 2002. *Alaska: An American Colony*. Seattle: University of Washington Press.

Hinrichs, C. Clare. 2003. "The Practice and Politics of Food System Localization." *Journal of Rural Studies* 19(1): 33–45.

Hinrichs, Clare. 2016. "Fixing Food with Ideas of 'Local' and 'Place.'" *Journal of Environmental Studies and Sciences* 6(4): 759-764. doi: 10.1007/s13412-015-0266-4.

Hodges Snyder, Elizabeth, and Ken Meter. 2015. "Food in the Last Frontier: Inside Alaska's Food Security Challenges and Opportunities." *Environment: Science and Policy for Sustainable Development* 57(3): 19–33. doi: 10.1080/00139157.2015.1002685.

Holtzman, Jon D. 2006. "Food and Memory." *Annual Review of Anthropology* 35(1): 361–378. doi: 10.1146/annurev.anthro.35.081705.123220.

Jessen, Cornelia M. 2011. "Refugees and Healthcare Providers in Anchorage, Alaska: Understanding Cross-cultural Medical Encounters." MA thesis, University of Alaska, Anchorage.

Kancewick, Mary, and Eric Smith. 1991. "Subsistence in Alaska: Towards a Native Priority." *University of Missouri Kansas City Law Review* 59(3): 645-673.

Kenner, Robert. *Food, Inc.* DVD. Directed by Robert Kenner. New York: Magnolia Pictures, 2009.

Komarnisky, Sara V. 2009. "Suitcases Full of Mole: Traveling Food and the Connections between Mexico and Alaska." *Alaska Journal of Anthropology* 7(1): 41–56.

Lappe, A. 2010. *Diet for a Hot Planet: The Climate Crisis at the End of Your Fork and What You Can Do About It*. New York, NY: Bloomsbury.

Law, Lisa. 2001. "Home Cooking: Filipino Women and Geographies of the Senses in Hong Kong." *Ecumen* 8(3): 264–283.

Lee, Molly. 2002. "The Cooler Ring: Urban Alaska Native Women and the Subsistence Debate." *Arctic Anthropology* 39(1–2): 3–9.

Lee, Molly. 2003. "'How Will I Sew My Baskets?' Women Vendors, Market Art, and Incipient Political Activism in Anchorage, Alaska." *American Indian Quarterly* 27(3–4): 583–592.

Lewis, Carol E., and Roger W. Pearson. 1990. "Three Development Models for Alaska's Agricultural Industry." *Yearbook of the Association of Pacific Coast Geographers* 52(1990): 109–124. doi: 10.1353/pcg.1990.0015.

Loo, Clement. 2014. "Towards a More Participative Definition of Food Justice." *Journal of Agricultural and Environmental Ethics* 27(5): 787–809. doi: 10.1007/s10806-014-9490-2.

Loring, Philip A., and S. C. Gerlach. 2009 "Food, Culture, and Human Health in Alaska: An Integrative Health Approach to Food Security." *Environmental Science and Policy* 12(4): 466–478. doi: http://dx.doi.org/10.1016/j.envsci.2008.10.006.

Mares, Teresa M. 2012. "Tracing Immigrant Identity through the Plate and the Palate." *Latino Studies* 10(3): 334–354. doi: 10.1057/lst.2012.31.

Meter, Ken, and Megan Phillips Goldenberg. 2014. *Building Food Security in Alaska*. Anchorage, AK: Alaska Department of Health and Social Services, with collaboration from the Alaska Food Policy Council.

Minkoff-Zern, Laura-Anne, Nancy Peluso, Jennifer Sowerwine, and Christy Getz. 2011. "Race and Regulation: Asian Immigrants in California Agriculture." In *Cultivating Food Justice: Race, Class, and Sustainability*, edited by Alison Hope Alkon and Julian Agyeman, 65–85. Cambridge, MA: MIT Press.

Mintz, Sidney. 1985. *Sweetness and Power: The Place of Sugar in Modern History*. New York: Viking.

Nestle, M. 2013. *Food Politics: How the Food Industry Influences Nutrition and Health*. Berkeley, CA: University of California Press.

Ohnuki-Tierney, Emiko. 1994. *Rice as Self: Japanese Identities through Time*. Princeton, NJ: Princeton University Press.

O'Malley, Julia. 2015. "Most Diverse Neighborhood in US Welcomes You in Alaska." *Al Jazeera America*, August 9, 2015.

Ostrom, Marcia. 2006. "Everyday Meanings of 'Local Food': Views from Home and Field." *Community Development* 37(1): 65–78. doi: 10.1080/15575330609490155.

Pollan, Michael. 2006. *The Omnivore's Dilemma*. London, UK: Penguin Publishing Group.

Pollan, Michael. 2008. *In Defense of Food: An Eater's Manifesto*. London, UK: Penguin Publishing Group.

Pothukuchi, K., and J. L. Kaufman. 1999. "Placing the Food System on the Urban Agenda: The Role of Municipal Institutions in Food Systems Planning." *Agriculture and Human Values* 16(2): 213–224.

Saleeby, Becky M. 2010. "Anchorage, Alaska: City of Hope for International Refugees." *Alaska Journal of Anthropology* 8(2): 93–102.

Slocum, Rachel. 2006. "Anti-racist Practice and the Work of Community Food Organizations." *Antipode* 38(2): 327–349.

Stevenson, Kalb T., Lilian Alessa, Andrew D. Kliskey, Heidi B. Rader, Alberto Pantoja, Mark Clark, and Nicole Giguére. 2014. "Sustainable Agriculture for Alaska and the Circumpolar North: Part I. Development and Status of Northern Agriculture and Food Security." *Arctic* 67(3): 271–295.

Taylor, Dorceta E., and Kerry J. Ard. 2015. "Detroit's Food Justice and Food Systems." *Focus* 32:13–19.

Tsong, Nicole. 2004. "New Start." *Anchorage Daily News*, November 14, 2004.

US Department of Agriculture (USDA). 2016. "Local Foods." USDA National Agricultural Library. https://www.nal.usda.gov/aglaw/local-foods.

Valiente-Neighbours, J. M. 2012. "Mobility, Embodiment, and Scales: Filipino Immigrant Perspectives on Local Food." *Agriculture and Human Values* 29(4): 531–541. doi: 10.1007/s10460-012-9379-5.

Watson, James, and Melissa L. Caldwell. 2005. *The Cultural Politics of Food and Eating: A Reader*. Malden, MA: Blackwell.

White, Monica M. 2011a. "D-Town Farm: African American Resistance to Food Insecurity and the Transformation of Detroit." *Environmental Practice* 13(4): 406–417.

White, Monica M. 2011b. "Sisters of the Soil: Urban Gardening as Resistance in Detroit." *Race/Ethnicity: Multidisciplinary Global Contexts* 5(1): 13–28.

II Labor: Fields and Bodies

Of all the topics within the immigrant-food nexus, immigrant farm laborers have been given the most attention by the alternative food movement and critical food scholarship. As we discussed in our introduction, films, books, and nonprofit campaigns on the agricultural industry's reliance on and marginalization of predominantly Latinx migrants have gained increasing visibility throughout the last two decades.

This critique of marginalized labor has been criticized by many as limited and secondary to environmental and health critiques within food conversations. Even still, the literature that is available remains relatively one-dimensional: the narrative of immigrants' labor within the food system is one of Latinx (largely male) workers experiencing harmful working conditions with little agency to change or leave these conditions. While this narrative is an accurate description of a portion of immigrants' experiences within food labor, it is by no means the full picture.

Part II of this book includes four chapters that broaden the scope of inquiry into immigrant labor within the food system, and within the immigrant-food nexus more generally. These authors address the historic and ongoing marginalization of migrant labor but take this as only the starting point for a much wider discussion of the many forms and faces of immigrant labor within the food system. Immigrant laborers are not only farmworkers; they are farmers and business owners. They are not only men; they are women who command vital roles both behind the scenes and on the front lines. They are not stand-ins for technological farming solutions; the social and scientific worlds of crop production are becoming increasingly wound together to necessitate hands in the field. They are not only carrying out manual labor; they are mobilizing transnational intellectual and cultural knowledge to run and grow farming enterprises.

It is only by recognizing the multiplicity of immigrant food labor experiences that we may begin to understand the multiplicity of actors and policy arenas that intersect in the immigrant-food nexus. These four chapters are not separate stories. They are interlinked pieces of the larger labor project being enacted through and acted against daily by immigrant bodies.

6 Labor and the Problem of Herbicide Resistance: How Immigration Policies in the United States and Canada Impact Technological Development in Grain Crops

Katherine Dentzman and Samuel C. H. Mindes

Introduction

In order to maintain its system of productivist industrialized agriculture, the United States relies heavily on migrant farm laborers (Bonanno 2015; Guan et al. 2015; Zahniser et al. 2018). With domestic workers unwilling to take such low-status, physically demanding, and low-paying jobs (Bonanno 2015; Zahniser et al. 2018), it has become necessary to depend on migrants. The United States gets the majority of its migrant farm labor from Mexico, with most of these workers coming seasonally to harvest horticultural and other specialty crops (Martin 2009; Zahniser et al. 2018). The reliance on these workers, however, has significant risks attached to it. With current political and social trends in the United States, migrant labor shortages have become a serious problem (Fan et al. 2015; Zahniser et al. 2018). This has the potential to devastate sectors of US agriculture that cannot function without the readily available and low-cost hand labor that Mexican migrant farm laborers provide.

The impact of foreign workers' availability on horticulture—and in particular its correlation with mechanization—has been well documented (e.g., Bellenger et al. 2008; Fan et al. 2015; Martin 2013). However, the effect on commodity field crops has been deemed largely irrelevant because of already high levels of mechanization in these systems (Martin 2013). Indeed, in the United States, only about 13% of migrant farmworkers are employed in field and grain cropping systems, whereas 84% work in fruit, vegetable, or nut farming systems (Hernandez, Gabbard, and Carroll 2016). This leads to the assumption that labor shortages will have minimal impact on the viability of US commodity grain crops (e.g., Zahniser et al. 2018). However, the problem of herbicide-resistant (HR) weeds is challenging this assumption. Many weed species have become resistant to popular herbicides, and no new herbicidal mode of action appears to be forthcoming (Livingston et al. 2015). Thus, controlling herbicide resistance in commodity grain crops increasingly requires removing weeds through manual labor—a role traditionally filled by Mexican immigrants.

Herbicide-resistant weeds are the result of reliance on an herbicide-only weed management system, and they present a significant threat to the profitability and continuity of commodity grain crop farms in the United States. Such weeds can devastate crop yields and cause the cost of production to skyrocket—in particular, the cost of hiring hand labor to remove resistant weeds from the field (Norsworthy et al. 2012). An alternative to herbicide-only weed management, integrated weed management (IWM), involves the integration of preventative, mechanical, cultural, chemical, and biological weed management techniques. This combination provides effective weed management while also avoiding development of herbicide resistance (Harker 2013). However, such programs often rely on the availability of hand labor, necessitating the use of foreign migrant laborers (Harker 2013). Therefore, the trend in availability of such laborers will likely have a meaningful impact on how commodity grain crop growers control their weeds.

Using focus group data with farmers from Minnesota, Iowa, Arkansas, and North Carolina, we address the issue of US foreign farm laborer dynamics in relation to herbicide resistance and commodity grain crops. Within our focus groups, labor shortages emerged as a relevant barrier to HR weed control, leading to significant challenges and concerns. Drawing from theory developed in a study of labor relations in the lettuce industry in California (Friedland et al. 1981), we investigate how this decrease in immigrant labor, specifically from Mexico, is impacting reliance on technological innovation in the herbicide industry and creating a counterpart to labor struggles and mechanization trends in horticulture.

Additionally, we assess two potential futures for HR weed management as a function of how national policies, programs, and social attitudes are impacting labor availability. We accomplish this via a comparison between our firsthand data from the United States and theoretical analysis of Canada's policies for temporary agricultural workers. Similar herbicide-resistance problems (see Heap 2017) and comparable reliance on Mexican temporary farmworkers (see Bonanno 2015; Weiler, McLaughlin, and Cole 2017a) in the United States and Canada provide a controlled context in which to test the impact of these political and cultural differences on future HR weed management outcomes, highlighting the importance of national immigration policies in determining systems of management in agriculture. This follows important work by Brandt (2002) documenting the work of Mexican women farmworkers along the tomato processing route between Mexico, the United States, and Canada. Our analysis integrates the social science area of immigrant labor issues and the physical science area of crop management—two areas that are not typically conceptualized as connected yet are becoming increasingly so in the lived experiences of farmers and laborers. Through this approach, this research will lend itself to new ways of discussing crop science and social issues as conjointly constituted, emphasizing the necessity for an interdisciplinary perspective in fields from food system studies, to agricultural science, to immigration politics.

Theoretical Background

When discussing labor shortages, the most common recommendation in horticulture is to mechanize (e.g., Guan et al. 2015; Halle 2007; Taylor, Charlton, and Yúnez-Naude 2012; Zahniser et al. 2018). Even when mechanization is not necessarily recommended, it is the common outcome of decreasing farmworker availability (Martin 2013). This trend reflects Friedland and colleagues' (1981) study of manufacturing in the lettuce and tomato industries. In their research, they posited that the elasticity and control of the labor supply plus the economic structure of an industry determines the type and extent of technological change in production.

The context for Friedland and colleagues' study was California in the 1960s, just before the end of the Bracero Program, which had been providing a reliable source of inexpensive (i.e., underpaid), legal immigrant labor. To deal with the anticipated labor shortage, the tomato industry threw substantial funding behind the development of mechanical harvesting technology. In contrast, the lettuce industry was less coordinated and slower to develop mechanization to replace migrant labor. Soon it became apparent that supplies of underpaid immigrant labor would continue unabated, largely because of undocumented immigrants, and that mechanization was no longer necessary. The lettuce industry ceased its research into mechanics, and throughout the 1970s, mechanization in the lettuce industry was limited by the efficiency of labor-intensive harvesting processes and the continued availability of underpaid labor. Friedland and colleagues (1981) proceeded to hypothesize three conditions that could theoretically induce mechanization in the lettuce industry: (1) blockage of the labor supply through limited migration and increased union organization, (2) reduction in the production cost of mechanical solutions, and (3) relocation of the lettuce industry from California to Florida, which would limit the number of Mexican migrants available to do harvesting. These conditions would essentially reproduce the context that led to significant mechanization in the tomato industry: a prolonged labor shortage combined with the relatively rapid development of cost-efficient mechanical laborsaving technologies.

In order to predict whether commodity grain growers in the United States and Canada will turn to mechanization for HR weed control, Friedland and colleagues (1981) suggest that we must first understand the two countries' migrant labor policies. Therefore, if there were significant differences between temporary migrant farmworker policies in the United States and Canada, we would likely see differences in how they propose to manage HR weeds. We discuss the literature on migrant farmworker programs in the United States and Canada to explore their respective effectiveness and problems.

Agricultural Labor Shortages and Farmworker Programs in the United States

The systems of US immigration and agricultural labor have long been interconnected (Devadoss and Luckstead 2011). Approximately 68% of all hired farm laborers in the United States are Mexican citizens, and 47% of all farmworkers are not legally authorized to work in the United States, according to self-reports from a 2013–2014 survey (Hernandez, Gabbard, and Carroll 2016). Since this is a self-reported statistic, the reality is that the percentage is likely much higher. Indeed, the organization Farmworker Justice puts the estimate of unauthorized migrant farmworkers at over 70% (Boudreau 2016), and the American Farm Bureau estimates between 50% and 70% (Zahniser et al. 2018). As unauthorized laborers, this population is particularly vulnerable to the industry's control and abuse.

Media and political rhetoric in the United States have historically held a negative attitude toward undocumented immigration (Martin 2015), and national immigration policies increasingly reflect this. Tougher border enforcement, the expansion of the US-Mexico border fence, and crossing conditions that are more dangerous and costly have drastically decreased the number of agricultural workers making the trip north since the late 1990s (Devadoss and Luckstead 2011; Fan et al. 2015; Martin 2013). Additionally, increased spending on domestic enforcement against undocumented workers—for instance, in the form of I-9 audits[1] and state-mandated E-Verify[2] programs to check workers' legal status—has been shown to have decreased undocumented farm labor employment and caused locally based labor shortages in the agricultural sector since 2006 (Devadoss and Luckstead 2011; Martin 2013). A 2018 report by the US Department of Agriculture concurs that the number of unauthorized Mexican immigrants has declined substantially, contributing to the tightening of the farm labor market in the United States (Zahniser et al. 2018). The Trump administration has aggravated this issue, as the president made migration a central issue in his 2016 campaign. Since taking office, he has enacted plans to build a wall on the Mexico-US border, increase the number of Border Patrol and Immigration and Customs Enforcement agents, and reduce the admission of foreigners more generally, resulting in a substantial decline in illegal migration between Mexico and the United States (Martin 2019).

In an attempt to better control the hiring of foreign farmworkers, the Immigration Reform and Control Act introduced section H-2A in 1986. Although a multitude of reforms to this program have been proposed since then—including the H-2C program championed by former Virginia representative Bob Goodlatte in 2018—none have passed (Martin 2019). This makes H-2A farmers' current best option for hiring foreign workers legally (Rickard 2015). H-2A hiring of foreign workers is allowable only if there are not sufficient domestic workers available and if hiring foreign workers will not negatively affect local wages (Bellenger et al. 2008). Producers must submit a request for workers to the Department of Labor (DOL) 60 days before they think they will need them. Once the DOL verifies that there is a need for immigrant farmworkers in the area, the request is transferred to US Citizenship and Naturalization Services, which recruits the appropriate workers. Although this process is supposed to take only 60 days, some farmers have complained of delays of up to an additional five weeks (Boudreau 2016). This program's intention is to alleviate the labor shortage; however, it is time consuming and difficult to use, discouraging both producers and migrants from employing it (Devadoss and Luckstead 2011). These conditions likely encourage the use of undocumented laborers, as employers see little advantage to hiring through H-2A versus hiring undocumented laborers, and anticipate clear advantages to their bottom line by using this less-protected undocumented labor population (Danger 2000).

In addition to the bureaucratic problems H-2A presents for employers, some legal experts compare worker conditions under H-2A to slavery or indentured servitude (e.g., Guerra 2004; Hall 2001). Indeed, conditions are often deplorable and workers' rights nearly nonexistent. Immigrant laborers hired under H-2A can only work for the grower that recruited them, locking them into a potentially poor work environment and leaving them susceptible to exploitation by employers who have the power to refer them to the authorities for deportation or blacklist them from future H-2A employment (Danger 2000; Guerra 2004; Hall 2001; Simms 2000). Additionally, workers do not see their contracts until they arrive in the United States, cannot choose their employer, cannot negotiate their wages, are excluded from the Migrant and Seasonal Agricultural Worker Protection Act, cannot sue their employer, and often face substandard housing conditions and very low wages (Danger 2000; Guerra 2004; Hall 2001; Simms 2000). Furthermore, H-2A reforms that benefit workers do not appear to be forthcoming, not in the least because of farmers' opposition to such reforms, as the current system clearly favors the employers over the employees (Norton 2016). Lobbying efforts to reform H-2A have focused on making it easier for farmers to bring in temporary workers, specifically lobbying for the removal of preferences for US laborers, for the reduction of temporary farmworker wages, and against the general sluggishness of the program (Tomson 2015; Francis 2018; Wheat 2018).

Despite these issues, farmers' H-2A requests are growing as border enforcement increases and immigration reform seems far off (Martin 2019). Specifically, H-2A requests are rising in response to the increasing labor shortage and the Trump administration's tougher enforcement of immigration laws, which is expected to decrease the availability of undocumented migrant agricultural workers (see chapter 1 of this volume for Kimberley Curtis's analysis of migrant farmworkers and increased border militarization in Arizona). In the first three months of 2017, 69,272 H-2A positions had been requested, a 36% increase from the same period in 2016 and by far the biggest jump in recent years (Charles 2017). Roughly 200,000 H-2A positions were certified in 2017 (91% of which went to Mexicans), and the first three quarters of 2018 suggest that certifications will jump another 21% (Zahniser et al. 2018). Ironically, President Trump himself employs H-2A farm guest workers at his Virginia vineyard (Martin 2019). Despite this increasing use of H-2A, which may be indicative of the labor shortage farmers are experiencing (Zahniser et al. 2018), experts anticipate that unless the program is reformed, labor shortages will continue to rise, with concomitant drops in agricultural productivity and value (e.g., O'Brien, Kruse, and Kruse 2014; Simms 2000; Zahniser et al. 2018). In the current US political climate, such enforcements, barriers, and stagnant policies are likely to continue to magnify unprecedented labor shortages (Guan et al. 2015; Martin 2019; Taylor et al. 2012). In particular, labor shortages will have a newly significant impact on farm industries traditionally thought of as not being highly labor dependent. This is likely to increase competition with the horticulture industry for laborers, further complicating migrant labor shortages. The assumption by most scholars that labor is not an issue on commodity grain farms is faulty and needs to be challenged in order to effectively address these issues.

Agricultural Labor Shortages and Farmworker Programs in Canada

The struggles with domestic labor availability that Canada faces are similar to those facing the United States (Preibisch 2010). However, in contrast to the US H-2A program, Canada's Seasonal Agricultural Worker Program (SAWP) is consistently cited as a model temporary agricultural worker program for its circularity—meaning workers who come to Canada rarely overstay their visas, instead returning to their home country and getting a new visa to return to Canada the following year (e.g., Castles, de Haas, and Miller 2014; Hennebry and Preibisch 2012; Massey and Brown 2011). Developed in 1966, the program allows agricultural employers in specific commodity sectors to recruit and hire workers who are citizens of Mexico or certain Caribbean countries to do work related to farm labor for a maximum of eight months (Castles, de Haas, and Miller 2014; Government of Canada 2016; Massey and Brown 2011). The Canadian government expressly stipulates the role of the foreign country to recruit and maintain a pool of qualified workers, ensure that workers have the proper documentation, and appoint representatives to assist foreign workers in Canada (Government of

Canada 2016). Because of the nature of the program, many of the participants are repeat migrants (Massey and Brown 2011).

The SAWP is extremely effective for employers looking to hire migrant laborers (Hennebry and Preibisch 2012), but it is far from ideal in terms of migrant rights. Castles, de Haas, and Miller (2014) specifically note its failures in terms of migrants' social and political rights as a result of the controlled circumstances under which migrants work, themes likewise highlighted by other critics of the program (e.g., Basok 2007; Preibisch 2010). While Canada presents itself as a very multicultural society, touting that over one in five people in Canada is foreign born (Chui and Flanders 2013), the origin of this migrant population is different from the countries of origin specified in the SAWP, suggesting that the connection between Canada and SAWP countries is purely economic. This point is further demonstrated by a December 2017 report by Canada's House of Commons Standing Committee on Agriculture, which offers 21 food policy recommendations, none of which relate to the livelihood of foreign workers, who are instead discussed as a commodity Canada needs (Finnigan 2017). Investigative pieces have also highlighted the struggles of SAWP laborers, demonstrating how many experience food insecurities, poor working conditions, and minimal pay (Weiler, McLaughlin, and Cole 2017b; MacLean Wells and McLaughlin 2018). The SAWP amplifies the power disparity between bosses and workers, as workers are afraid to complain because they might get fired and deported (Weiler, McLaughlin, and Cole 2017b). Though the SAWP has received widespread praise, the major benefits of the program are limited to the economic gains of employers.

As a result of the SAWP, Mexican workers in particular have become a major source of agricultural labor in Canada, with the number of Mexican agricultural workers granted authorization to work in Canada jumping 112% between 1994 (when the North American Free Trade Agreement, NAFTA, came into effect) and 2001. As of 2012, approximately 27,000 Mexican workers come to Canada through the SAWP every year. Although this population is significantly less than in the United States, it still makes up a large part of Canada's agricultural labor force (Weiler, McLaughlin, and Cole 2017a).

Stunningly, 98.5% of SAWP workers finish their contracts each year, 80% are repeat hires, and almost none stay illegally after their contract expires (Mueller 2005; Preibisch 2007). Compared with an estimated 70%

of Mexican agricultural workers in the US without legal documentation, Preibisch (2007) found in a 2004 study that only about 15% of rural agricultural workers in Ontario were working without visas. Part of the reason for these trends is the characteristics of workers recruited to the SAWP with regard to marriage, number of children, education, and experience (Massey and Brown 2011). However, a better explanation for this trend is the benefits the migrant laborers receive from the SAWP.

Although there are certainly clear workers' rights issues—such as forced saving schemes, no path to permanent residency, and an enforced lack of mobility (Hennebry and Prebisch 2012)—Mexican farmworkers in Canada have significant advantages over their counterparts in the United States. Temporary SAWP workers in Canada earn 9.3% more than legal temporary agricultural workers in the United States (Massey and Brown 2011). For undocumented workers in particular, the gap is even wider: legal SAWP workers in Canada earn about 29% more than undocumented workers in the United States (Massey and Brown 2011). Additionally, the SAWP partially pays for workers' transportation costs, provides free housing, and virtually guarantees job contracts (Basok 2000). In early 2019, the Canadian government announced a plan allowing immigration officers to issue open work permits to temporary foreign workers who can prove they have been abused by their employer, addressing the serious issue of workers being tied to one employer who can take advantage of them and terminate their work permits (Keung 2019). For all these reasons, Massey and Brown (2011, 130) find that "once embedded in the Canadian system, [Mexican agricultural workers] tend to return to Canada and do not look upon the United States as an alternative.... In other words, Mexicans appear to shift from American to Canadian migration but not vice versa." This is clear evidence that Mexican agricultural workers find the Canadian temporary agricultural worker program more favorable than that of the United States, even with its own drawbacks.

All this suggests that the United States is much more likely than Canada to experience labor shortages related to temporary migrant labor, and Mexican labor in particular. Indeed, the increasing political and social threats migrant workers face in the United States may divert some of these workers into the Canadian system. We might therefore expect to see different trends in HR weed management given the difference in the two countries' relative access to the hand labor of temporary Mexican agricultural workers. To investigate this trend, and specifically US farmers' experiences, opinions, and perceptions, we analyzed data from 10 focus groups with commodity crop farmers from four states.

Methods

Data analyzed in this chapter were collected as part of a larger research project funded through USDA-Agriculture and Food Research Initiative grant no. 122422 on "Integrating Human Behavioral and Agronomic Practices to Improve Food Security by Reducing the Risk and Consequences of Herbicide-Resistant Weeds." Between February and May 2015, 10 focus groups with corn and soybean growers were conducted. This population has traditionally had minimal reliance on hired labor; however, the corn and soybean industries are particularly reliant on herbicides and thus are threatened by HR weeds. Two focus groups took place in Arkansas, four in Iowa, two in Minnesota, and two in North Carolina. Hereafter, focus groups are referred to using the state abbreviation followed by a number signifying chronological order (e.g., MN1 signifies the first Minnesota focus group).

Focus groups had between 6 and 10 participants each, as recommended by Morgan (1997), for a total of 64 participants in the study. Table 6.1 presents a summary of participant demographics by state. While the sample misses some of the diversity of US farmers—such as small farmers and minority farmers—it does reflect the majority experience of US commodity grain crop growers. After an icebreaker question, three consistent lines of open-ended questioning were employed: (1) How should/would a farmer react to HR weeds on their own farm? (2) How should/would a farmer react to HR weeds on their neighbor's farm? (3) Is herbicide resistance a short-term or long-term problem?

Following Saldaña's (2015) technique, an emergent coding process was used to identify themes arising naturally from the focus group discussions. Coding of focus group data was completed in three phases following transcription. First, emergent themes were identified by in-depth reading of each transcript. Second, themes were coded into categories while noting representative quotations in the form of statements and conversations. Finally, the transcripts were recoded with the main themes in mind—one of which was labor. Other main themes included techno-optimism and

Labor and the Problem of Herbicide Resistance

Table 6.1
Demographics of focus group participants by state

	Minnesota	Iowa	North Carolina	Arkansas
Number of participants	16	25	11	12
Males	16	23	11	12
Females	0	2	0	0
Age	32–68	33–77	44–79	24–73
White	16	25	8	11
Nonwhite	0	0	3	1
Acres managed	200–6,200	110–6,000	0–1,600	2,200–9,200
Acres owned	0–2,600	0–1,500	0–900	40–2,200
Farms with partner	12	18	5	10
Grows corn	16	25	8	9
Grows soybeans	16	25	10	12
Grows cotton	0	0	0	8

individualism, which are addressed elsewhere (Dentzman, Gunderson, and Jussaume 2016; Dentzman and Jussaume 2017; Dentzman 2018). Subthemes and representative quotations were then identified.

Findings

Although we did not specifically ask focus group participants about labor struggles related to HR weeds, they consistently brought up this theme on their own. Specifically, the farmers mentioned how difficult it is to find laborers, the increasing need to employ temporary migrant workers to hand weed, and the high cost of employing these workers. In Iowa, growers mentioned labor as a significant issue impacting their ability to control weeds physically.

IA2

PARTICIPANT: And then, you've got the issue about labor, which is a major problem here in Iowa for agricultural endeavors.

Specifically, focus group participants in Iowa mentioned their inability to find laborers to hoe HR weeds out of their fields. Additionally, several growers in North Carolina witnessed fellow farmers hire migrants to pull weeds when they had never done so in previous years, a change they attributed to growing problems with HR weeds such as pigweed (*Amaranthus palmeri*).

NC2

PARTICIPANT 1: They was—beans were probably about that tall [holds hands roughly 6 inches apart]. And the pigweed's about that tall [holds hands roughly 12 inches apart]. And they had a group of Mexicans going through and pulling them up. It was right there on the 38, before you get to Springville. But that had to be labor intense, but—

PARTICIPANT 2: —at least he was pulling them up.

PARTICIPANT 3: Looking at bean process, it don't take many hours pulling weeds before it becomes, I guess, economically inefficient. —

PARTICIPANT 1: Cost prohibitive.

PARTICIPANT 3: —having to go out there and pay somebody to pull weeds up.

PARTICIPANT 1: But that's the first time I'd ever seen them do it. But this farmer, he was doing it.

In this discussion, growers directly referenced the newness of seeing Mexican laborers in a soybean field pulling HR weeds. Before the widespread occurrence of these weeds, such laborers would have been unnecessary. Now, even though they were sure it must be costly, growers recognized the increasing importance of foreign temporary agricultural workers to soybean production as a result of herbicide resistance. For the first time in a long time, successfully growing soybeans required hiring migrant workers.

When discussing the specifics of labor shortages, our focus group participants frequently concentrated not only on the *need* for migrant workers but also on the *availability* of migrant labor for HR weed control. In the following discussion in Minnesota, several growers explain how the supply of migrant labor is dwindling.

MN2

MODERATOR: Do you think that's going to be an increasing problem?

PARTICIPANT 1: Labor—

PARTICIPANT 2: Sure seems like it is.

PARTICIPANT 3: —always is.

PARTICIPANT 1: Yeah.

PARTICIPANT 2: Sure seems like it's always a...I mean, we went from—we went from...We would have probably 25 part-time laborers, you know, before round up sugar beets that came up—I mean, they moved—they came up from Texas or Mexico or wherever they all came from, and that—they're not coming anymore. Those people aren't out there to do that.

This discussion reflects the reality that migrant laborers are decreasing in availability in the United States right as commodity grain crop growers are starting to need to hire them (Charles 2017; Fan et al. 2015). In addition to this issue of availability, growers in our focus groups were also concerned with the cost of hiring migrant workers.

NC1

PARTICIPANT 1: Sitting on the porch, you make more than you can out there pulling weeds. I mean it's just so expensive to hire people to pull the weeds. I mean, I do it in tobacco, but that's terribly expensive.

PARTICIPANT 2: Yeah.

PARTICIPANT 3: It cost me $25,000 one year, pulling weeds, just pulling weeds.

AR2

MODERATOR: How much y'all spend each year on hoeing [HR weeds in corn]?

PARTICIPANT: I don't know. To be honest, he says every year, we're stopping at $100,000, and we've chopped about a month past that every year.

These two examples show just how serious a cost it is for growers to hire laborers to hand remove HR weeds. Although the participants believed that the cost was in a sense "worth it," it is doubtful that such an expense will be sustainable for the majority of farms in the long run. Both the availability and cost of labor look to be increasingly significant barriers to physical management of HR weeds. According to Friedland and colleagues (1981), this should result in a movement toward technological and mechanical innovation in weed control.

In contrast to Friedland's theory, mechanized innovations to alleviate labor shortages related to HR weeds have been minimal in the United States. Growers instead tend to pin their hopes on chemical technology solutions.

MN1

PARTICIPANT: I think we're going to have to find chemicals.

IA2

PARTICIPANT: I think we're all hoping somewhere in that chemistry, there's something that comes around that's a new version. Yeah, stall long enough, maybe they'll figure something out, give us another product. That's about it.

Rather than mechanization, growers were hoping for future chemical innovation to provide labor savings in the face of herbicide resistance. However, chemical solutions are becoming few and far between. It takes up to 10 years to develop and release a new herbicide mode of action, and experts believe that there are few new herbicides left to discover (Livingston et al. 2015). This leaves commodity grain crop growers in a paradox: labor is becoming increasingly necessary while simultaneously becoming more scarce and expensive, yet the producers of the chemical herbicides they have come to rely on to avoid labor problems are offering no new solutions.

Discussion

Our focus group discussions suggest that migrant labor shortages *are* increasingly relevant to US commodity field crop growers because of the increasing prevalence of HR weeds that must be hand pulled. Although not all focus group participants had hired foreign laborers to pull weeds, they were widely aware of other farmers who had done so. They also had justified concerns over the decreasing availability and increasing cost of such labor. This prompted a continued hope for the development of new chemical herbicide technologies. However, such new technologies have not appeared on the market in over 20 years and are unlikely to do so in the near future (Beckie 2014). This corresponds with Friedland and colleagues' second condition for mechanization in a commodity sector—the economic and opportunity costs of chemical solutions are becoming high enough that mechanical solutions have a reduced cost in comparison. Additionally, their first condition for mechanization is already being met in the United States, as stringent immigration policies and an ineffective temporary agricultural worker program are blocking the temporary migrant agricultural worker supply (Zahniser et al. 2018). This suggests that as chemical herbicide development continues to lag and become less effective, the US commodity grain industry will undergo a mechanical transformation in HR weed control.

While the US situation seems primed for mechanization of weeding, the temporary agricultural workers program in Canada suggests a different trajectory for its grain industry. Though Canada is also facing the lag in chemical innovation, Canadian farmers are significantly less impacted by shortages of foreign agricultural workers. Following the theory of Friedland and colleagues, we would therefore expect Canadian farmers to continue relying on the relatively stable supply of inexpensive temporary agricultural workers from Mexico and put off a movement toward mechanization. When labor is inexpensive and readily available, a significant systemic change in agriculture, such as the mechanization of weed control, is highly unlikely (Friedland et al. 1981).

In the future, we expect a divergence in HR weed control techniques in the United States and Canada as a result of the two countries' differing attitudes toward and policies for Mexican temporary agricultural laborers. The answer to what comes next in the fight against HR weeds therefore depends on individual countries' policies toward migrant labor. In the United States, the tandem need for and shortage of migrant labor is pushing US grain crop growers toward reliance on herbicide-based innovation for weed control. Such a "solution" does not challenge the current structure of the agricultural system or larger migration policies in the United States. In this sense, it is the simple solution, because it requires no significant departure from current practices. However, this system is breaking down in the face of arrested chemical innovation, and systematic social change—in terms of migration policies and the structure of agriculture and its labor practices—is not being considered as a viable solution. Given the current political climate and the oppressive focus on productivist agriculture (McDonagh 2014), the conditions in the United States seem incompatible with such changes. If the structures of US agriculture and immigration policies are not reformed, the next solution growers reach for may well be mechanical weed management, which would slightly alter agricultural practices through scientific change yet continue to ignore needed social reform.

In Canada, we are unlikely to see a similar movement toward mechanical weed control. While Canadian commodity grain crop growers also face serious herbicide-resistance threats and a lack of new herbicide development—satisfying Friedland and colleagues' second condition for mechanization—they do not fulfill the first condition. That is, the migrant labor supply is not being blocked or reduced. Thus, the path of least resistance is to avoid undergoing any type of change, whether agricultural or social. However, just because social change is not *necessary* for growers to combat herbicide resistance does not mean that it should be disregarded. The serious and often brushed aside problems with Canada's SAWP present a moral imperative that cannot be ignored. Temporary foreign agricultural workers are not just a commodity, and they deserve the same treatment and opportunities as native-born Canadians. If Canada truly wants to have a model temporary migrant agricultural worker program, it has the opportunity to lead the way by reforming its program with a focus on the human rights of workers. The problem of HR weeds, however, is unlikely to provide an impetus for such change.

Conclusions

According to our focus groups, migrant labor shortages are not only a horticulture problem. Instead, they are increasingly becoming a problem for commodity grain crop growers as HR weeds become a serious issue and herbicide innovation ceases. Although more research is needed, it appears that Friedland and colleagues' theory of labor determining technological change in an agricultural sector is useful for predicting the future of HR weed control in countries with differing migrant labor policies. It also suggests that controlling HR weeds will not be an issue of only agriscientific or only social-scientific change. Rather, the two are inextricably intertwined in the determination of HR weed control practices.

Taking this into account, we would expect a future in which reliance on chemicals is no longer the solution to labor shortages and HR weeds. Instead, we would predict continued reliance on migrant labor in Canada and the development and adoption of a mechanized weed control technology in the United States. However, the potential mechanization of weed control is currently only in its nascent stage. The only successful mechanical weed management innovation to date is the Harrington Seed Destructor (HSD) in Australia, which attaches to a combine and pulverizes weed seeds in a rotating cage mill (Walsh, Harrington, and Powles 2012). It is not a stand-alone weed management solution but rather can be combined with chemical weed control to delay resistance (Walsh, Harrington, and Powles 2012). Although demonstrably effective at destroying weed seeds, the HSD costs about $240,000 and only two were sold between 2012 and 2014—both in Australia (Jacobs and Kingwell 2016; Walsh, Harrington, and Powles 2012). The HSD is currently unavailable in the United States, although it is predicted to become commercially available in 2019 (Jason Norsworthy, personal communication, July 8, 2017). At this point, it is unclear whether mechanical weed control such as the HSD will be effective enough—and immediate enough—to keep US farmers profitable in the face of rising costs associated with HR weeds. If mechanical solutions are ineffective, this could open the door for farmers and other agricultural stakeholders to start considering systematic social change as a more viable solution to controlling HR weeds.

Both the US and Canadian cases provide opportunities to highlight the problems with their temporary migrant agricultural worker programs. These problems are not only practical but also moral, as with the serious human rights abuses in the programs of the United States and Canada. At the intersection of technology and social forces, new conversations about national immigration policies and the structure of agriculture can potentially flourish. Particularly in the United States, if mechanization of weed control does not provide the hoped for solution to HR weeds, farmers and other stakeholders may find a growing incentive to lobby for change in how agricultural labor is managed. This underlines the fact that addressing herbicide-resistance management is increasingly becoming a technological issue inseparably intertwined with social justice and migration policies. In this light, the future livelihoods of farmers and agricultural workers, as well as the future of agriculture as a whole, depend on our ability to employ an interdisciplinary lens to a multitude of issues in agriculture. Such a perspective must be used to analyze technological and social issues as conjointly constituted determinants of both agricultural problems and possible solutions.

Notes

1. I-9 audits became preferable to workplace raids in the Department of Homeland Security (DHS) following the 2008 US presidential election (Martin 2013). During I-9 audits, DHS agents review employment records and notify employers of any unauthorized workers.

2. E-Verify is a federal program that requires all US employers, regardless of employment sector, to electronically submit employees' information to check their legal status against government databases. It was developed to ensure that only those who are legally permitted to work in the United States can do so.

References

Basok, Tanya. 2000. "Migration of Mexican Seasonal Farm Workers to Canada and Development: Obstacles to Productive Investment." *International Migration Review* 34 (1): 79–97.

Basok, Tanya. 2007. *Canada's Temporary Migration Program: A Model Despite Flaws*. Migration Policy Institute. https://www.migrationpolicy.org/article/canadas-temporary-migration-program-model-despite-flaws.

Beckie, Hugh J. 2014. "Herbicide Resistance in Weeds and Crops: Challenges and Opportunities." In *Recent Advances in Weed Management*, edited by Bhagirath S. Chauhan and Gulshan Mahajan, 347–364. New York: Springer.

Bellenger, Moriah, Deacue Fields, Kenneth Tilt, and Diane Hite. 2008. "Producer Preferences for Migrant Labor and the Wage, Hours, and Gross Sales Effects in Alabama's Horticulture Industry." *HortTechnology* 18(2): 301–307.

Bonanno, Alessandro. 2015. "The Political Economy of Labor Relations in Agriculture and Food." In *Handbook of the International Political Economy of Agriculture and Food*, edited by Alessandro Bonanno and Lawrence Busch, 249–263. Cheltenham: Edward Elgar Publishing.

Boudreau, Catherine. 2016. "Looming Crop Losses as Farmers Face Labor Shortages." *Politico*, Morning Agriculture, April 22, 2016. http://www.politico.com/tipsheets/morning-agriculture/2016/04/looming-crop-losses-as-farmers-face-labor-shortages-clinton-sanders-sharply-disagree-on-soda-tax-uk-ag-minister-its-gonna-be-lonely-213910.

Brandt, Deborah. 2002. *Tangled Routes: Women, Work, and Globalization on the Tomato Trail*. Aurora, ON: Garamond Press.

Castles, Stephen, Hein de Haas, and Mark J. Miller. 2014. *The Age of Migration: International Population Movements in the Modern World*. 5th ed. New York: Guilford Press.

Charles, Dan. 2017. "Government Confirms a Surge in Foreign Guest Workers on U.S. Farms." National Public Radio, May 18, 2017. https://www.npr.org/sections/the salt/2017/05/18/528948143/government-confirms-a-surge-in-foreign-guest-workers-on-u-s-farms.

Chui, Tina, and John Flanders. 2013. *Immigration and Ethnocultural Diversity in Canada: National Household Survey, 2011*. Ottawa: Statistics Canada. https://www12.statcan.gc.ca/nhs-enm/2011/as-sa/99-010-x/99-010-x2011001-eng.pdf

Danger, Cecilia. 2000. "The H-2A Non-immigrant Visa Program: Weakening Its Provisions Would Be a Step Backward for America's Farmworkers." *University of Miami Inter-American Law Review* 31(3): 419–438.

Dentzman, Katherine. 2018. "'I Would Say That Might Be All It Is, Is Hope': The Framing of Herbicide Resistance and How Farmers Explain Their Faith in Herbicides." *Journal of Rural Studies* 57 (January): 118–127.

Dentzman, Katherine, Ryan Gunderson, and Raymond Jussaume. 2016. "Techno-optimism as a Barrier to Overcoming Herbicide Resistance: Comparing Farmer Perceptions of the Future Potential of Herbicides." *Journal of Rural Studies* 48 (December): 22–32.

Dentzman, Katherine, and Raymond Jussaume. 2017. "The Ideology of US Agriculture: How Are Integrated Management Approaches Envisioned?" *Society and Natural Resources* 30(11): 1311–1327.

Devadoss, Stephen, and Jeff Luckstead. 2011. "Implications of Immigration Policies for the US Farm Sector and Workforce." *Economic Inquiry* 49(3): 857–875.

Fan, Maoyong, Susan Gabbard, Anita Alves Pena, and Jeffrey M. Perloff. 2015. "Why Do Fewer Agricultural Workers Migrate Now?" *American Journal of Agricultural Economics* 97 (3): 665–679.

Finnigan, Pat. 2017. *A Food Policy for Canada: Report of the Standing Committee on Agriculture and Agri-food*. Ottawa: House of Commons, Canada.

Francis, Janae. 2018. "Farmers Lobby for Immigration Reform to Address Labor Shortages." *Christian Science Monitor*, February 23, 2018. https://www.csmonitor.com/USA/2018/0223/Farmers-lobby-for-immigration-reform-to-address-labor-shortages.

Friedland, William H., Geoffrey Dunn, Amy E. Barton, and Robert J. Thomas. 1981. *Manufacturing Green Gold: Capital, Labor and Technology in the Lettuce Industry*. New York: Cambridge University Press.

Government of Canada. 2016. "Hire a Temporary Worker through the Seasonal Agricultural Worker Program—Overview." https://www.canada.ca/en/employment-social-development/services/foreign-workers/agricultural/seasonal-agricultural.html.

Guan, Shengfei, Feng Wu, Fritz Roka, and Alicia Whidden. 2015. "Agricultural Labor and Immigration Reform." *Choices* 30(4): 1–9.

Guerra, Lisa. 2004. "Modern-Day Servitude: A Look at the H-2A Program's Purposes, Regulations, and Realities." *Vermont Law Review* 29(1): 185–214.

Hall, Mary Lee. 2001. "Defending the Rights of H-2A Farmworkers." *North Carolina International Law and Commercial Regulation* 27(3): 521–538.

Halle, Sara R. 2007. "Proposing a Long-Term Solution to a Three-Part American Mess: US Agriculture, Illegal Labor, and Harvest Mechanization." *Drake Journal of Agricultural Law* 12(2): 359–390.

Harker, K. Neil. 2013. "Slowing Weed Evolution with Integrated Weed Management." *Canadian Journal of Plant Science* 93(5): 759–764.

Heap, Ian. 2017. *The International Survey of Herbicide Resistant Weeds*. www.weedscience.org.

Hennebry, Jenna L., and Kerry Preibisch. 2012. "A Model for Managed Migration? Re-examining Best Practices in Canada's Seasonal Agricultural Worker Program." *International Migration* 50(S1): e19–e40.

Hernandez, Trish, Susan Gabbard, and Daniel Carroll. 2016. *Findings from the National Agricultural Workers Survey (NAWS) 2013–2014: A Demographic and Employment Profile of United States Farmworkers*. Report 12, December. Washington, DC: US Department of Labor, Employment, and Training Administration, Office of Policy Development and Research.

Jacobs, Ashley, and Ross Kingwell. 2016. "The Harrington Seed Destructor: Its Role and Value in Farming Systems Facing the Challenge of Herbicide-Resistant Weeds." *Agricultural Systems* 142 (February): 33–40.

Keung, Nicholas. 2019. "Ottawa Proposes Open Permits for Migrant Workers Who Are Abused." *Toronto Star*, January 11, 2019.

Livingston, Michael, Jorge Fernandez-Cornejo, Jesse Unger, Craig Osteen, David Schimmelpfennig, Tim Park, and Dayton M. Lambert. 2015. "The Economics of Glyphosate Resistance Management in Corn and Soybean Production." Economic Research Report ERR-184. Washington, DC: US Department of Agriculture.

MacLean Wells, Donald, and Janet McLaughlin. 2018. "The Cruel Trade-off at Your Local Produce Aisle." *The Conversation*, January 16, 2018.

Martin, Philip. 2009. *Importing Poverty? Immigration and the Changing Face of Rural America*. New Haven, CT: Yale University Press.

Martin, Philip. 2013. "Immigration and Farm Labor: Policy Options and Consequences." *American Journal of Agricultural Economics* 95(2): 47–75.

Martin, Philip. 2015. "Immigration and Farm Labor: Challenges and Opportunities." *AgBioForum* 18(3): 252–258.

Martin, Philip. 2019. "Trump, Migration, and Agriculture." *Border Crossing* 9(1): 19–27.

Massey, Douglas S., and Amelia E. Brown. 2011. "New Migration Stream between Mexico and Canada." *Migraciones Internacionales* 6(1): 120–144.

McDonagh, John. 2014. "Rural Geography II. Discourses of Food and Sustainable Rural Futures." *Progress in Human Geography* 36(6): 838–844.

Morgan, David L. 1997. *Focus Groups as Qualitative Research*. Qualitative Research Methods16. Thousand Oaks, CA: Sage Publications.

Mueller, Richard E. 2005. "Mexican Immigrants and Temporary Residents in Canada: Current Knowledge and Future Research." *Migraciones Internacionales* 3(1): 32–56.

Norsworthy, Jason K., Sarah M. Ward, David R. Shaw, Rick S. Llewellyn, Robert L. Nichols, Theodore M. Webster, Kevin W. Bradley et al. 2012. "Reducing the Risks of Herbicide Resistance: Best Management Practices and Recommendations." *Weed Science* 60(sp1): 31–62.

Norton, Dean. 2016. "NY Farm Bureau President: Farmworker Union Will Put Family Farms Out of Business (Commentary)." *The Post Standard*, June 3, 2016. http://www.syracuse.com/opinion/index.ssf/2016/06/ny_farm_bureau_president_farmworker_union_will_put_family_farms_out_of_business.html.

O'Brien, Patrick, John Kruse, and Darlene Kruse. 2014. *Gauging the Farm Sector's Sensitivity to Immigration Reform via Changes in Labor Costs and Availability*. Washington, DC: American Farm Bureau Federation.

Preibisch, Kerry. 2010. "Pick Your Own Labor: Migrant Workers and Flexibility in Canadian Agriculture." *International Migration Review* 44(2): 404–441.

Preibisch, Kerry L. 2007. "Local Produce, Foreign Labor: Labor Mobility Programs and Global Trade Competitiveness in Canada." *Rural Sociology* 72(3): 418–449.

Rickard, Bradley J. 2015. "On The Political Economy of Guest Worker Programs in Agriculture." *Food Policy* 52 (April): 1–8.

Saldaña, Johnny. 2015. *The Coding Manual for Qualitative Researchers*. London: Sage Publications.

Simms, Theodore C. 2000. "A Fighting Chance: An Examination of Farmers' New Freedoms and Familiar Problems under the H-2A Guestworker Program." *Drake Journal of Agricultural Law* 5(2): 501–519.

Taylor, J. Edward, Diane Charlton, and Antonio Yúnez-Naude. 2012. "The End of Farm Labor Abundance." *Applied Economic Perspectives and Policy* 34(4): 587–598.

Tomson, Bill. 2015. "Farmers: Trump 'Terrible for Agriculture.'" *Politico*, September 1, 2015. https://www.politico.com/story/2015/09/donald-trump-2016-farmers-fear-argriculture-213201.

Walsh, Michael J., Raymond B. Harrington, and Stephen B. Powles. 2012. "Harrington Seed Destructor: A New Nonchemical Weed Control Tool for Global Grain Crops." *Crop Science* 52(3): 1343–1347.

Weiler, Anelyse M., Janet McLaughlin, and Donald C. Cole. 2017a. "Food Security at Whose Expense? A Critique of the Canadian Temporary Farm Labour Migration Regime and Proposals for Change." *International Migration* 55(4): 48–63.

Weiler, Anelyse M., Janet McLaughlin, and Donald C. Cole. 2017b. "Helping Migrant Workers Must Be Part of New Food Policy." *Toronto Star*, December 22, 2017.

Wheat, Dan. 2018. "H-2A Rule Changes May Come." *Capital Press*, January 18, 2018. www.capitalpress.com/Washington/20180118/h-2a-rule-changes-may-come.

Zahniser, Steven, J. Edward Taylor, Thomas Hertz, and Diane Charlton. 2018. *Farm Labor Markets in the United States and Mexico Pose Challenges for US Agriculture*. United States Department of Agriculture Economic Research Bulletin 1476-2018-8188, November.

7 Labor and Legibility: Mexican Immigrant Farmers and Resource Access at the US Department of Agriculture

Laura-Anne Minkoff-Zern and Sea Sloat

Introduction

Following a US Department of Agriculture (USDA) staff member in her white sedan with government plates, we drove our own unmarked rental car along a winding country highway. We passed corn and soybean fields, farmhouses, and a small downtown with a few local businesses. We drove up a gravel driveway and parked behind the USDA car. Trailing the staff member, a white female soil conservationist, we walked unannounced onto a farm with a few acres of diverse vegetables, a farmhouse, a shed, and a hoop house. The hoop house had been financed through a grant from the USDA's Natural Resources Conservation Service (NRCS), giving the staff member rights to visit and inspect the structure and property randomly for the first three years in order to validate that it is code compliant and being used properly.

The farm we visited is owned and operated by a Mexican immigrant farmer, one of a small number of immigrant farmers who directly participate in a USDA-funded program. USDA staff in the Northern Neck of Virginia promote the hoop house, or "high tunnel," installation program to local vegetable farmers. These tunnel-shaped greenhouses allow farmers to start their seeds and get their crops to market earlier in the season. The USDA covers the entire cost of the hoop house. In exchange, the farmer must agree to keep it in production for a minimum of three years, maintain meticulous records of their growing practices and finances, and allow USDA officials onto their property unannounced. This program is one of a variety of financial assistance opportunities for small- and medium-scale fruit and vegetable farmers through the USDA's NRCS and Farm Service Agency (FSA) (Farm Service Agency 2015).

Despite the fact that Latino/a farmers are a growing presence among new farmers in the United States, they have a low rate of inclusion in USDA programs nationally.[1] According to official USDA agricultural census data, self-defined Latino/a farmers utilized USDA loans and other direct assistance programs at about one-third to one-half the rate of white farmers. The number of farms with principal operators of "Spanish, Hispanic, or Latino origin" grew from 50,592 in 2002 to 55,570 in 2007. In 2012, the number increased again to 67,000 farms, a 21% increase over five years, with Latinos making up 3% of all principal operators.[2] Of those 67,000 Latino/a farm operators, the vast majority (64,439) were the primary farm business owners as well. In contrast, during the same period, the population of white principal operators fell 5% and overall the number of farmers dropped 4% (USDA 2014). As many Latino/a farmers transition from working as laborers in others' fields to positions as farm owners and operators, they, along with other farmers of color, represent the new face of a flourishing generation of farmers.

This chapter addresses why immigrant farmers are so unlikely to participate in USDA direct financial assistance programs, despite immigrant farmers' growth as a new group of farmers and particularly as a group that the USDA declares they want to support. We contend that the standardization of practices and bureaucracy inherent in receiving USDA assistance stands in stark opposition to the agrarian norms and practices of immigrant farmers and hinders their participation in USDA opportunities. The requirements of standardization help to maintain a racialized class boundary in US agriculture today, playing a large role in preventing immigrant farmers from moving up the agricultural ladder. While monitoring and recording farmers' activities is necessary at some level for the USDA to assure that funds are used appropriately, the extent to which farmers are asked to track activities and comply with standardization is impossible for most immigrant farmers. Furthermore, if their different practices and limited literacy and linguistic abilities are not considered, these farmers will never be able to take full advantage of the programs they so desperately need to succeed.

Between 1997 and 2000, four separate lawsuits targeted the USDA for racial and gender-based discrimination, particularly in FSA loan programs. In response to these suits, the US secretary of agriculture during the Obama administration, Thomas J. Vilsack, proclaimed a "new era of civil rights" in a memorandum to all USDA employees. In this memo, he announced an overhaul of the equal employment opportunity, civil rights, and program delivery processes at the agency, with the intent to "ensure fair treatment of all employees and applicants" (Vilsack 2009).[3]

Despite this proclamation and the fact that their numbers are growing, immigrant farmers to this day are not extended the same opportunities as other farmers, because their practices are often incompatible with the standardization and bureaucracy required to be properly acknowledged and supervised by the USDA. Their direct market approach, planting of diverse crops, reliance on family labor, and lack of record keeping stand in contrast to the dominant model of US industrial agriculture.[4]

It is not simply the size or scale of their farms that bars them from accessing USDA resources, although that certainly limits what is available to them. The farmers in this study have limited formal education, literacy, and English-language skills, and are therefore exceptionally daunted by the paperwork necessary for government grants, loans, and insurance applications. Additionally, it is not routine for immigrant farmers to record and track their own farming progress and decisions in writing. In contrast, their farming knowledge tends to be documented and disseminated through word of mouth. As has been the case for other farmers who do not replicate state-sanctioned or dominant forms of farming, these practices and forms of agrarian knowledge sharing may be interpreted as "unscientific" or "illegible" to the state and therefore not deemed worthy of acknowledgment (Scott 1998) or, in this case, acceptable for funding. Many small-scale diversified crop and vegetable farmers run up against the same challenges when looking for government resources, yet for the immigrant farmers in this study, the expectation for standardized practices is compounded with the above-mentioned lack of formal education, literacy, and language abilities. These barriers are made worse by workers' distrust of US government agencies as a result of their immigration experiences.

There is a growing body of geographical, anthropological, and sociological research on farm labor that critically engages with the politically produced vulnerability and exploitation of the immigrant body. This literature contributes to our understanding of historical and modern-day labor conditions in the agrifood system, which is necessary to gain a comprehensive picture of the political economy of food production and advocate for workers' rights throughout the food system. In particular, this work investigates the relationship between the immigrant worker and the state, providing a nuanced analysis of how US national policy and immigration agencies reinforce unjust working conditions and a racialized workforce (see Allen 2008; Brown and Getz 2008; Guthman and Brown 2015; Gray 2013; Holmes 2013; Mitchell 1996; Sbicca 2015; and many others). However, critical analysis of Latino/a workers thus far does not include the possibility that some immigrant workers are in fact advancing in this agrarian class system. Furthermore, there has been almost no comprehensive inquiry into how immigrant farm owners are experiencing state apparatuses. The research presented in this chapter makes this needed intervention, exploring how immigrant farmers interact with the state through their engagement, or lack thereof, with the USDA.

This chapter is based on semistructured interviews between 2011 and 2016 with over 70 immigrant farmers in Washington, Minnesota, California, Virginia, and New York, as well as 47 interviews with staff in government and nonprofit programs who work with immigrant farmers. Almost all farmers emigrated from Mexico, and all identify as Latino/a or Hispanic. They are a mix of resident aliens, naturalized citizens, and undocumented immigrants who have been in the United States for a range of 4 to 25 years. Most speak limited English, and Spanish is their first language, although for some even Spanish is a second language.[5]

All farmers in this study own their farm business, differentiating them from a farm laborer working under an employer. The farmers have been operating their own farms for 2 to 20 years. They all farm on a relatively small scale, on plots ranging from 3 to 80 acres, with most between 10 and 20 acres. The majority practice some form of integrated pest management with low chemical input or organic cultivation, growing diverse crops using mostly family labor. Most farmers prioritized direct sales, specifically farmers markets. Some, particularly in California, could not enter into direct markets because of market saturation and had no option but to sell to produce brokers. Almost all farmers interviewed expressed a desire to maintain this farming style and to remain living on or near the land they cultivate.

These practices contrast with the dominant industrial model most commonly promoted by the USDA. The industrial agriculture model has long been problematic for smallholder farmers as well as more diversified growers, regardless of race, ethnicity, or citizenship status. Earl Butz, secretary of agriculture in the Nixon administration, was known for his mantra, "Get big or get out" (Scholar 1973). Butz's policies, and those of the USDA leadership since, have focused on supporting the large-scale production of commodity crops, corn and soy in particular, mainly through commodity price support and crop insurance programs. These decisions are not made only at the agency level. US agricultural policy is largely set by the United States Farm Bill, which is voted on by Congress every five years. By setting priorities and outlining fiscal parameters, the Farm Bill contributes to the prioritization of large-scale industrial production and deprioritizes the needs of smallholders, "specialty" crop growers (mainly fruit and vegetable producers), and other diversified growers (see Ahearn, Yee, and Korb 2005; Clapp and Fuchs 2012; Dimitri, Effland, and Conklin 2005; DuPuis 2002; among others).

In what follows, we discuss how particular USDA practices, programs, and expectations are unsuited to immigrant farmers' ways of cultivation. In addition to linguistic and cultural norms related to bureaucracy, paperwork, and communication, their farming practices are not typical of most commercial farmers in the United States, as they fit what might be deemed a more alternative farming approach. From the ways they plan for their season to the specific crops they grow, our research has shown that Latino/a immigrant farmers are not producing food in a way that conforms to the industrial agrarian model understood by the USDA, therefore making state resources inaccessible and limiting farmers' potential economic success.

Citizenship, Race, and Legibility

The United States has a long history of constituting citizenship—and related rights to land and resources—through whiteness. Racial formations, which occur through a process of "historically situated projects in which human bodies and social structures are represented and organized" (Omi and Winant 1994, 55–56), are imposed and reinforced via power relations within the US food and agriculture system. Previous groups of immigrants and farmers of color have been excluded from full citizenship rights in the United States because of state-sanctioned policies, which are reinforced through daily experiences of racialized exclusion. Nonwhite immigrant farmers have been explicitly dispossessed of land and capital, in many cases because of their racial and citizenship status (Chan 1989; Foley 1997; Matsumoto 1993; Minkoff-Zern et al. 2011; Wells 1991, 1996). These processes have succeeded in creating agricultural racial formations, resulting in the ownership and operation of US farms remaining under primarily white control.

The unjust and uneven consequences of agricultural racial formations are not limited to immigrants; there is a long and well-recorded history of discrimination against US-born farmers of color, particularly African American and Native American farmers (see Clearfield 1994; Daniel 2013; Gilbert, Sharp, and Felin 2002; Grim 1996; Payne 1991; Ponder 1971; Simon 1993; and many others). This discrimination has ranged from overtly racist treatment at local and federal USDA offices to deficient literacy assistance, legal counsel, and advertisement of available opportunities to help nonwhite farmers access and maintain their land and markets (Gilbert, Sharp, and Felin 2002).

Daniel (2013) draws on Scott's legibility argument to explain USDA discrimination against black farmers in the civil rights era, providing historical context within which to understand USDA policy and practice today. African American farmers in the United States, like Mexican and other immigrant farmers of color, have been displaced from their livelihoods many times over. This displacement occurred historically through the capture and enslavement of their ancestors from their homelands, and more recently as landowners and tenant farmers who faced systematic discrimination by the USDA. During the New Deal era, large farms and gridlike orderly homesteads were idealized as the form for spreading modern agricultural technologies. Black farming operations did not fit this model of efficiency and modernism, and therefore were not considered for subsidies and grants, contributing to the 93% decline in the number of black farmers from 1940 to 1974 (Daniel 2013).

Conversely, scholars have argued that the USDA has a history of democratic planning and resource distribution, as shown in the work of many agency leaders and other individuals who have worked explicitly with farmers of color, African American farmers in particular (Couto 1991; Gilbert 2015). These arguments directly conflict with Scott's monolithic description of the state. As such a large government agency, there is no single consistent way staff or leadership interacts with the public. Despite the generally industrial focus of USDA funds, there are USDA opportunities for small-scale farmers as well as for those who have been deemed sustainable or socially disadvantaged by the agency. These include the Sustainable Agriculture Research and Education (SARE) program and other research and development related to local food initiatives, such as farmers markets, which are the primary markets for the immigrant farmers included in this study.

Additionally, in our research we encountered USDA staff who are actively engaged with farming communities of color and some who specifically focus on immigrant and/or Latino/a farmers. Unfortunately, these practices were not the norm, and the staff who actively pursue opportunities to work with Latino/a immigrant or socially disadvantaged farmers expressed that there was a lack of structural support from the agency in that pursuit. Although there are USDA programs targeted to sustainable or diverse growers, this information cannot reach the farmers if they are not on the radar of the state in the first place.

The existence of immigrant farmers is often unknown or overlooked in day-to-day, on the ground USDA operations. In beginning our research with immigrant farmers, the first author made cold phone calls to USDA regional headquarters in five states across the United States: Virginia, New York, California, Minnesota, and Washington. In each case, when the author first called and asked to speak to someone who works with "immigrant farmers," the person on the end of the line responded as if the caller had asked about *farmworkers*, not farm business owners. The author consistently had to explain, "I am looking to speak with someone in your office that might work with immigrant *farmers*, as in farm business owners, not laborers." Even in regions where immigrant farmers exist in significant numbers, it took substantial explanation to start a conversation where USDA staff understood the specific group of farmers the author was interested in discussing. Staff were either unaware that Mexican immigrant farmers existed in their region or were so accustomed to thinking of Latino/a immigrants as agricultural workers that they disregarded their encounters with immigrant farmers until probed directly.

Even when Latino/a or other farmers of color do succeed in making it in the door of a USDA office, they have experienced rampant discrimination based on their racial identity, as evidenced by several lawsuits against the agency. In 1999, a class action lawsuit was settled by black farmers alleging racial discrimination by the USDA between 1981 and 1996 while applying for farm loans and assistance. In 2000, another class action suit was filed against the USDA on behalf of Hispanic farmers and ranchers who were discriminated against from 1981 to 2000, also while applying for USDA loans.

The USDA admitted to discrimination, and this case is currently being settled via a claims process where farmers are eligible to receive from $50,000 to $250,000 (Hispanic and Women Farmers and Ranchers Claims and Resolution Process 2012; Martinez and Gomez 2011). According to our contact with the Office of General Counsel at the USDA, the claims administrator received over 50,000 claims. The USDA approved 14.4% of the claims, while the rest were rejected. The USDA provided a one-line explanation to farmers whose claims were not accepted: "You failed to provide sufficient documentation, or the documentation that you provided was not sufficient to meet the requirements under the Framework" (Zippert 2015). As we will discuss, this statement reflects many immigrant farmers' general lack of standardization and documentation practices, which, we argue, are necessary in order to be deemed legible in the eyes of the USDA.

As is demonstrated by the growing numbers of immigrant farmers, those under pressure to conform often continue to create alternative agrarian spaces. Research by Wells (1996) on the struggle of Mexican immigrants in California agriculture in the 1970s and 1980s illustrates the ways Mexican farmers' practices have been persisting in this context. Her study reflects our own findings that Mexican immigrants prefer to make their farming decisions independently and find technical advice from governmental outsiders unsuitable to their own experiences and practices. Additionally, Wells observes that immigrants' lack of material resources and formal education to invest in their farm businesses leads them to be more dependent on their personal social networks and previous farm experience, which differentiates them from white farmers, who are more likely to learn from university and marketing guidelines.

This chapter thus advances literature on immigration and racial discrimination in agriculture, shedding light on how the USDA's processes are promoted as universally accessible or color-blind while they in fact maintain racial and ethnic divides in agriculture. Applying the notion of illegibility to the practices of immigrant farmers, we explore how government expectations of modernization largely function as gatekeepers to agricultural development and growth, despite individual and structural efforts to create inclusivity. In the case of immigrants, farmers marginalized by state authorities are still rising in number and drawing on their own agrarian knowledge and norms to preserve their agrifood traditions and lifestyles. These farmers are cultivating in a way that contributes to local economies

Labor and Legibility

and ecosystems, as well as creating a more culturally diverse populace of US farm owners. Although they are currently making their businesses work, many function on the edge of economic stability. Without government support and acknowledgment of these differences in agrarian practice, their livelihoods and farm businesses may not survive in the long term.

Time, Labor, and Spatial Control

If a visitor knows where to look, they might be able to tell an immigrant's field from their neighbor's. In contrast to the monocrop, uniform rows of wheat and corn that line most of the side of country highways in the Northern Neck of Virginia, Latino/a immigrants' fields tend to include huge varieties of produce, each row different from the next. Among the cultivated crops, plants such as *purslane* (also known as *verdolaga* or pigweed)—seen as a common weed by US-born farmers—are left to grow between the rows. Farmers in this region harvest such plants for their Latino/a customers and themselves to consume in soups and stews. Juxtaposing the perfectly managed rows of grain grown by midscale white farmers and kept meticulously free of wild plants by regular doses of pesticides and pest-resistant genetically modified seeds, the immigrant farmers' fields show signs of agroecological variety.

All farmers interviewed saw starting their own farm as a way to regain independence in their daily lives and labor in the face of their limited material wealth and political standing. In contrast to their experience as farmworkers, they have the ability to choose when to rise, what to plant, and how to pick their crops, as long as they operate a productive farm. Cultivation using practices that reflect their own experience reasserts immigrant farmers' control over their own labor. To protect this autonomy, many of the farmers we spoke with shied away from interactions with the state where they may be subjected to standardizing their practices to match a particular form of farming.

Each farmer interviewed has a unique story, but they all share the common experience of previously working as farm laborers. One Mexican farmer living in Virginia recounted his journey of starting his own business, which provides insight into why immigrant farmers place such importance on maintaining independence:

> When I decided to work for myself, I was working for someone else. I saw that after I worked for him for about five years, and he was becoming successful, making a lot of money. And I stayed the same, earning six dollars an hour.... One day

I said to him, "To start, this is good. But now I see that you're just there doing nothing, and I don't make anything. I don't make money. I'm the only one working." Because I was the only employee he had. ... He had at least two hundred, five hundred thousand dollars in earnings that I had made for him. And I said, "No, I'm killing myself for you. It's over. I'm going to start my own business." And that's how it happened.

This farmer, without access to standard bank loans because of his lack of a well-documented income history and related low credit score, started a farm by saving his small earnings. This was mirrored by all other farmers in this study, whose access to loans was scant. Beginning by renting a small plot and slowly saving enough to buy land, they started with almost nothing in terms of capital investment and depended on their experience, knowledge, self-exploitation, and family labor to advance their business.

Immigrant farmers' personal histories of exploitation as workers motivate them to seek more control over their daily activities and decision-making power concerning their land. All the farmers we spoke with relayed the physical and emotional challenges of farming: consecutive months of intensive labor, often 12 hours a day, seven days a week. They expressed that not being assured a paycheck at the end of the week is a precarious way to live. One farmer explained, "Here we live just from the land. There's no one paying us $8 an hour. There's no one paying us." As independent business owners, they are subject to the unpredictability of the market. As farmers, they are additionally vulnerable to uncertain weather and climate conditions. Overwhelmingly, though, the satisfaction that comes with making their own decisions keeps them farming, regardless of the struggles. As one farmer shared, "I feel happy that it's *my* business, that we can make our own decisions." Even in the most difficult times, the desire to maintain control over one's labor and growing practices transcends the daily obstacles of small-scale farming.

On their farms and in their businesses, farmers avoid cultivation systems imposed on them by outsiders, be they wholesalers who would tell them what to plant and how much (in order to secure a market) or government officials whose programs require particular crops and techniques to qualify for assistance, such as in the cover crop and hoop house programs. All the farmers interviewed plant diverse fruits and vegetables, an important strategy for selling directly to customers at farmers markets, their primary outlet for sales. Some noted that they sold to their extended community as well, as part of a more informal market. Rarely did we hear of them selling to restaurants or local stores; luxury crop buyers usually go with more socially connected and better-marketed white farmers, and contracts with large grocery chains go through a wholesale purchaser, requiring larger quantities than they grow. In most regions, they are able to avoid selling through a middleman or outlets that would require reducing their diversity or standardizing their practices. Growing diverse crops also often reflects their previous farming experience in Mexico and Central America, although climate, markets, crop varieties, and other resource availability differ greatly.

Farmers' diverse crops range from standard farmers market produce such as kale and heirloom tomatoes to less common products such as peanuts and purple potatoes. In addition to ones well known to American customers, they also plant Latin American crop varieties. Many farmers grow and sell herbs such as *pápalo* and *chipilín, pipián* (a squash variety), *tomatillos*, and hot *chiles*, which are hard to find in many parts of the United States. However, these choices to cultivate diverse crops, which work well for direct markets and reflect their own experience as farmers preimmigration, are not typically supported by the USDA programs made available to them in their local offices. For example, the regional office in the Northern Neck region of Virginia offers a cover crop assistance program, subsidized through state funds. But as the staff member from the local NRCS office told us, this program is not tailored to their needs as diversified fruit and vegetable farmers:

> I also offer this cover crop program for them. That program is through...it's a state program. But most of them—the cover crop has to stay on the land, between certain planting dates and certain dates that you have to destroy. And that date, the destroyer date is after. Because they start planting around February first: the beginning of February they start discing their land, preparing their land. And that cover crop has to stay on there until the middle of March. And that's not good for vegetable farmers at all because they need that time, they need that land. When it's ready to go, they're ready to go.

"So the cover crops work better for the grain farmers?," we asked:

> Yes. I have offered several times. I go out there and just try to push the program. And they say no, it's just not good for them because of the rules and regulations of the cover crop program.

This example of poor seasonal fit with available NRCS programs could be equally true for any fruit or vegetable farmer in the region. Yet, for immigrant farmers, who have fewer farming options because of their limited access to capital investment, land, and markets, this misalignment reinforces an existing inequality for already disenfranchised farmers.

Another example of this misalignment is the hoop house program. In addition to being subject to random visits and having to provide a detailed log of what is planted, how much was spent, and how much profit was made, farmers must also plant particular crops according to USDA guidelines in order to participate. Farmers must prepare and adhere to an operation and maintenance plan that includes particular instructions as to proper irrigation and planting practices and erosion control. This plan has to be reviewed and approved by an NRCS official. One farmer who chose to participate in the hoop house program conveyed both gratitude and frustration: "We were planting tomatoes, because they're very particular. They [the USDA] want certain stuff. You can't go ahead and do anything you want with them [the hoop houses]. ... And it's good help. I'm not saying it doesn't help, but we've managed to come so far on our own."

While the farmer expressed appreciation for the financial assistance, she also questioned whether the planting restrictions were worth the support. The requirement for standardization feels like a relinquishment of some part of her agrarian autonomy or the ability to make all farming decisions as she wishes. Even for those who succeed in securing state resources, they seem unsure about the decision to work within certain rules and regulations.

As can be expected from any government institution, the USDA requires extensive paperwork before, during, and after taking advantage of their loans, grants, or insurance options. When farmers were asked what they think the greatest challenge is for Latino/a farmers accessing USDA programs, most mentioned the paperwork. Although white farmers may also be resistant to paperwork and general bureaucracy, the fact that most farmers we interviewed did not have an education past middle school means they are lacking the literacy skills necessary to fill out the required paperwork in any language. Because of intimidation, most will never enter the door of the USDA to inquire about opportunities. For others, it may be the ultimate reason they stall in the process and fail to obtain the grant, loan, or insurance package.

Most farmers never looked into USDA programs like these, because of their suspicion of the government and government officials. This discomfort was compounded by their inability to navigate state bureaucracy, compounded by the language barrier. Regionally based USDA staff are often aware of the Latino/a farmer presence in their areas and lack of participation in programs. They discussed with us the ways they attempted to outreach to them, yet they were limited if they did not have a Spanish-speaking staff in their offices. In Virginia, a USDA staff member told us that there must be 10% participation in USDA programs in the region in order for bilingual forms to be made available. However, it is unlikely that there will ever be more than 10% participation if the paperwork is not made available in Spanish in the first place. This catch-22 is one example of the ways in which raced and classed inequality is structurally maintained at the USDA, aggravating the already tenuous history of USDA discrimination.

Without Spanish-speaking outreach abilities, most farmers never hear about the programs available. When asked about the USDA, most farmers interviewed were unaware of opportunities accessible to them. USDA FSA loans are designed for farmers who struggle with traditional bank loans and are meant to be a farmer's first line of credit. Although many farmers interviewed told us they were unable to get access to credit from regular banks, they were unaware that USDA loan programs existed for this reason specifically. Even those who spoke nearly perfect English found the forms intimidating. One immigrant farmer, who has obtained US citizenship, told us, "I tried in the past to get a small operating loan. And I didn't feel confident enough to fill out the application by myself because there were a lot of questions I didn't know." Since attempting to apply for her first USDA loan, this farmer has since applied for another loan, which she successfully secured with the assistance of the local FSA staff. Yet the level of confidence needed to walk into a government office where a huge stack of paperwork awaits is unrealistic for most. This expectation is especially difficult when understood in the context of the tense relationship between most rural Latino/a immigrants and the state, given their histories of immigration and the current rise of anti-immigrant sentiment in US politics.

The fact that paperwork and the related language barrier is the greatest impediment to aid for immigrant farmers is well understood by USDA staff in these counties. One local USDA staff member explained, "Most of our [Latino/a] producers used to come in the office. They don't come in anymore. I think it's English. Because we had one that couldn't speak English, and he would always bring his son in here. And then the forms. We have some forms that are in Spanish, but most of our forms aren't. I think it's... where they're used to dealing with more cash than a lot of paperwork. I think they find the paperwork a little overwhelming."

In addition to noting that the written forms themselves are a technical challenge, she highlighted that immigrant farmers are not accustomed to operating in bureaucratic environments. Even if the forms were in Spanish, their limited formal education makes the process of filling out paperwork extremely daunting. Ultimately, the paperwork and related language barrier is a reflection of the broader structural challenge of fitting immigrant farmers' diverse and nonconforming agrarian practices into standardized boxes. As we discuss, for immigrant farmers to move beyond these barriers and thrive in the challenging world of US farming, historically racist legacies and present-day racialized forms of exclusion must be accounted for, and government agencies must begin significant and structural change.

Toward a New Era of Inclusion

Under Vilsack's guidance, the USDA took several steps working toward a new vision of equality at the federal level. Since 2009, it has provided civil rights training to employees, established the Office of Advocacy and Outreach to aid beginning and socially disadvantaged farmers, and claims to be working toward resolving civil rights lawsuits inherited from previous administrations. The department also vowed to be an equal opportunity employer and create a workforce that "represents the full diversity of America" (USDA 2015).

This was all under President Obama's administration. As we were completing fieldwork, Donald Trump was elected president and Sonny Perdue, an agribusiness executive who took a strong anti-immigrant stance as the governor of Georgia, was sworn in as the secretary of agriculture. On a national level, we are seeing massive cuts in government spending, meaning further cuts on a regional and local scale in funding to university extension offices, grants, staff, and staff training, such as funds that could be used to improve racial inclusion of immigrant farmers in institutions, opportunities, and programs. While it is too early to know exactly how such reforms under the new administration will unfold for farmers in the long run, projections do not look positive. After the election, we followed up with staff at the USDA to inquire about what they thought the new administration would mean for immigrants and other farmers of color. Requests for feedback were either declined or not answered. While our inquiry was not exhaustive, we can imagine most staff still employed by the USDA are not looking to critique the administration from their current positions.

Unfortunately, even during the Obama administration, we found that despite claims of increased racial equality from the federal offices of the USDA, little change on the ground was being made in local and regional offices to directly help Latino/a immigrants overcome obstacles in order to transition from the role of farmworker to farmer in the United States. The processes of monitoring and standardization, as currently required by USDA programs, exacerbate the racial exclusion of immigrant farmers from state programs and ultimately from the advantages other farmers receive. This uneven rural development must be understood in the context of the historical relationship between Latino/a immigrants and the state as well as through the lived experiences of those struggling within a system where their practices are not deemed readable. Today's Latino/a immigrant farmers follow in this pattern of racialized others being left out of a system where predominantly white practices are deemed legible, and therefore legitimate, and predominantly brown and black practices are not.

As previously mentioned, programs that are developed for the specific needs of diversified fruit and vegetable, or specialty crop, growers already exist within the USDA. There are also microloan programs available through the FSA that are designed for "nontraditional" farmers. These require less paperwork and could be greatly helpful for Latino/a immigrants as they transition to farm ownership. Additionally, an Office of Minority and Socially Disadvantaged Farmers Assistance (MSDA) has been established within the FSA with the express purpose of assisting farmers such as those who participated in this study. These programs are a great start toward making government-supported programs available to immigrant growers. Regrettably, because of social divides and language and educational barriers, these programs are unknown to those most in need of assistance.

We do not claim that the USDA is the only institutional boundary for immigrant farmers or the only place where improvements can and should be made. Immigrant farmers struggle with access to capital, outreach and access to markets, general business skills, and many other management practices. While there are many entrepreneurial and nonprofit ventures that focus on advancement for and training of small farmers, farmers of color, and immigrant farmers, they are often working on shoestring budgets, with varying levels of accountability to their clients, and have limited access to resources and markets themselves. The USDA is the only state institution that claims to provide economic opportunities for rural communities and agricultural producers of the United States.

Immigrant farmers are challenging the historical racialized legacies of farming in the United States despite the odds and are persisting in new markets and climates that are seemingly unattainable. This chapter asks researchers and critical theorists to better recognize the perseverance of nonwhite farmers in order to build on our understanding of agricultural transitions and racial formations. As critical food scholars, we recommend that other food scholars and scholar-activists reframe their research and writing concerning immigrants of color from work that focuses on their immigrants' victimization to work that emphasizes their active role in creating sustainable food system change (in this volume, see Schmid, chapter 8; Passidomo and Wood, chapter 12; and Situational Strangers, chapter 14).

The USDA has the opportunity to support immigrant farmers' growth, but in order for programs and funding to reach the most financially disadvantaged beginning farmers, the agency must do more to recognize the challenges immigrant farmers experience in the current system. The state and civil society are by no means separate entities, and many within the USDA are actively working to create reforms to address its history of racism, but until these institutional norms are challenged, farmers of color and immigrant farmers in particular will continue to struggle with agrarian class mobility and with land and food-producing industries that remain primarily in white hands.

Notes

This chapter has been adapted from an article published in the journal *Agriculture and Human Values* 34(3): 631–643 (2017). Parts of this chapter can also be found in *The New American Farmer: Immigration, Race, and the Struggle for Sustainability* by Laura-Anne Minkoff-Zern (Cambridge, MA: MIT Press, 2019).

1. Although we completely support the transition to using the gender-inclusive Latinx instead of Latino/a, we use Latino/a, as that was the term most often used by participants in identifying themselves.

2. These numbers do not tell us how many are first-generation immigrants. The number of operators that were also owners before 2012 is not available.

3. Lawsuits include the *Pigford v. Glickman* and *Brewington v. Glickman* class action lawsuits for African American farmers, the *Keepseagle v. Vilsack* settlement for Native American farmers, and the Hispanic Farmers and Ranchers and Female Farmers and Ranchers claims processes.

4. We are not claiming that family labor is inherently a better system or more equitable, only that it is evidence of a particular form of farming (see Feldman and Welsh 1995; Reed et al. 1999; Riley 2009).

5. In Washington and California, many farmers interviewed identify as Triqui or Mixteco (indigenous to Mexico).

References

Ahearn, M. C., J. Yee, and P. Korb. 2005. "Effects of Differing Farm Policies on Farm Structure and Dynamics." *American Journal of Agricultural Economics* 87(5): 1182–1189.

Allen, P. 2008. "Mining for Justice in the Food System: Perceptions, Practices, and Possibilities." *Agriculture and Human Values* 25(2): 157–161.

Brown, S., and C. Getz. 2008. "Privatizing Farm Worker Justice: Regulating Labor through Voluntary Certification and Labeling." *Geoforum* 39(3): 1184–1196.

Chan, S. 1989. *This Bittersweet Soil: The Chinese in California Agriculture, 1860–1910*. Berkeley: University of California Press.

Clapp, J., and D. Fuchs. 2012. *Food*. Cambridge: Polity Press.

Clearfield, F. 1994. "Reaching Out to Socially Disadvantaged Farmers: Rural Development and a Changing USDA." In *Proceedings of the 51st Annual Professional Agricultural Workers Conference*, edited by N. Baharanyi, R. Zabawa, W. Hill, and A. Parks, 135–142. Tuskegee, AL: Tuskegee University.

Couto, R. A. 1991. "Heroic Bureaucracies." *Administration and Society* 23(1): 123–147.

Daniel, P. 2013. *Dispossession: Discrimination against African American Farmers in the Age of Civil Rights*. Chapel Hill: UNC Press Books.

Dimitri, C., A. B. Effland, and N. C. Conklin. 2005. *The 20th Century Transformation of US Agriculture and Farm Policy*, vol. 3. Washington, DC: US Department of Agriculture, Economic Research Service.

DuPuis, E. M. 2002. *Nature's Perfect Food: How Milk Became America's Drink*. New York: NYU Press.

Farm Service Agency. 2015. *Actively Engaged in Farming*. Farm Service Agency Programs and Services. www.fsa.usda.gov/programs-and-services/payment-eligibility/actively_engaged/index.

Feldman, S., and R. Welsh. 1995. "Feminist Knowledge Claims, Local Knowledge, and Gender Divisions of Agricultural Labor: Constructing a Successor Science." *Rural Sociology* 60:23–43.

Foley, N. 1997. *The White Scourge: Mexicans, Blacks, and Poor Whites in Texas Cotton Culture.* Berkeley: University of California Press.

Gilbert, J. 2015. *Planning Democracy: Agrarian Intellectuals and the Intended New Deal.* New Haven, CT: Yale University Press.

Gilbert, J., G. Sharp, and M. S. Felin. 2002. "The Loss and Persistence of Black-Owned Farmland: A Review of the Research Literature and Its Implications." *Southern Rural Sociology* 18(2): 1–30.

Gray, Margaret. 2013. *Labor and the Locavore: The Making of a Comprehensive Food Ethic.* Berkeley: University of California Press.

Grim, V. 1996. "Black Participation in the Farmers Home Administration and Agricultural Stabilization and Conservation Service, 1964–1990." *Agricultural History* 70(2): 321–336.

Guthman, J., and S. Brown. 2015. "I Will Never Eat another Strawberry Again: The Biopolitics of Consumer-Citizenship in the Fight against Methyl Iodide in California." *Agriculture and Human Values 33(3)*: 1–11.

Hispanic and Women Farmers and Ranchers Claims and Resolution Process. 2012. https://www.farmerclaims.gov.

Holmes, Seth. 2013. *Fresh Fruit, Broken Bodies: Migrant Farmworkers in the United States.* Berkeley: University of California Press.

Martinez, J., and R. E. Gomez. 2011. *Identifying Barriers That Prevent Hispanic/Latino Farmers & Ranchers in Washington State from Participating in USDA Programs and Services.* Yakima, WA: Rural Community Development Resources (RCDR) Center for Latino Farmers.

Matsumoto, V. J. 1993 *Farming the Home Place: A Japanese American Community in California, 1919–1982.* Ithaca, NY: Cornell University Press.

Minkoff-Zern, Laura-Anne, Nancy Peluso, Jennifer Sowerwine, and Christy Getz. 2011. "Race and Regulation: Asian Immigrants in California Agriculture." In *Cultivating Food Justice: Race, Class, and Sustainability,* edited by Alison Hope Alkon and Julian Agyeman, 65–85. Cambridge, MA: MIT Press.

Mitchell, Don. 1996. *The Lie of the Land: Migrant Workers and the California Landscape.* Minneapolis: University of Minnesota Press.

Omi, M., and H. Winant. 1994. *Racial Formation in the United States.* New York: Routledge.

Payne, W. C., Jr. 1991. "Institutional Discrimination in Agriculture Programs." *Rural Sociologist* 11(1): 16–18.

Ponder, H. 1971. "Prospects for Black Farmers in the Years Ahead." *American Journal of Agricultural Economics* 53(2): 297–301.

Reed, D. B., S. C. Westneat, S. R. Browning, and L. Skarke. 1999. "The Hidden Work of the Farm Homemaker." *Journal of Agricultural Safety and Health* 5(3): 317–327.

Riley, M. 2009. "Bringing the 'Invisible Farmer' into Sharper Focus: Gender Relations and Agricultural Practices in the Peak District (UK)." *Gender, Place and Culture* 16(6): 665–682.

Sbicca, J. 2015. "Food Labor, Economic Inequality, and the Imperfect Politics of Process in the Alternative Food Movement." *Agriculture and Human Values* 32(4): 1–13.

Scholar, H. H. 1973. "Federal Farm Policies Hit." *Reading Eagle*, October 23, 1973.

Scott, James C. 1998. *Seeing Like a State: How Certain Schemes to Improve the Human Condition Have Failed*. New Haven, CT: Yale University Press.

Simon, M. F. 1993. "Addressing the Problems of Agency Utilization by Kentucky's Black Limited Resource Farmers: Case Studies." In *Challenges in Agriculture and Rural Development, Proceedings of the 50th Annual Professional Agricultural Workers Conference*, edited by R. Zabawa, N. Baharanyi, and W. Hill, 129–134. Tuskegee, AL: Tuskegee University.

US Department of Agriculture (USDA). 2014. *Census of Agriculture*. www.agcensus.usda.gov.

US Department of Agriculture (USDA). 2015. *Secretary Vilsack's Efforts to Address Discrimination at USDA (OASCR, Vilsack's Efforts to Address Discrimination at USDA)*. http://www.ascr.usda.gov/cr_at_usda.html.

Vilsack, T. J. 2009. "Memo to All USDA Employees: A New Civil Rights Era for USDA." USDA Department of the Secretary. http://www.usda.gov/documents/NewCivilRightsEra.pdf.

Wells, M. J. 1991. "Ethnic Groups and Knowledge Systems in Agriculture." *Economic Development and Cultural Change* 39(4): 739–771.

Wells, M. J. 1996. *Strawberry Fields: Politics, Class, and Work in California Agriculture*. Ithaca, NY: Cornell University Press.

Zippert, J. 2015. "USDA Approves only 14% of Completed Claims in the Hispanic and Women Farmers and Ranchers Discrimination Settlement." *Greene County Democrat*.

8 Enterprising Women of Mexican American Farming Families in Southern Appalachia

Mary Elizabeth Schmid

Introduction

> We were just agriculturalists and now we do agribusiness. Our parents were campesinos. Our husbands are agriculturalists, but we have turned our work into agribusiness.
> —Liliana

Liliana has lawful permanent residency (LPR, or "green card") status in the United States and Mexican citizenship, much like other adult members of her binational kin group. She is in her late forties, seems to never stop moving, and always knows the local "going rate" for a 25-pound box of tomatoes. She and her family grow fresh-market tomatoes as their principal cash crop in the Carolina hill country of southern Appalachia. Their enterprise is part of a kin-based group of allied enterprises in southern Appalachia and the Mexican Bajío.[1] Members of this binational kin group maintain multiple political-economic identities, collaborate on farming enterprise projects, and contribute to the global agrifood system from multiple regions in North America.

In a number of ways, these Mexican American farming families defy the stereotype that shapes the American public's imagination of who migrant farmworkers are, what they know, and how they contribute to food economies. They are part of a recent 21% increase in Latinx[2] owned and operated farms in the United States from 2007 to 2012—a growth concentrated in the western and southern states (USDA 2012). However, in the contemporary southeastern US context, this phenomenon is obscured by the social category of "farmworker," which serves as a "racialized code word"

(Torres, Mirón, and Inda 1999, 10) for Latinx and is often conflated with the national identity of Mexican. Much of the popular and even academic literature on Mexico-US migration and the contemporary labor process in southeastern US agriculture follows the reductionist assumption that Mexicans and Mexican Americans are marginalized, temporary, (mostly) male, and migratory individuals who contribute to the food economy only as farmworkers. This framing does not take workers' enterprises into account or shed light on the ways in which persons from this racialized group gain access to land, labor, resources, and the multiple elements necessary to exercise the power to produce and distribute agrifood commodities in the contemporary, globally restructured context. This chapter challenges this framing and counterconstructs Latinas' roles in agriculture in southern Appalachia to support a reconceptualization of these (typically racialized) women as part of binational farming families who contribute to the North American agrifood system in diverse, significant ways.

Through ethnographic examples, this chapter shows how women from these families negotiate temporalities, manage resources, organize tasks, and collaborate in and across their farming enterprises. I argue that through creative socioeconomic practices they cultivate *comercio* mētis: practical commerce skills and wisdom that they apply to their work in the fresh fruit and vegetable (FFV) industry in southern Appalachia. To do this, I discuss *marketeras* (female marketers at terminal produce markets, also known as regional farmers markets), their *comercio* mētis in the FFV industry in southern Appalachia, and their perspectives on their farming enterprise practices and kin-based collective strategy. I conclude by indicating how the cultivation of *comercio* mētis—as exemplified by these binational families—is contributing to the revitalization of sustainable small to mid-level farming and a diverse community economy in southern Appalachia.

Family FFV Farming Enterprise Women: *Comercio* Mētis and Alternative Visions

Managing time and resource flows is essential to the success of farming fresh fruits and vegetables (FFV). In this context, "women's work" constitutes activities like these, which create value for enterprises in both the short and long terms. This work is both "intellectual" and "manual" and manifests in fields, packinghouses, farm offices, and marketplaces, where women complete FFV enterprise activities such as negotiating sales and maintaining records. According to many FFV industry actors, the management of enterprise flows—in the fields, farm office, and in-between—is critical to maintaining contemporary FFV farming enterprises. Robert, a tomato farmer in southern Appalachia, whose wife does the enterprise record keeping, believes, "The financial part is just as big as the work. More so nowadays. That is the single thing that will get a farmer: He can't handle money. If he can't handle the finances, it will get him on anything. He can do the work. Used to, you could make a living with your back. If you don't manage it now, you can't."

The farming enterprise women I spoke with agreed with Robert's assessment and believe that it takes a team to run an FFV farming enterprise. Yaneli links her arrival in the United States to the beginning of her family's farming enterprise in southern Appalachia:

> Yo me vine en los 90's y él ya había hecho una aplicación por mí, e hizo el proceso y todo. Por entonces, ya mi esposo empezó a traer más dinero y entonces decidimos hacer una compañía. Cuando vino, él no [lo] hacía por su propia cuenta, no hasta que yo llegue. Después fue cuando el empezó a hacer sus cosas por su propia cuenta, sembró él solo y yo empecé a manejar el negocio.

> (I came in the 90s and he had already done an [immigration, LPR] application for me and did the process and everything. By then, my husband had started to make more money so we decided to create a company. When I came, he wasn't working for himself, not until after I arrived. That was when he began to do things by himself, he grew things by himself and I started to manage the business.)

Notice the critical phrase "we decided." Yaneli, together with her husband, developed their farming livelihood enterprise strategy. As I show later in the chapter, Yaneli and her kin act collectively to organize their farming enterprises through coordination and gendered work realms but do not consider their relations hierarchical.

To make sense of the enterprise work that women like Yaneli do, I use the term *comercio* mētis to refer to the practical knowledge and skills that agrifood system actors develop while managing distribution and marketing for their kin's farming enterprises. Mētis is a Greek term introduced by Aristotle to differentiate "know-how" from "know that." This term points to the significant difference between practical wisdom and skills learned from experience in the community, in contrast to technical expertise learned through application and memorization of scientific rules. Mētis is practical wisdom, meaning it requires insight and ethical consideration. Mētis affords the ability to make decisions that consider long-term accomplishments and immediate tasks.

Following other anthropologists, I suggest that mētis is culturally produced and critical to the functioning of agrifood systems. For anthropologists like James Scott, mētis is a kind of "local knowledge" that "is contextual and practical" and explains how peasants survived despite destructive state intervention (Scott 1998, 320). Susanne Freidberg recognized the mētis of global green bean trading women in Burkina Faso who manage critical segments of neocolonial "commodity paths" (Freidberg 2004). In the context of global sushi trading in Tokyo's Tsukiji marketplace, Theodore Bestor found that "women appear mostly in seemingly supporting roles, as clerical workers in traders' back offices and cashiers sequestered in tiny booths" (Bestor 2004, 84). Though this work seems to be done in the "back room," the *comercio* mētis needed for (and produced through) these creative socioeconomic activities is what sustains enterprises and agrifood systems. In key ways, this work—which is gendered and often goes unremunerated—is like the provisioning work that women do for their farmworking community members, as Fabiola Ortiz Valdez discusses in chapter 9 of this volume. As such, this work and these women contribute to the total work process necessary for the circulation of food as commodities within agrifood systems. For these reasons, these women and their work should be neither underestimated nor understudied.

Family farming enterprise women like Liliana are typically not recognized as innovators, long-term food economy contributors, or creators of *comercio* mētis—which they are. This is because, as Latinas in southern Appalachia, they are racialized and stereotyped as "migrants"—symbolically marked as short-term "others" (Fabian [1983] 2002)—and reified as "cheap labor" of and for the globalized FFV industry (Collins 1995). As Minkoff-Zern and Sloat show in chapter 7 of this book, some Latinx farming families experience barriers to accessing USDA programing resources—at least in part because of the inaction of local and state agency offices despite federal initiatives aimed at reconciling civil rights issues.

Though these Mexican American women face discrimination in various arenas—whether it be at the Department of Motor Vehicles or in the grocery store line—they are just some of many Latinx who have gained socioeconomic mobility in agricultural communities (Du Bry 2007) and made places for themselves in the southeastern United States (Peacock, Watson, and Matthews 2005; Smith and Winders 2007) and southern Appalachia (Barcus 2007; Margolies 2012) over the last three decades. They persist with their rural livelihoods despite the structural violence of anti-immigrant policy that has spread since the 1990s from the US West Coast to eastern states (Kingsolver 2010) and is linked to localized federal immigration policing methods that threaten the mobility of Latinx, such as 287(g), which sanctions the racial profiling of drivers (Stuesse and Coleman 2014).

Farming families are realizing an alternative vision of and for themselves as Mexican Americans in the agrifood system through kin-based enterprises. Practices of kin relatedness matter in modern political economies in that they organize economic behavior and influence industrial systems (McKinnon and Cannell 2013). Kin-based enterprises are sites where people develop capitalist (Yanagisako 2002) *and* noncapitalist ethics, logic, and practices that represent "alternative visions" that are possible when people have "lived experience or recent memories of alternative practices on which to build alternative visions" (Rothstein 2007, 11). These women show how exchange relations and enterprise strategies can defy the negative, short-term exchange logic implicit in contemporary US market capitalism. They see their family-based FFV farming enterprises as collective projects, where both women and men make decisions about the enterprise and the family group. In this way, they apply an alternative logic that may prove to be more sustainable for community economy development (Gibson-Graham 2006).

As such, these farming enterprise women—and their collective strategies, cooperative exchange relations, and contributions to the agrifood system—represent a model for an alternative approach to US agrifood livelihoods. Though they do not carry the "alternative food movement" banner, they manifest "diverse livelihood possibilities" that "have always coexisted" with dominant capitalist modes of industry (Kingsolver 2016, 38). They are reconfiguring forms of relatedness from the Bajío in southern Appalachia through their farming enterprise endeavors.

Marketeras "Moving" Fresh-Market Produce

The term "fresh" did not gain its power in the US food market until after WWII. During this era, agriculture was restructured in places across the globe so that wholesale fruit and vegetable suppliers could increase their market share by "meeting"—though some might argue producing—US consumers' year-round demand. William Friedland refers to the resulting order of this transition as the "global fresh produce system" (Friedland 1994). In the food industry, "fresh-market" is a term for fruits and vegetables sold—usually in lesser quantities—at regional farmers markets by farming enterprises, to buyers who use them for household consumption, distribute them to local restaurants, or resell them to corner stores or at roadside stands. This type of exchange supports a multiscalar, community food economy. Though these regional supply chains and family-based FFV farms were once the norm, most of the fruits and vegetables sold in the United States today are sold through wholesale brokerage firms as part of a globalized food system. In the few regional wholesale FFV marketplaces that are still active, immigrants bolster the agrifood system and create the potential for socioeconomic sustainability.

Timing is a critical factor in "fresh"-food production, distribution, and marketing at every scale of commerce. The southern Appalachian region has a vegetable and tomato harvest season that runs from late June to early November. This is referred to as their "window of opportunity"—or period when production is possible. These temporal "windows" set up the regional industry rhythms and organize people's year-round work schedules. This is one reason why, for instance, urbanites in Nashville eat broccoli grown by Mayans in Guatemala (Fischer and Benson 2006).

Time can be viewed as both economic and political, since time is significant to making and claiming resources (Ferry and Limbert 2008). This is especially obvious in the food industry, where coordinated timings of circulations (e.g., food, cash, and information) can make or break an enterprise. The most obvious coordinated timings are maturation cycles of fruits and vegetables. The quicker you sell "fresh" produce, the more easily you can create profit. The older the fruit or vegetable, the less value it commands in the marketplace. For this reason, *marketeras* develop and rely on *comercio* mētis to "move" the waves of produce that are "coming off" the fields and into the marketplaces.

The pace at which food is circulated and exchanged matters greatly to the sustainability of FFV farming enterprises. This is why people in the FFV industry talk about "moving" fresh-market produce and why particular *marketeras* excel in the farmers marketplaces. This aspect of *comercio* mētis of these farming enterprise women—and the *comercio* mētis that allows them to manage timings and exchange relations—is important to their family farming enterprises and to the food economy in southern Appalachia.

Andrea and her 65-year-old mother, Beatrice—known by everyone in the market as *tia* (or aunty)—are successful Mexican American *marketeras* in the Carolina hill country of southern Appalachia. They manage sales and other enterprise logistics for their family's produce enterprise. During the harvest season, they rent three stalls at the regional (terminal) spot market. They have 15 years of experience, and their enterprise now grows on approximately 100 acres of leased land. They sell from this farmers market produce stall and from their packinghouse about an hour south of the market. As far as Andrea knows, her family grew crops in Mexico but mostly to eat and feed to their animals rather than to sell commercially as they do today. When I asked Andrea if farming was a family tradition, she responded, "I guess. Brought up doing it."

As the birds sang from the rafters above in the metal shed, Andrea drove the forklift, rearranged the pallets, stacked eight high, and moved them from one section of the stall to another. Meanwhile, Beatrice opened white boxes of bright yellow straight-neck squash. Once the boxes of yellow squash were opened and set up at an angle so that buyers could see inside them without bending over, she organized the boxes of plump, red tomatoes, which—Andrea told me later—they needed to sell soon or they would lose all their market value. Beatrice spent much of her time in the marketplace culling the produce "with age" (or taking out the decaying pieces from the boxes), while Andrea managed the exchange relations and transactions.

Marketeras like Beatrice are skilled produce graders. Grading (also called sorting and "quality control") is an economic practice that creates value for a produce farming enterprise. Tomato and vegetable harvesters pack produce boxes in the trucks in the fields while harvesting or in the packing shed after harvesting. There are various packing strategies, such as grouping produce by "age," size, shape, and color. The *marketeras* must maintain their perishable merchandise as best they can, since their income depends on it. Culling is one way to do this. The longer a box of produce sits in the marketplace, the more often a *marketera* must cull the decaying produce pieces, since that decay quickly spreads, rendering the whole box worthless. Once this happens, the enterprise "takes a complete loss"—contributing to the already narrow profit margins of small and midscale, capital-intensive conventional FFV farming enterprises. If the *marketeras* cull too much produce, the enterprise also loses money. Grading produce is a key example of *marketeras' comercio* mētis, a practical skill guided by industry knowledge, which improves with experience.

FFV industry actors—such as agricultural extension agents, produce buyers, and marketplace staff—admire the proclivity of *marketeras* like Beatrice and Andrea for managing their inventory and "reading" marketplace circulations to quickly "move" (or continuously sell) their perishable products at viable market prices. *Marketeras* "read" the "movement" of goods in the marketplace to find out the "going rate" and "line up" buyers. The fruits and vegetables that their family enterprise produces enter the marketplace at different times during the season, because the enterprise attempts to plant and harvest its crops sequentially. It does this so that it can continuously supply its buyers throughout the season and improve its chances at a reliable positive revenue flow.

Communicating with the harvesters and customers is key to a *marketera*'s job. Selling perishable commodities is not possible without cooperation and the practice of reliable communication and trust between the buyer and seller. *Marketeras* must consistently manage supply and demand, coordinating timings of deliveries and shipments. As one farmer in southern Appalachia said, "You have to have it sold before you put the seed in the ground." This means that *marketeras* must arrange their harvest season sales months before harvesting and throughout harvest season. Their customers come to the marketplaces in person to assess the "fresh" merchandise and/or to pay for and pick up orders. While selling produce quickly is important, trustworthiness (reputation) and cooperative practices are also central to *marketeras' comercio* mētis and critical to maintaining FFV farming enterprises.

Collective Strategy: Cooperative Practices of Farming Families

The noteworthy role of cooperation in food provisioning is implicit in global commodity chain studies but not fully acknowledged. Cooperation takes place when people overcome the "collective action dilemma," meaning "situations in which the production of some group benefit is limited or prevented by the temptation [of an individual] to free ride," or take without equitably giving (Cronk and Leach 2013, 9). The recognition of potential social cooperation is even more important in our current era, when "insidious capitalist individualism" seems to be such a powerful organizing force for US society (Davis 2016). As a community of allied producers, these farming enterprise women act as innovators, creating value (Gudeman 2001) by using cooperation strategically within the capitalist global fresh produce system to accumulate wealth and generate employment for themselves, their kin, and their food economy communities. In this section, I address this collective strategy as an aspect of these farming enterprise women's *comercio* mētis.

Managing circulations and cycles as a group of allied farming families is not a simple task. Women and men work together as couples to maintain their farming enterprises by working with allied kin group members whose enterprises produce and market the same perishable commodities. While some people may imagine these kin simply as competitors in the same perishable commodity market, these women do not. Instead, they see their kin as allies and organize coordination among their farming enterprises to compete better as a group within the industry. The trust and cooperation that allows for this collective strategy depends on metaknowledge, or the cultural knowledge that lies beneath the "coalition identities [that] help shape individuals' social lives in important ways, and individuals, in turn, contribute to the strength of coalitions" (Cronk and Leech 2013,122). These farming families work as a kin-based coalition of FFV enterprises—not exactly as a cooperative or as a corporation.

These enterprise women created their kin-based coalition of nine produce farming families as a scale-making project, serving as more than just a safety net. It is an innovative collective strategy within an industrialized, neoliberal capitalist system. The women share certain machinery, information, and resources. In this sense, *comercio* mētis illustrates solidarity in practice. Silvia, one member of the coalition, explained how this works:

> El propósito de la cooperativa es marketing, vender el producto. El comprador que nos recibe casi todo el producto. Nosotros comenzamos trabajando con los dos compradores que nos reciben y hemos ido incorporando. Nos van pidiendo más producto. Entonces todos pueden beneficiarse. Todo el producto se vende por el medio de la cooperativa. Somos socios.

> (Marketing is the cooperative's objective, to sell the product. There is a buyer who receives almost all our product. We [as a group of enterprises] started to work with two buyers that received our product and we have been incorporating. They have been asking us for more product. So, we can all benefit. All the product is sold through the cooperative. We are partners.)

They market themselves as a group of Latina-led immigrants with family-based farming enterprises. They manage the paperwork for their multiple enterprises together, which enables them to sell a large percentage of their produce as a group under one brand. As one member, Juana, said, "Se da publicidad porque somos hispanas y somos mujeres. Minorías en tres partes: mujeres, hispanas, y agricultores." ("It gives publicity [to our buyers] because we are Hispanic, and we are women. Minorities in three parts: women, Hispanic, and agriculturalists.") Later, Juana told me that women in agriculture today must look for these kinds of opportunities, just as their female relatives in Mexico sought out subsidies from the Mexican government to buy tractors for their family grain farming enterprises. They understand a key point about value creation: "Gains large and small can be built into the linking of one scale to another" (Guyer 2004, 94).

These farming enterprise women understand that there would be too much risk involved if they pooled their perishable products, as Guadalupe explained:

> Cada quien tiene su propio establecimiento, pero todos van al mismo comprador. Somos juntos, pero no revueltos. Ella trae su tráilero. Yo traigo el mío. Ella y su esposo lo lleva. Al mismo lugar, no se "mixta" (mezcla). Entonces cada quien lo suyo y le paga depende de la clase o la calidad de lo que mandes.
>
> (Each person has their own establishment but everybody goes to the same buyer. We are grouped together but not mixed up. She brings her trailer (guy). I bring mine. She and her husband take theirs in. To the same place, [the product] is not mixed together. So, for each their own and [the buyer] pays each person depending on the type or quality of what you send [to the buyer].)

Juana later adds to this explanation in English, "Because if they went together they could both be rejected instead of splitting them up and not risking that chance. Each person brings their product and each person receives their check. I just put it all in QuickBooks for everyone and reconcile the bill at the end of the year." Through this model, they are leveraging their resources as a collective, yet maintaining their capitalist objective.

These women gain production power and buying power through their collective strategy. Some FFV brokers (and sometimes wholesale buyers) offer to purchase some of the inputs needed at the start of the season for the FFV farming enterprises that supply them with a large percentage of their product. This is important because farming enterprises often buy these inputs on credit, which they pay back after they receive payment for their harvest. If the production materials (like seeds and plastic mulch) are paid for by their brokers or buyers, these women and their farming enterprises do not have to pay as much interest at the end of the season. As a group, they were able to access this advantage.

The coalition is a scale-making project (Tsing 2005). An "economies of scale" logic supports the widely held belief that increasing scale is the key to surviving in farming today. The more volume a farming enterprise produces for market, the more power the enterprise has when negotiating price and accessing markets, or customers such as grocery stores, who buy most of their produce from wholesale FFV brokerage firms, sometimes called "aggregators." This is like a new corporate FFV brokerage firm organizational model called a "grower network," which is becoming more prevalent in the US organic produce industry. In this case, the broker sells produce from various midscale farming enterprises under one brand, and each enterprise is paid a percentage of the final sale for the quality and quantity of produce they "turn in" to the grower-network brokerage firm. The percentage that the brokerage firm claims varies but seems to range between 10% and 25% of each sale.

Like this corporate model, the coalition of binational farming families discussed here pools their reputation, not their product. This grower-network aggregator strategy requires corporate-level investment that these families are unable to access. These farming family women create and enact alternative modes of enterprise—such as this coalition—at least in part because owning and operating farming enterprises in the contemporary neoliberal US agrifood system is otherwise unattainable because of the exorbitant expenses involved in starting and maintaining fresh-market produce farming enterprises.

The coalition sells most of its produce to two wholesale brokerage firms (also referred to as buyers). These buyers are not always trustworthy, and in these instances the farming enterprise "takes the hit" (in industry terms). This puts their next season (and livelihoods) in jeopardy. Laura offered an example of this:

> Los compradores que nosotros tenemos nos dicen que nos van a pagar two mil pesos. Nosotros sabemos que estos two mil pesos van a llegar. No es que no nos paguen. Es una de las cosas más importantes, tener [un] comprador, pero tener [un] comprador que te pague two mil pesos. Tú ya puedes hacer planes con esos two mil pesos porque tú sabes que no matter what, esos dólares te los van a pagar.

> Y hay otras personas que venden y no saben o venden a consignment. Y siembran sin saber... Es una inversión, bien grande.

> (The buyers we have tell us they are going to pay us two thousand pesos [meaning dollars]. We know that those two thousand pesos are going to arrive. It isn't that they wouldn't pay us. To have a buyer is one of the most important things, but [you must] have a buyer who would pay you the two thousand pesos. You can already make plans for those two thousand pesos because you know that, no matter what, they are going to pay you those dollars. And there are other people that sell and do not know, or they sell on consignment. And they plant without knowing.... It is a big investment, quite big.)

Selling together as a coalition means that together these FFV family farming enterprises also have more power when demanding payment if they are not paid in a timely manner by the buyer. Payment in the "fresh" produce industry is not as simple as one might think. Buyers often keep a line of credit with the farming enterprises, and some never pay their debt. Fresh produce brokers can (and often do) claim that the produce was "bad" (or decaying) when it arrived at their warehouse, often using cell phone photos of a few opened boxes as proof. According to farmers, however, it is common knowledge that brokers may store your crop for too long because they are unable to sell it quickly. When this happens, the farming enterprise must "swallow" the loss because brokers will not pay a farmer for produce that decays in the brokerage firms' coolers. Shipments of even one tractor-trailer "load" can be worth tens of thousands of dollars. As one coalition member told me, "They pay you within the month or maybe six months later, depending on the buyer. My husband had to drive down to Florida last season to demand [a $30,000] payment. They ignored our calls and everything."

To develop the *comercio* mētis to judge whether a buyer is or is not trustworthy is clearly a critical skill in developing a successful fresh-market fruit and vegetable enterprise. Outsiders to the FFV industry may wonder why these women representing low- to mid-volume farming enterprises sell to wholesale brokerage firms at all. The reason is that large-scale buyers—like grocery store chains and fast-food restaurant distributors, who follow an "economy of scale" logic—purchase almost exclusively from brokerage firms instead of directly from farming enterprises. These women use their coalition strategy to access these markets and negotiate the risk as a group because they know that marketplaces like where Andrea sells do not currently offer reliable prices or comparable sales opportunities when it comes to large quantities of perishable commodities that have a definite shelf life linked to their market value. By aggregating their enterprise efforts, they can access marketing options that are more advantageous and create enterprise practices that are more sustainable.

Another critical cooperative practice that this collective strategy formalizes is informal money lending. Each of these women values this practice and described it as one of the reasons their enterprises still exist today. I asked one of the members, Diana, why they organized this group and what the most important aspect of the group was, and she said:

> Que estar unidos. Que ayudarse uno a otro, aunque estemos separados. Si por decir yo me atoró. Yo no tengo bolsillo para el pago de tractor este ano. Vamos a suponerlo. No tengo. Me fue súper mal. Ellas me ayudan a pagarlo. Ya sea de dinero de uno o si no tiene tampoco con el dinero otro.
>
> (To be united. To help each other even though we are separate. Let's say I got myself into a jam. I do not have money to make my tractor payment this year. We are going to suppose this. I do not have the money. [The harvest season] was really bad for me. They help me pay it. It could be money from one or, if she doesn't have it either, then with the money from another [member].)

With the endless financial risks involved in growing fresh-market produce (e.g., flooding, bacteria, decay, and volatile market prices), this cooperative resource flow management practice offers these women a lifeline when the season or month goes poorly for their enterprise.

Members of this group rely on each other and engage in a mixture of positive, neutral, and negative exchange relations within the group and as a group within the larger FFV industry. Though space does not permit full coverage here, it is important to note that these binational kin also collaborate with their relatives who farm basic grains in the Mexican Bajío, where most of these women and men were born. They organize farming enterprise agreements during their annual trips to their *rancho* (village) over the holiday season. These cooperative agreements among kin include the option of becoming *medieros*.

When people act as *medieros* in this context, they practice crop sharing. They cooperatively create and operate a farming enterprise as partners who contribute different pieces that equate to halves of the farming enterprise equation. In this binational farming kin group case, kin members who work in (and have legal status in) the United States collaborate with kin members in Mexico on grain growing enterprises, focusing on white *maíz* (corn), wheat, and sometimes barley. Often, the kin member who mostly resides in the United States—and works in fresh-market produce there—contributes half the inputs necessary for production. Though enterprise arrangements differ, the contributed *media* (half) of the binational *mediero* centers on the financial burdens, such as the cost of agrichemicals and the payment of workers. These costs are covered by the binational *mediero* though she or he is not present in the fields in Mexico during the production phase. The other *mediero* contributes labor—including the time, energy, and physical work involved in the fields and the paperwork—as well as the remaining costs incurred during production, such as tractor maintenance.

The binational status of *medieros* and other farming family members who largely reside in the United States—and make their income in dollars—allows for investment and ease of mobility between the United States and Mexico, which is necessary to coordinate this binational kin-based farming strategy. While men dominate the grain farming labor in their *rancho*, these are family-based exchange relations, and women play active roles in organizing these collective strategies among kin members and in the everyday work of sustaining their multiple farming enterprises. Collective, kin-based strategies like these present a real alternative to the practices that dominate the now inequitable globalized agrifood system, and it would seem that it could—if paired with supporting programming and government policies—introduce possibilities for greater equity and sustainability.

Long-Term Agrifood Systems: Family Farming Enterprises and Community Economy

Women are powerful actors in agrifood systems in the most local, regional, and global ways. With their *comercio* mētis and transnational knowledge of agricultural markets and exchange relations, Mexican American farming enterprise women are changing the agricultural landscape and its potential for sustainable small to mid-level farming in southern Appalachia. Though the popularized image of Appalachia is a homogenous site of extraction and poverty, the region constitutes a world of political-economic possibilities, as many Appalachian scholars have shown (Fisher and Smith 2012). These Mexican American women illustrate this point through their collective creation of wealth and employment for their kin, the agrifood community, and regional economies.

With their *comercio* mētis, they organize logistics, maintain exchange relations, and contribute to rural economies in immeasurable ways in the United States. As women most of whom were born and raised in a rural *rancho* in the Mexican Bajío, they have a bifocal lens, meaning that their experience navigating two political economies has afforded them the ability to assess both contexts (Rouse 1991) and their positionality in each (Zavella 2011). This ability to analyze multiple political-economic contexts and envision multiple simultaneous political-economic identities for themselves (Gibson-Graham 2006) and their kin is key to their success as founding members of fresh-market fruit and vegetable farming families in the United States.

These Mexican American women's experiential knowledge and skills—manifesting in the marketplace and in their collective strategies—offer examples of alternative farming practices that are effective and sustainable in small to midscale FFV farming enterprises in southern Appalachia. This study shows that localized state-based and federally funded programming—like agricultural cooperative extension programming—needs to be reevaluated to recognize gender and the roles of Spanish-speaking contributors to the agrifood system. These contributors need to be considered and consulted as programming and policy are designed to create sustainable and diverse community economies. These findings also point to a need for those within the alternative food movement and food scholarship to be critical of where and to whom we look for alternative and innovative practices.

Assumptions about class categories within the US food economy realm obscure the power of existing alternative approaches to value creation—alternatives that can teach us about the class process (Resnick and Wolff 1987) and create viable pathways toward a transition to a more just food economy. Instead of thinking of the US food economy solely in terms of its deficits, we must shift toward scholarship and policy that recognizes and empowers the industry actors who are already employing collective strategies that could transition our current neoliberal, globalized agrifood system from corporate owned to collectively led. These enterprising farming family women and their *comercio* mētis prove that alternative practices and exchange relations are already happening. They are possible when US capitalist logic is challenged and collective enterprise is valued.

While some of the reasons for their success have a cultural basis, as I have indicated, these women's stories may also be studied and applied in many parts of the United States—as well as in and between other countries—where strong kinship ties provide the trust necessary for collective action. Cooperative practices support the development of diverse community economies, which prove to be more sustainable and reliable as a source of long-term livelihoods than what is available to the majority in the current globalized food industry. Though the power of kin-based cooperation can be harnessed for neoliberal projects—as was the case with the privatization of oil in Argentina (Shever 2012)—kin-based cooperative practices and collective, long-term livelihoods are developed by families across the world. These families own and operate their enterprises and innovate through a mixture of capitalist and noncapitalist practices, even in neoliberal policy environments (Rothstein 2007). These families need programming and policy support to fortify their multigenerational livelihoods and rural economies.

Farming families in the United States now produce "global crops"—meaning that they circulate globally and are simultaneously defined by local production relations (Kingsolver 2011). Global crops are produced and distributed through various scales (i.e., global, local, and regional) of cooperative enterprise arrangements, which enterprises use to navigate food production in a political-economic environment largely shaped by global economic market integration. Most farming family enterprises encounter inequitable access to market share even in their local communities, because of agricultural policies that exclude low- to mid-volume food producers from a chance at a viable long-term livelihood. Farming communities in the United States experience "social trauma" as a result of these altered agricultural policies and the linked volatile global crop prices (Dudley 2000).

Latinx contributions to the food economy in southern Appalachia are diverse and significant. However, current anti-immigrant policy that supports deportation tactics such as raids and traffic stops are saturating rural communities in the southeastern United States with fear of separation of family and devastation of agricultural industries. A diverse community food economy will not be achievable without clear and humane immigration policies that promote long-term sociopolitical belonging. *Comercio* mētis of immigrant farming families and immigration reform are inextricably linked to the well-being of the FFV industry and farming communities in the rural United States. Inclusive agricultural and immigration policies—that recognize and support those already living out productive ways of participating in the food system—are necessary for the United States to create the conditions that will generate new long-term political-economic stability and equitable agrifood systems for generations to come.

Notes

1. The Bajío is an industrialized agricultural region of west-central Mexico that includes parts of the states of Aguascalientes, Jalisco, Guanajuato, and Querétaro.

2. Following activists in southern Appalachia, I use the term "Latinx" instead of Latin@ or Latino/a. I use Latinx when I am referring to both men and women, which is often glossed with the usage of Latinos when referring to women and men.

References

Barcus, Holly R. 2007. "The Emergence of New Hispanic Settlement Patterns in Appalachia." *Professional Geographer* 49(3): 298–315.

Bestor, Theodore C. 2004. *Tsukiji: The Fish Market at the Center of the World*. Berkeley: University of California Press.

Collins, Jane. 1995. "Transnational Labor Process and Gender Relations: Women in Fruit and Vegetable Production in Chile, Brazil and Mexico." *Journal of Latin American and Caribbean Anthropology* 1(1): 178–199.

Cronk, Lee, and Beth L. Leach. 2013. *Meeting at Grand Central: Understanding the Social and Evolutionary Roots of Cooperation*. Princeton, NJ: Princeton University Press.

Davis, Angela Y. 2016. *Freedom Is a Constant Struggle: Ferguson, Palestine and the Foundations of a Movement*. Chicago: Haymarket Press.

Du Bry, Travis. 2007. *Immigrants, Settlers, and Laborers: The Socioeconomic Transformation of a Fanning Community*. New York: LFB Scholarly Publishing.

Dudley, Kathryn Marie. 2000. *Debt and Dispossession: Farm Loss in America's Heartland*. Chicago: University of Chicago Press.

Fabian, Johannes. (1983) 2002. *Time and the Other: How Anthropology Makes Its Object*. New York: Columbia University Press.

Ferry, Elizabeth Emma, and Mandana E. Limbert, eds. 2008. *Timely Assets: The Politics of Resources and Their Temporalities*. Santa Fe, NM: School for Advanced Research Press.

Fischer, Edward F., and Peter Benson. 2006. *Broccoli & Desire: Global Connections and Maya Struggles in Postwar Guatemala*. Stanford, CA: Stanford University Press.

Fisher, Stephen L., and Barbara Ellen Smith. 2012. *Transforming Places: Lessons from Appalachia*. Urbana: University of Illinois Press.

Freidberg, Susanne. 2004. *French Beans and Food Scares: Culture and Commerce in an Anxious Age*. Oxford: Oxford University Press.

Friedland, William H. 1999. "The New Globalization: The Case of Fresh Produce." In *From Columbus to ConAgra: The Globalization of Agriculture and Food*, edited by Alessandro Bonanno, Lawrence Busch, William H. Friedland, Lourdes Gouveia, and Enzo Mingione, 210–231. Lawrence: University Press of Kansas.

Gibson-Graham, J. K. 2006. *Postcapitalist Politics*. Minneapolis: University of Minnesota Press.

Gudeman, Stephen. 2001. *The Anthropology of Economy*. Malden, MA: Blackwell.

Guyer, Jane. 2004. *Marginal Gains: Monetary Transactions in Atlantic Africa*. Chicago: University of Chicago Press.

Kingsolver, Ann. 2010. "Talk of 'Broken Borders' and Stone Walls: Anti-immigrant Discourse and Legislation from California to South Carolina." *Southern Anthropologist* 35(1): 21–40.

Kingsolver, Ann. 2011. *Tobacco Town Futures: Global Encounters in Rural Kentucky*. Long Grove, IL: Waveland Press.

Kingsolver, Ann. 2016. "When the Smoke Clears: Seeing Beyond Tobacco and Other Extractive Industries in Rural Appalachian Kentucky." In *The Anthropology of Postindustrialism: Ethnographies of Disconnection*, edited by Krista Harper, Seth Murray, and Ismael Vaccaro, 38–55. Milton Park: Routledge.

Margolies, Daniel S. 2012. "Taquerias and Tiendas in the Blue Ridge: Viewing the Transformation of Space in a Globalized Appalachia." *Appalachian Journal* 39(3/4):: 246–268.

McKinnon, Susan, and Fenella Cannell, eds. 2013. *Vital Relations: Modernity and the Persistent Life of Kinship*. Santa Fe, NM: School for Advanced Research Press.

Peacock, James L., Harry L. Watson, and Carrie R. Matthews, eds. 2005. *The American South in a Global World*. Chapel Hill: University of North Carolina Press.

Resnick and Wolff. 1987. *Knowledge and Class: A Marxian Critique of Political Economy*. Chicago: University of Chicago Press.

Rothstein, Frances Abrahamer. 2007. *Globalization in Rural Mexico: Three Decades of Change*. Austin: University of Texas Press.

Rouse, Roger. 1991. "Mexican Migration and the Social Space of Postmodernism." *Diaspora* 1(1): 8–23.

Scott, James C. 1998. *Seeing Like a State: How Certain Schemes to Improve the Human Condition Have Failed*. New Haven, CT: Yale University Press.

Shever, Elana. 2012. *Resources for Reform: Oil and Neoliberalism in Argentina*. Stanford, CA: Stanford University Press.

Smith, Barbara Ellen, and Jamie Winders. 2007. "'We're Here to Stay': Economic Restructuring, Latino Migration, and Place-Making in the U.S. South." *Transactions of the Institute of British Geographers NS 33*: 60–72.

Stuesse, Angela, and Mathew Coleman. 2014. "Automobility, Immobility, Altermobility: Surviving and Resisting the Intensification of Immigrant Policing." *City and Society* 26(1): 51–72.

Torres, Rodolfo D., Louis F. Mirón, and Jonathan Xavier Inda. 1999. "Introduction." In *Race, Identity, and Citizenship: A Reader*, edited by Rodolfo D. Torres, Louis F. Mirón, and Jonathan Xavier Inda, 1–16. Malden, MA: Blackwell.

Tsing, Anna Lowenhaupt. 2005. *Friction: An Ethnography of Global Connection*. Princeton, NJ: Princeton University Press.

US Department of Agriculture (USDA). 2012. *Census of Agriculture*, vol. 1. "Table 58. Spanish, Hispanic, or Latino Origin Principal Operators—Selected Farm Characteristics 2012 and 2007." https://www.nass.usda.gov/Publications/AgCensus/2012/Full_Report/Volume_1,_Chapter_1_US/st99_1_059_059.pdf.

Yanagisako, Sylvia Junko. 2002. *Producing Culture and Capital: Family Firms in Italy*. Princeton, NJ: Princeton University Press.

Zavella, Patricia. 2011. *I'm Neither Here nor There: Mexicans' Quotidian Struggles with Migration and Poverty*. Durham, NC: Duke University Press.

9 Gender, Food, and Labor: Feeding Dairy Workers and Bankrolling the Dairy Industry in Upstate New York

Fabiola Ortiz Valdez

Introduction

Many dairy farms in New York State are located in predominantly pro-Trump counties, where "Hillary for Prison" signs line the streets along with the occasional Confederate flag. Before the 2016 presidential election, I was aware that dairy workers faced many challenges, including mobility in and out of the farms for fear of law enforcement harassment, but after the election, this already troubling landscape dramatically intensified. Two seemingly quaint little towns, the sites of my field research, increasingly became, at least to my eyes, overtly uninviting, threatening, and isolating places for immigrants to live.

Generally, dairy workers work up to 12 hours per day, performing tasks such as milking, feeding, herding, caretaking, and driving, as well as supervising other workers. Given the limited free time outside work that workers have to meet their basic needs, one of my main roles during my fieldwork was to take workers to the local Wal-Mart for groceries. Before the election, those grocery trips could last up to four hours. After the November 2016 election, the same trips would take a rushed 30 minutes because of the fear of being targeted by immigration enforcement. Nevertheless, I kept seeing the same amount of food in the houses and the same dishes being prepared, the workers were adapting to this change, becoming more self-sufficient by canning tomatoes, growing a larger amount of beans, and raising more chickens. Although these activities were present before the election, they increased substantially after it.

New York State is a leader in the US dairy industry and is the nation's largest producer of yogurt. A large segment of the workers who produce the required milk are undocumented men from Mexico and Guatemala, living in farm housing provided by their employer and located on the farm premises (Fox et al. 2017; Hamilton and Dudley 2013). The data presented in this chapter are part of my larger work on labor relations and labor organizing among dairy workers in central New York. I started preliminary fieldwork in 2013, working with labor and advocacy organizations focusing on undocumented farmworkers. Through my advocacy and grassroots organizing work, I was able to conduct ongoing visits to over two dozen dairy farms for a period of three years. In early 2016, I selected two dairy farms as the focus of my research, where I conducted interviews with workers and farm owners. I've lived in the United States for almost ten years—for the last five years that I have been in central New York, I have spent countless hours looking for traditionally Mexican products and have failed at efforts to reproduce my *abuela*'s (grandmother's) cooking. On a personal level, it was fascinating how I was suddenly able to visit geographically and socially isolated dairy farms and eat traditional Mexican food, food that never failed to make me homesick and make me feel at home at the same time.

Because of grueling work schedules and geographic isolation, male workers often are not able to take time to cook for themselves. Instead, they rely on the women living on the farms to fulfill that need, often in exchange for pay. Although the number of women who live and work at the farms is small, their presence is significant; it quickly became apparent that women's labor was a key aspect of the whole dairy production. For the last six months of my fieldwork, I shifted my focus to interviewing and conducting participant observations with as many women as I could. This chapter examines the ways in which women who live on two secluded dairy farms assume roles, primarily in the reproductive space (i.e., farm housing), that greatly impact the production space. Drawing on these six months of ethnographic research at two farms, I illuminate the role of female immigrants in providing not only food but also cultural sustenance for the workers, while at the same time filling a vital role in sustaining the functionality of dairy production more broadly by keeping workers fed and fulfilled.

I begin this chapter with an overview of the dairy industry in New York State, particularly the growth in production and manufacture of dairy products thanks to the Greek yogurt boom. I give special attention to the overall labor and living conditions of workers on dairy farms. I introduce my two field sites in terms of demographics, living arrangements, and productivity.

I then look at the ways in which women use strategies to mitigate the challenges in providing traditional food for workers as well as the pressures to provide food from workers' home countries. I provide snapshots of the lives of three women, arguing that through traditional food practices, immigrant women make it possible for workers to maintain a connection to their life back home. Simultaneously, these women are creating a new home place where workers create communities and sanctuaries away from law enforcement. Throughout, in order to illustrate how women explore different ways to re-create traditional food, I refer to Marte's concept of "food maps" as "perceptual models of how people experience their boundaries of local home through food connection" (Marte 2008, 47). I conclude that the foodways that these women provide serve as out-of-farm programs through which farmworkers can access food at no cost to the farmer (Minkoff-Zern 2014), and that these foodways serve as one of the main forces driving productivity at the farms. Under current conditions, male dairy workers decide to keep their jobs in part because of their access to traditional and comforting food, or at the very least because those foodways make the job tolerable.[1] Even though women are not referred to by anyone as workers, in reality their domestic labor is as critical as any other type of labor in making the farm function.

The argument that capitalism has benefited from workers' social reproduction outside of the boundaries of the working day is certainly not new (see Kasmir 2008). However, this chapter aims to make an empirical contribution to how this phenomenon presents itself in dairy farms in upstate New York. This research echoes other scholars' rejection of the assumption that the kitchen, or farm housing for that matter, is merely a place of women's oppression (Abarca 2006). However, the gendered politics of domesticity mean that at times the women themselves do not see themselves as "true" workers (i.e., on the payroll) and often think of farmers as *buena gente* (good people) who let them live on the farm for free. I hope that the insight into these women's lives shows the ways they exercise agency in providing traditional cooking and helps us understand their labor within the food system in ways that resist the gendered manual/domestic labor hierarchy that I, along with fellow scholars Mary Elizabeth Schmid (chapter 8) and Catarina Passidomo and Sara Wood (chapter 12) in this volume, are complicating.

It's Official, New York Loves Dairy

The National Agricultural Workers Survey estimated that in 2014 at least 47% of farmworkers in the United States were unauthorized immigrants and 73% were foreign born (US Department of Labor 2016). This dependence is the outcome of a combination of factors, including neoliberal policies in the United States and Central America, international trade agreements, and historical and contemporary state-sponsored violence (Green 2011). These factors have produced both low-wage agricultural jobs in the United States and a surplus of cheap and expendable labor from Latin America (Green 2011, 368). This is a workforce that encounters daily conditions of intense and growing precariousness, including low wages, lack of legal protections, intensified work hours, and inferior-quality housing. Studies show that in farm camps across the country, migrant farmworkers tend to reside in houses marked by overcrowding, unsanitary conditions, and remoteness from health clinics, grocery stores, and public transportation (Benson 2008, 2011; Cartwright 2011;Holmes 2011, 2013: Quesada, Hart, and Bourgoi 2011; Saxton 2013).

A decline in consumption of fluid milk (Farm Credit East Knowledge Exchange 2015) and an increase in consumer demands for other dairy products have created a "yogurt gold rush" in New York State, increasing overall milk production and making New York the nation's largest producer of yogurt (Office of Governor Andrew M. Cuomo 2013). New York's yogurt industry started growing in 2007, when companies such as Chobani opened plants across the state (Cornell Cooperative Extension Harvest New York 2015).

Approximately 47,000 migrant farmworkers and their families arrive in New York State annually, some of them to work on dairy farms (BOCES Geneseo Migrant Center 2012). This labor force toils under precarious conditions, including exclusion from important legal protections because of their occupational and documentary status, 12-hour work shifts, dilapidated housing, exposure to toxins, surveillance, separation from their home communities, and harassment by federal and state police.[2] Because dairy farms adhere to a nonseasonal 24/7 production process, the precarious conditions faced seasonally by other migrant workers are encountered year-round by most dairy workers.

Producing and Reproducing on a Dairy Farm

Farm housing is a term and condition of employment. Housing farmworkers for work that is mostly seasonal (e.g., harvesting apples, potatoes, and berries) ensures that the farms are provided with an adequate supply of labor (Occupational Safety and Health Administration 2014). Even though dairy production is not seasonal agriculture, most dairy farmers provide housing for their workers (Fox et al. 2017). Farm housing is subject to regulation and inspection under the 2011 Agriculture and Markets Law and must also meet the Department of Health codes for migrant worker housing (New York Department of Health 2011). However, inadequate housing conditions are typical, including unsanitary conditions, lack of basic utilities, and overcrowding (NFWM 2014).

From over three years of fieldwork, I can attest to the overall poor and hazardous living and working conditions of dairy workers in New York that other scholarly works have explored (Gray 2016; Sexsmith 2017). Workers live near their workplace, with no need for transportation to get to their jobs. All the dairy farms I visited are located far from health clinics, churches, and grocery stores. The lack of public transportation leads workers to depend on private car services to leave the farms at least once a week to buy groceries, go to church, and so forth. Isolation, fear of deportation, not being able to speak English, harassment by immigration and police officers, and having no means of transportation accentuate the segregation and confinement of this population. Dairy workers report high anxiety and depression symptoms because of, among other things, their social isolation (Fox et al. 2017; Sexsmith 2017), yet leaving the farm is risky behavior. There have been several cases of workers detained by the State Police while they were passengers in private transportation vehicles.[3]

Dairy farm owners in New York have reported challenges and struggles themselves (Hamilton and Dudley 2013). Finding local labor has become more and more difficult, and smaller farms have started consolidating.[4] To maintain and accelerate production, farmers continue to hire immigrant farmworkers willing to work under these conditions.[5] In order to keep their labor force stable, farmers have to resort to different strategies. I argue that one of these strategies is to allow women to live on the farm, often without working in the milking parlors, so they can clean the houses, occasionally cover shifts, serve as company for the workers, and cook. According to the women on the farms, this practice is relatively new; workers' anecdotes suggest that it started in the late 1990s. Providing male workers access to traditional food serves two purposes: it serves as a tool to keep them from quitting their jobs, and it keeps them safe by not having them take the risk of leaving the farm to get food.

Fox Farm and Valley Farm are similar in many ways but significantly different in others. Both are located in central New York, have between 1,200 and 1,500 cows, and employ 11 to 15 immigrant workers—mostly men from Guatemala and Mexico. Both provide overcrowded farm housing with a high ratio of workers to utilities and do not charge rent or utilities to workers, not an uncommon practice among dairy farms. Both farms have hazardous working conditions, such as lack of protective equipment when working with chemicals, exposed wires in the barns, and other dangers, and a working day of 10 to sometimes 16 hours. Finally, workers at both farms must abide by certain living restrictions that make it hard to access food, imposed by the farmer in order to protect their workforce (e.g., not leaving the farm in groups of three or more and not being allowed to have a car), and by the changing political environment, which has made cooking on-site all the more essential.

Feeding Dairy Workers

By focusing on the narratives of three women living on the farms, I offer examples of the limitations and challenges that these women must overcome to fulfill their roles. On both farms, male workers pay the women for the food they cook for them[6]; however, the farm owners do not compensate the women. Through these narratives, I illustrate how farm owners derive significant stability in their workforce in exchange for the labor of the women.

Bertha

Bertha[7] used to work at Fox Farm. There are four separate living spaces shared by a total of 12 children,[8] including two babies and one passing as an adult and thus able to work. Fox Farm is also home to nine women: two full-time workers, three part-time workers, and four spouses or family members who stay at home and don't work at the parlor.

Gender, Food, and Labor

Bertha is originally from Guatemala. She worked at Fox Farm for over five years before she married a white US-born supervisor from that same farm. After they married, Bertha stopped working on the farm and moved to a town 20 minutes away, but she started working in one of the farmhouses, cooking for at least 12 workers every day. Bertha and her two small children arrive at the farm at 6:00 a.m., when her husband starts his shift. Bertha spends almost 12 hours at her sister Belen's house. Belen is one of the nine women living at Fox Farm. She has three kids, and her husband works at the farm. Bertha takes care of the four-year-old child of Manuela, one of the two women who work in the barns. In addition, whenever Valentina, the other female worker, is called into work, she drops her newborn baby off with Bertha for the day. In a given day, Bertha must take care of five children, sometimes with the help of her sister, and cook food for at least 12 workers, enough for 36 plates.[9]

Each worker pays Bertha $150 to $180 per week for three daily dishes. They arrived around nine in the morning to get their first meal. Some workers stay in Bertha's kitchen and have their food there, others bring it to the farm, and others just take it home for whenever they get a chance to eat. I often participated in the eating as well. At the beginning, I felt guilty for eating her food because she never accepted any payment from me, but it quickly became clear that not doing so would be disrespectful.

Bertha often struggles to think of different dishes to cook. She said to me, "I just don't know what else to cook. I already did mole, I already did menudo...what else can I do?" I asked her, "Does it really matter what you cook? Isn't it good enough that you make Mexican and Guatemalan food?" Every time she answered me with the same gravity: "Can you imagine if after working all this time they come and get the same food that I gave them yesterday? They pay too good money for that."

Many times, I ate with one or two workers at the table while Bertha started cleaning up. I asked workers what they would do without Bertha, given their long work hours. Their answers didn't vary much. Armando, who works the night shift, said laughing, "I probably wouldn't eat [otherwise]....[It] is nice to wake up and come to the kitchen and have food that tastes like it used to." Another worker, Antonio, said, "One day I'll take you to the farm and you'll see the crap that the *gueros* have in that fridge, it's all Hot Pockets and blue drinks! That's all what the Americans eat! I mean, I eat that sometimes, but only when I'm really hungry. I wouldn't be able to *durar* (to last) without Doña Bertha." At various times I saw Salvador, who lives next door, come and get his weekly tortillas from Bertha. He pays her $60 per week and also provides a kilo of *maseca* for her. He almost never talked to me, but every time I saw him pick up his order, Bertha told me, "Imagine, how can he work *a gusto* if he doesn't have his tortillas?"

To a large extent, Bertha is able to be self-sufficient. She has her own house with her husband 20 minutes from Fox Farm. There she is able to make these foods by growing corn, tomatoes, onions, and beans, raising chickens, and she has a couple of goats. She also had approximately 50 jars with tomatoes to make her own salsa. These independent food-growing practices constitute Bertha's foodmap. Every week she asked me to bring her meat in bulk from the stores in Syracuse. I always asked her to come with me, but she never did, saying, "I don't have time. Where would I leave the kids?" Bertha and her family have their own car, so she does not hire *raiteros*, the people (usually US citizens) who drive workers in and out of the farm to the store or different appointments[10]; however, a trip to Syracuse would definitely set back Bertha's day.

In terms of economic gain, Bertha doesn't earn much from the food. According to Bertha, if she sold a plate for $15, her earnings would only be $5. "I don't do it for the money.... I do other things for money, like childcare, selling phone cards.... But the food is a different thing, the workers need the food," she explained. I asked Bertha if she knew what the farm owner thought of her selling food to his workers. She answered, "I know he likes it. He tells my husband. It is a good thing I do it, because he is very good people and lets us be here all day, so my husband doesn't have to worry about it. He also lets my siblings work here.... He lets us [the women and children who do not work at the farm] live here for free, so the least I can do is to feed people." I asked Bertha what would happen if she didn't show up one day. She laughed it off and said, "The whole farm would shut down." This is certainly a valid prediction; without Bertha, up to four workers[11] would have to stay home taking care of their kids and wouldn't be able to show up to work. What seems to be even more alarming to Bertha, however, is that the workers would not eat properly: "Many workers will have to eat whatever leftovers they had from the day before, or to eat the awful food the *gueros* buy at the Dollar Store nearby."

Bertha's main role, then, is to prepare homemade food for the workers and to care for the workers' children, two paramount activities that in many ways ensure not only the dairy production but also the safety of the labor force. There thus emerges a clear contradiction in Bertha's description of her labor. According to her, the farmer likes that she cooks, and sees the benefit to his workers from this, but at the same time he is doing them (particularly Bertha's husband) a favor by letting Bertha stay on the farm all day for free and even letting her and her family benefit from selling her food. In a way, Bertha's perception devalues her own labor while at the same time recognizing that the farm owner relies on women like her to keep the business afloat. Because this contradiction is not unique to Bertha, I'll return to address it in-depth later in the chapter.

Camila and Estela

Camila and her husband, Eber, live at Fox Farm. Eber is one of the "best workers at the farm," according to Camila. They have their own living space, which they share with their three children, her sister Estela, and Estela's daughter. Camila and Estela are both undocumented, and neither works at the farm. Camila's family is not related to anyone in the other houses, so they often feel isolated from everyone else. They say, "It's just us here. We don't cause problems. Sometimes people here don't like us, but we try to keep to ourselves."

After their children leave for school, Camila prepares Eber's breakfast. A typical breakfast includes beans, eggs, and chorizo. As soon as Eber leaves, Camila and Estela start cooking again to prepare food either for the five children in the house, for themselves, to sell to two workers on the farm, for selling at a church event, or to take a trip to the nearby farms to sell to workers there.

I asked Camila why she wasn't regularly selling food to the workers on this farm like Bertha was. She said that Bertha was already doing that and that she and Estela did not want to cause any problems. Camila and Estela do sell food to the workers on the day that Bertha doesn't cook. When Estela first arrived at the farm (approximately two years ago), she started cooking for Victor, one of the workers, but had to stop because of a disagreement regarding money. "He said that he paid me, but he didn't, and he sometimes came to get his food when he knew I was alone. I don't like that," she explained. Estela's sentiment echoes the larger issues of power dynamics between men and women at the farms, as well as the sexual harassment that women experience. Power dynamics and harassment are issues that are an intrinsic part of women's everyday life at the farm, dynamics that while not focused on within this chapter, warrant significant attention in current food and labor scholarship.

As part of this research, I was able to conduct fieldwork in Mexico and to visit Camila's and Estela's families in Guatemala. It was during that trip that I met Estela, before she arrived in the United States. I stayed long enough to observe the ways in which she and her family cooked. We made tamales and tortillas, and killed and deboned chickens, the same as I did with Camila and Estela at the farm in the United States. The similarities of these food practices make them truly transnational, even though neither woman ever articulated them as such. Whether in Guatemala or New York, they followed their set steps when killing chickens: first, pulling the neck out really hard so you would kill it with one stroke; second, dipping it in hot water for exactly five seconds, just enough to make the feathers soft without beginning to cook the meat. Through multisited ethnography, I was able to observe how culturally specific practices carried across nations are a way to experience and re-create their home (i.e., foodmaps), which are an example of what Mares calls "powerful ways to enact one's cultural identity and sustain connections with families and communities who remain on the other side of the border" (Mares 2012, 35).

As opposed to Bertha, who never let me help her, Camila was very willing to receive help. Once, Estela and Camila were preparing chickens they sold to a Pentecostal church. I offered to come and help. I plucked a total of three chickens in the same time each of them plucked ten. Given that I slowed the food production every time, I often quit the process and instead sat down, drank coffee, and asked questions.

One of their main sources of income is to go to the nearby farms and sell tamales. When I asked Camila why tamales, she laughed and answered, "What am I going to do? Hamburgers? They would throw it in my face." She also said the tamales are easy to carry and that they remind the workers of home, so they will pay good money for them.[12] When I asked whether the farmers let her come in, she said, "Of course they let me in, they have no women there."

I attended a church retreat hosted at Valley Farm,[13] where Estela and Camila provided 50 chickens and several pounds of beans, plus hundreds of tortillas for approximately one hundred people, a large proportion of the central New York immigrant dairy workers and their families. One of the pastors told me that they usually hire women from dairy farms to do the cooking. He continued, "I don't know what we would do without them. A lot of people see the church in general as their family, so these events are really important. We need to make people feel like they are back in Guatemala, and the food helps." It is clear that the food here is more than simple convenience or satisfying a craving; Camila and Estela's cooking plays a central role in the community's yearning for home and in shaping their lives in the United States (Mares 2012).

I was often amazed to find things like chorizo, queso fresco, and corn husks in Camila's kitchen. I asked her, "Where did you get this stuff? I always look for them in Syracuse and I never find it." She laughed and said, "Wal-Mart! You are not really looking, Fabiola." She was right. I started to pay more attention to the "international aisle" at the Wal-Mart and started to see mole, *maseca*, and other products that other stores in Syracuse do not carry.

Camila and Eber have a truck that they bought under the farmer's name. They occasionally use it to get groceries, but often I was the one who brought Camila to the nearest Wal-Mart, to save time. Before the 2016 election, it was rare for her to ask me to drive her, but whenever I did we would spend up to three hours in the store, walking around, comparing prices, and running errands for other people. After January 2017, Camila started asking me more often to bring her to the store because she didn't feel comfortable driving herself anymore, saying, "I think there is a lot more police now after Trump won.... I saw on Facebook that ICE is in the Wal-Mart parking lot every now and again." As a result of this fear that both Camila and Estela experience because of their risk of being deported, the trips to the store went from taking three hours before the election to sometimes as quick as 30 minutes since then.

According to Camila, she "borrows" the space outside her house to raise chickens and to plant beans. Not unlike Bertha, Camila and Estela have some access to self-sustained food practices; however, nearby stores are their main source of food. Not unlike Bertha, neither Estela nor Camila consider the farm owner as benefiting from their cooking or their presence. "The farmer is very nice," Camila said. "He is very good to us. He lets us live here for free. He lets us have our kids here. He lets us sell food to the workers whenever workers need it [meaning when Bertha takes a day off]. How does he benefit from that? In no way at all. Eber goes to work happy because he has his belly full of beans and knows that he is going to come back to some nice food and a clean house." I asked her once what would happen if she wasn't allowed to live on the farm. She said, "Eber [is] the farmer's best worker. He would quit."

Inez

Inez is one of the two worker's spouses who live at Valley Farm. She lives there with her husband, Martin, and their two kids. In contrast to Fox Farm, there are only six women and three children living at Valley Farm. Two women work full-time caring for calves, two of them are spouses, and one is the mother of a worker. There are a total of 11 immigrant workers, including the two female workers, living in three buildings, separated into four living spaces.

Inez's six-year-old goes to school, and her three-year-old stays home with Inez while Martin goes to work at 6:00 a.m. She cooks breakfast at around 5:00 a.m. just for her family and then starts cooking for other workers. Inez is the only woman who sells food to workers. She cooks for seven workers who according to her "are too tired after work to cook for themselves." She doesn't have to cook for more because the other workers have either a spouse, mother, or sister living with them.

Valley Farm is in an area with a high presence of Border Patrol officers, where there have been several cases of State Police cooperation with immigration enforcement to detain people. By providing food for the workers, Inez solves the problem of food access while also providing a way for workers to claim a sense of place in a foreign and increasingly aggressive place (Mares 2012). I often saw a worker named Ruben coming and getting his food from Inez. One day he made a joke about having to eat the same thing he ate a few days ago, "Inez, again, *caldo de pollo*? Didn't you make this a couple of days ago?," he asked. Inez laughed and brushed it off with a hand gesture. I asked Ruben if he preferred hotdogs and hamburgers instead. He laughed and said, "No way! *Caldo de pollo* twice a week is not too bad." One night, I told Martin how lucky he was to have tortillas with every

Gender, Food, and Labor

meal—that *tortillerias* are one of the things I missed most about home. He responded, "To be honest with you, I don't even think about it. I mean, I like it, but it doesn't even occur to me that there will be no tortillas."

Every other day at around 8:00 a.m., a vendor named Manuel comes over with a van filled with Mexican and Goya[14] products. Manuel travels from farm to farm to sell products he gets from New York City or Boston; his mobile van is one of the main ways Inez is able to create her own food-map. There are some specialty products that the women, particularly Inez, buy from him. "You just can't get these anywhere," Inez told me often. Whenever I had the luck to be there when Manuel arrived, Inez usually treated me to a bag of Rancheritos chips and a Jarritos soda. I never felt comfortable taking them because of the high markup,[15] but Inez always insisted. I asked her, "Why do you buy stuff that you can find in other stores and that he sells so expensive?" She answered, "It beats leaving the farm. Can you imagine? What if they [border patrol/ICE] catch me? What if they take me away from my kids and my husband? Too much risk, that's why Manuel comes." I asked Manuel why he went to the trouble of visiting all these farms instead of opening a local store. He said, "[This practice] is better for business...two reasons why I come here: workers don't have the time to go to the store, they work a lot of hours; and workers are afraid to leave the farm." I did not ask Manuel directly about his immigration status, but given his comments, I assume that he does not run the same risk of deportation.

As opposed to Bertha and Camila, Inez and her family have no opportunity to plant their own crops and no means of transportation, so she relies heavily on *raiteros*, who charge her around $50 for a 20-minute trip to the store. Sometimes she would ask favors from the other workers who went to the store, although she would rarely ask me to bring her or to pick up things for her. "I don't want to bother you," she said many times.

> Inez was the only person who told me specifically how the farm owner depends on her to keep workers there. Inez, not unlike Bertha and Camila (who supports Estela), lives largely off her partner's income. Inez told me that she doesn't really make any money when she cooks for the workers. "Listen, I'm going to be honest with you. This does not make me any money," she said. I asked her why she does it then, and she responded, "The farm owner lets women stay in the houses so his workers have someone to cook for them.... He told me that the guys used to leave because they had nobody to cook for them. He started asking

his workers why so many people left. They told him that [it was] because they couldn't cook for themselves. So, since the eighties the farmer started letting women and spouses living here so the workers would stay. This farm cannot be without women.... That's how they hire people here. They look for young males, but even better if they have a wife."

I was surprised at how candid the farmer was with her, but Inez did not seem surprised by this. For Inez, the farmer's honesty was necessary for her to fulfill her role at the farm. I told Inez, "It's like you are a worker on the farm too." "I am not. I don't milk cows or anything," she responded.

Traditionally, farm labor organizers and scholarship on agricultural labor have overlooked the issues female farmworkers face. Though some authors have certainly highlighted the role of female farmworkers in the labor movement (see Rose 2008), there is still a large disparity within this literature regarding the roles of male and female labor leaders. Based on the stories presented in this chapter, it seems that a similar mistake occurs within the actual workplace: confining women to traditionally female-defined work, such as reproductive work, reinforces traditional gender relations that perpetuate vulnerability. According to Keiko Budech, studies have also shown that these women face a number of inequalities outside the domestic sphere, such as the gendered effects of pesticide exposure, sexual harassment in the fields, and domestic violence (Budech 2014,17).

As stated earlier, these gendered inequalities are not the main focus of this chapter; however, it would be a mistake not to mention that these women face several vulnerabilities that perpetuate the existing oppressive systems rooted in agricultural labor. It is not my intention to present these women as passive beings unaware of their own exploitative conditions. However, I believe that their gendered vulnerability plays a role in keeping them from seeing themselves as workers, or at least to talk about themselves as workers with me. Future research would benefit from gender-focused analysis of the reproductive work of women living on farms.

Challenges and Limitations: "It Is Hard for Us to Go and Buy Food"

Researchers who study the homemaking process adopted by undocumented immigrants identify limited income and access to quality affordable housing as the two main constraints for these populations (Hadjiyanni 2014). Access to traditional food, and even food in general, is an additional challenge for dairy workers trying to find a place to call home and create a sense of belonging.

Both farms discussed here have housing restrictions imposed by the farmer, such as not being allowed to drive together to buy groceries (or anywhere for that matter), at least not more than three workers together. One farmer reasoned, "If immigration stops the car, it would be worse if they take more workers instead of if they only take one or two." As troublesome as this sounds, the farmer's logic is not wrong: there have been numerous examples of local police detaining passengers on their way to church, on their way to and from a city park, and going to the store. During my research, I felt the tension every time I took people in my car. On those trips, I saw everyone as a possible threat. Anyone could call immigration at any time. Even though everybody living at the farm knows the risk they take by leaving the farm, often I felt that my role was to protect them, to use my green card and my driver's license as tools when interacting with law enforcement. I have extensive training on what to do if I get stopped by a police agent; however, every time I saw a state trooper or Border Patrol agent, those horror stories came to mind. Given my own immigration status, I often did not trust myself to handle a situation in which the workers were protected.

Relying on a *raitero* is also risky behavior. A few months ago, there was an incident at Fox Farm where a *raitero* got into an argument with Bertha and ended up calling the USCIS, causing ICE agents to arrive at the farm. When that happened, most workers ran to hide in the woods. Fortunately, ICE did not arrest anybody, and nobody was reprimanded, but as a result, that *raitero* is not allowed to come back to the farm, and farm owners have become stricter with their housing rules.

Navigating stores like Wal-Mart is not easy either. Bertha is the only participant discussed here with enough English proficiency to go to the store by herself. Camila and Estela have trouble with language and have low levels of literacy in their own language, not to mention the unwelcome looks from the local, primarily white population.

Even though women have little capital and resources, social or otherwise, including the limited availability of traditional ingredients, they still fulfill their roles. Women create their own foodmaps as they map out the multiple locations they need to navigate in order to feel at home (Marte 2008, 47). The women presented in this chapter are constantly recreating their own foodmaps by exploring and creating actual places to access ingredients and traditional foods, whether it is their own backyard, the mobile van, or the international aisle at Wal-Mart. These three women have different access to food, but all have to navigate their foodmaps in order to create senses of place and home for themselves and the workers. Somehow, within all the limitations they face, women at the farm get the ingredients they need to make things "taste like home."

Making Dairy Labor Cheap and Keeping the Labor Force Safe

Male workers constantly emphasize the significance of eating Mexican or Guatemalan food. A worker from Valley Farm told me, "You see, we cannot go home, right? We miss it.... It's good to eat food from home. It makes me happy, although sometimes it makes me sad. It makes me miss my family more, but I get over it." As the makers of this food, the women's stories captured in this chapter show how they themselves must overcome challenges and concentrated efforts to sustain "culturally meaningful practices" through food (Mares 2012, 36) and in doing so maintain dairy production moving forward.

The women do not find selling food to be a lucrative business, but they often expressed pride and a sense of duty in cooking for people, making it clear in their statements that they know the significance of their contributions to the farms. All the women knew that I thought the owner benefited a lot from their presence and labor. They all felt flattered by my assumptions, and sometimes they agreed with me and sometimes they did not, often within a single conversation. Following acknowledgment of the benefit they provided the farm, I often heard contradictory comments such as, "The farmer is really nice; he allows us to have our own *huertas*," "[we] live here for free even though we are not workers; he is doing us a favor," and "He lets us make business here by selling our food without charging us anything."

In order to have their labor force subsist, the farmers utilize the women's role in the reproduction space, an argument put to the forefront by scholars of the twentieth century as a way to develop further Karl Marx's limited analysis of the role of women in the production process (Federici 2010). However, in addition to the notion presented by Marx, in this case the farmer benefits in two ways. First, his workers not only subsist but are also happy reminiscing about home, which in turn makes them stay longer at the farm.

Gender, Food, and Labor

Second, without having to venture outside the farm to look for traditional food, his workers are safe and they get the food at home, costing the farm owner almost nothing.[16] The food that the women provide can be thought of as examples of what Minkoff-Zern calls "programs that allow farmworkers to access food," such as food banks or lunch programs, "which ultimately function to subsidize farmworker exploitation" (Minkoff-Zern 2014, 97).

These stories show how the ramification of women's role on the farms is dialectical: on the one hand, it subsidizes workers' exploitation, and on the other it protects workers from greater risk and provides them with not only comfort and community but also joy and emotional relief (Gimenes-Minasse 2016, 95) to both men and women who are inhabiting marginalized and isolated spaces. In some ways, these women are aware that the owners are engaged in a codependent exchange with them, creating the conditions of possibility (stoves, kitchens, where to live, etc.) for this re-creation of home and relative protection in exchange for their labor and sacrificed spatial independence. In times of increasing vulnerability, a complicated relationship is forming: culturally appropriate food offers farmworker women and men indulgence, opportunities to socialize, physical comfort, and nostalgia, and for the farmers, it offers a safety net.

Notes

1. Women play different roles at the farm, providing food as part of the reproductive labor is just one of them.

2. The Fair Labor Standards Act (FLSA) excludes industries for which the work has historically been performed by racial minorities, such as agricultural workers. In New York State, farmworkers do not have the right to overtime pay, weekly days off, or collective bargaining.

3. See http://progressive.org/dispatches/immigrants-face-the-death-of-a-dream-by-james-goodman-180206/ and https://www.democratandchronicle.com/story/news/2017/03/24/arrests-follow-overnight-protests-at-boarder-patrol-station-in-irondequoit/99572000/.

4. A trend that started in the 1980s across the country (Sexsmith 2017).

5. Dairy farms don't qualify for temporary visa programs like H-2A. However, at the time this chapter was written, there was a Senate proposal to give dairy farms access to such temporary work visas.

6. With the exception of the women's partners.

7. All farms and people mentioned in this chapter are given a pseudonym.

8. Children do not work at the farm but live there because they are sons and daughters of workers.

9. Three meals for 12 workers.

10. A 30-minute trip to Syracuse usually cost approximately $100.

11. Belen, Valentina, and Manuela.

12. Around $3 per tamale.

13. Fox Farm and Valley Farm are 30 minutes apart.

14. Brand of Latin American foods sold in many Latin American countries and in the United States.

15. For example, a 1.5-liter bottle of Jarritos costs around $3.50 at Wal-Mart, but Manuel sells it at $5.00.

16. Farmers do not pay the women for their work; however, by allowing them to live at the houses at no cost, they provide them with utilities and equipment needed for cooking (i.e., gas, electricity, stoves, cooking utensils, etc.).

References

Abarca, M. 2006. *Voices in the Kitchen: Views of Food and the World from Working-Class Mexican and Mexican American Women.* College Station: Texas A&M University Press.

Benson, Peter. (2011) Tobacco Capitalism: Growers, Migrant Workers, and the Changing Face of Global Industry. Princeton: Princeton University Press.

Benson, Peter. 2008. "El Campo: Faciality and Structural Violence in Farm Labor Camps." *Cultural Anthropology* 23(4): 589–629.

BOCES Geneseo Migrant Center. 2012. http://migrant.net/.

Budech, Keiko A. 2014. "Missing Voices, Hidden Fields: The Gendered Struggles of Female Farmworkers." Pitzer Senior Theses Paper 54, Pitzer College. http://scholarship.claremont.edu/pitzer_theses/54.

Cartwright, Elizabeth. 2011. "Immigrant Dreams: Legal Pathologies and Structural Vulnerabilities along the Immigration Continuum." *Medical Anthropology* 30(5): 475–495.

Cornell Cooperative Extension Harvest New York. 2015. http://www.dairyfoods.com/blogs/14-dairy-foods-blog/post/91019-new-york-loves-dairy-processors.

Farm Credit East Knowledge Exchange. 2015. "Challenging Times, Changing Markets: Northeast Milk Marketing in 2015." https://www.farmcrediteast.com/News-and-Events/News/20151029DairyReport.aspx.

Federici, Silvia. 2010. "The Reproduction of Labour-Power in the Global Economy, Marxist Theory and the Unfinished Feminist Revolution." Reading for January 27, 2009, University of California Santa Cruz seminar "The Crisis of Social Reproduction and Feminist Struggle."

Fox, Carly, Rebecca Fuentes, Fabiola Ortiz Valdez, Gretchen Purser, and Kathleen Sexsmith. 2017. *Milked: Immigrant Dairy Farmworkers in New York State*. Report by the Workers' Center of Central New York and the Worker Justice Center of New York. New York, NY.

Gimenes-Minasse, Maria Henriqueta Sperandio. 2016. "Comfort Food: Sobre Conceitos e Principais Característica." *Contextos da Alimentação—Revista de Comportamento, Cultura e Sociedade* 4(2): 92–102.

Gray, Margaret. 2016. "Dairy and Death in New York State." *Adelphi University Magazine*, Spring.

Green, Linda. 2011. "The Nobodies: Neoliberalism, Violence, and Migration." *Medical Anthropology: Cross-Cultural Studies in Health and Illness* 30(4): 366–385.

Hadjiyanni, T. 2014. "Transbodied Spaces: The Home Experiences of Undocumented Mexicans." *Space and Culture* 18(1): 81–97.

Hamilton, Emily, and Mary Jo Dudley. 2013. *The Yogurt Boom, Job Creation, and the Role of Dairy Farmworkers in the Finger Lakes Regional Economy*. Ithaca, NY: Cornell Cooperative Extension.

Holmes, Seth. 2011. "Structural Vulnerability and Hierarchies of Ethnicity and Citizenship on the Farm." *Medical Anthropology: Cross-Cultural Studies in Health and Illness* 30(4): 425–449.

Holmes, Seth. 2013. *Fresh Fruit, Broken Bodies: Migrant Farmworkers in the United States*. Berkeley: University of California Press.

Kasmir, Sharryn. 2008. "Dispossession and the Anthropology of Labor." *Critique of Anthropology* 28(1): 5–25.

Mares, Teresa M. 2012. "Tracing Immigrant Identity through the Plate and the Palate." *Latino Studies* 10(3): 334–354.

Marte, Lidia. 2008. "Migrant Seasonings: Food Practices, Cultural Memory, and Narratives of 'Home' among Dominican Communities in New York City." PhD diss., University of Texas at Austin, ProQuest Dissertations Publishing.

Minkoff-Zern, Laura-Anne. 2014. "Subsidizing Farmworker Hunger: Food Assistance Programs and the Social Reproduction of California Farm Labor." *Geoforum* 57:91–98.

NFWM. 2014. http://nfwm.org/2014/11/nfwm-selected-2014-top-rated-nonprofit/.

New York Department of Health. 2011. https://www.health.ny.gov.

Occupational Safety and Health Administration. 2014. https://www.osha.gov/.

Office of Governor Andrew M. Cuomo. 2013. "Governor Cuomo Announces New York State Is Now Top Yogurt Producer in the Nation." April 2013. https://www.governor.ny.gov/news/governor-cuomo-announces-new-york-state-top-yogurt-producer-nation.

Quesada, J., L. K. Hart, and Philippe Bourgoi. 2011. "Structural Vulnerability and Health: Latino Migrant Laborers in the United States." *Medical Anthropology: Cross-Cultural Studies in Health and Illness* 30(4) 339–362.

Rose, Margaret. 2008. *Traditional and Nontraditional Patterns of Female Activism in the United Farm Workers of America, 1962–1980*. Albuquerque: University of New Mexico Press.

Saxton, Dvera. 2013. "Layered Disparities, Layered Vulnerabilities: Farmworker Health and Agricultural Corporate Power on and off the Farm." PhD diss., American University.

Sexsmith, Kathleen. 2017. "'But We Can't Call 911': Undocumented Immigrant Farmworkers and Access to Social Protection in New York." *Oxford Development Studies* 5(1): 96–111

US Department of Labor. 2016. *Findings from the National Agricultural Workers Survey (NAWS) 2013–2014: A Demographic and Employment Profile of United States Farmworkers*. Report 12, December. Washington, DC: US Department of Labor, Employment, and Training Administration, Office of Policy Development and Research. https://www.doleta.gov/agworker/pdf/NAWS_Research_Report_12_Final_508_Compliant.pdf.

III Identity Narratives and Identity Politics

For all people, food is intimately intertwined with identity. We share experiences and build relationships over meals with family, friends, and strangers. Food scholars and food enthusiasts alike can attest to food's power to start a meaningful conversation. Immigrants perhaps forge this link between food and identity at a particularly intimate level. As we detailed in our introduction, food stands as an umbilical link between where one was and where one is today. It is no surprise, then, that a central theme within the immigrant-food nexus is the formation, development, and deepening of a politics of identity.

Part III of this volume includes five chapters telling stories of how food intersects with identity. This concept of "identity" is not limited to individuals: our authors investigate identities of the nation-state, the family, the organization, the kitchen, and even the "alternative food movement" itself. Within these chapters, immigrants find their identities at times included and at times excluded within the many borders and boundaries the previous chapters began to unravel. Yet at all times, these immigrants are themselves agents in molding their transnational identities and the meanings embodied in their translocal foods. They perform their identities through growing, cooking, and eating foods, some of which originate as recipes from their previous homes and some of which are developing through newfound kinship with fellow makers and eaters.

In these stories of identity formation and development, the power of narrative is clear. As narrative allows these immigrants to speak to the roles foodways play in their lives, it is clear that food holds incredible significance in daily life as well as larger community experiences, whether in moments of joy or fear, togetherness or tension.

10 The Canadian Dream: Multicultural Agrarian Narratives in Ontario

Jillian Linton

Introduction

In the lead-up to the 2011 Canadian federal elections, Michael Ignatieff, then leader of the Liberal Party, announced the party's plan for a national food policy. On the first page of the document highlighting the main policy goals, Ignatieff states, "Buying local is good for our farmers—who grow the world's highest-quality foods—for our families, and for our environment" (Liberal Party of Canada 2010). He went on to give campaign speeches from farm plots about the importance of home-grown Canadian food but was unable to sway voters and suffered a devastating loss at the polls (Swainson 2010). While the Liberal Party did not go on to win the election, this focus on creating local food policy gained increasing political support at the municipal, provincial, and national levels. Since then, local food procurement bills have been passed by several municipalities in Ontario, the province voted in favor of the Local Food Act in 2013, and supermarkets have increasingly sourced locally to meet growing consumer demand (Flavelle 2009). As different levels of government increasingly turn toward the promotion of localized production and consumption, it is worthwhile to analyze the rhetoric used to justify and support local food.

Discourse is intertwined with historical narratives and moral codes that outline the priorities and expectations of the society or social group in question (Fairclough 2003). Since discourse is often used to guide and justify political agendas, a close study of it can give insight into the underlying ideas driving policy, the power structures that are being endorsed, and the potential inequalities and exclusions being promoted (Razack 2002). To date, discourse on "local food" has been more thoroughly studied in the United States, where local food narratives have been associated with socially, and at times racially, exclusive language (Born and Purcell 2006; Hinrichs 2003). Researchers in different parts of the country, from California, to Iowa, to Washington, have noted that the imaginaries that organize alternative food movements[1] are often built around white identities and viewpoints (Ramírez 2015; Flora et al. 2012; Alkon and McCullen 2011; Guthman 2008; Slocum 2007). Although the exact organization of these food communities is different, they are frequently founded and centered on particular agrarian myths, such as the idea that white farm families made use of abundant empty land to grow food and create the fertile landscapes of today. Alkon and McCullen coined the term "white farm imaginary" based on their observations of California farmers markets. In their studies, they found that this white farm imaginary "romanticizes and universalizes an agrarian narrative specific to whites while masking the contributions and struggles of people of color in food production" (Alkon and McCullen 2011, 938). This white farm imaginary both follows and supports a nationalistic discourse that fails to recognize the historical exploitation of slave labor, existing racialized farmers (see Minkoff-Zern and Sloat, chapter 7, this volume), or the migrant farmworkers who do most of the cultivation. That being said, the white farm imaginary is not static but shifts according to its temporal and spatial contexts (McCullen 2008). While Canada and the United States exist on the same continent and their local food communities have transnational connections, it would be false to assume that Canadian farm imaginaries would work in the same way. There is a growing body of research that focuses on Canada's agricultural imaginaries and how Canada actively excludes Indigenous peoples and migrant farmers or workers (Rotz 2017; Wakefield et al. 2015) and the way that racialized boundaries exist in specific local food spaces such as Vancouver and Toronto (Gibb and Wittman 2013; Campigotto 2010). These researchers show that whiteness does play a role in agriculture north of the border but that it works differently because of Canada's divergent context, colonial history, and politics of multiculturalism vis-à-vis the United States.

Despite the government's official stance of celebrating diversity, Canadian multicultural policy has been criticized as a state-legitimizing tactic that reframes the long-standing problem of national unity, particularly indigenous sovereignty claims and French versus English division, into a society built on difference and diversity (Mackey 2002; Bannerji 2000).

Notwithstanding the country's rhetoric of inclusion, racialized immigrants continue to find that they are unable to achieve full status as Canadian, even upon obtaining citizenship. All the while, multicultural policy makes it difficult to name issues of systemic racism, as the policy's outward image obscures the continued racial exclusion of certain bodies from the Canadian national project (Bannerji 2000). Approaching these topics at the intersection of Canadian food politics and immigrant experience, this chapter conducts a discourse analysis of local food using data from the *Toronto Star* to determine the specific narratives and imaginaries that are used to support and justify local food in Canada. Considering Canada's policy of multiculturalism, immigrant-friendly rhetoric, and outward promotion of cultural diversity, its local food imaginary may also read as inclusive. I argue, however, that the *absences* in this rhetoric are most telling. The omission of indigenous perspectives, continued power imbalances, and historic inequality throughout these articles promotes a nationalistic ideal that is reductive and exclusionary.

Methods: The Changing Use of Local Food

To better understand the discourses around local food in the Greater Toronto area and the farm imaginaries at play, I completed an analysis of textual news media.[2] I chose the *Toronto Star* newspaper because it is a daily newspaper that is regionally focused yet has the highest readership of any newspaper in the country. It is a left-leaning publication, generally more liberal than any of the other daily papers with similar distribution levels. I conducted an online search of all articles in the ProQuest Toronto Star, Toronto Ont. Database (pubid—44892), which includes articles dating back to 1985. The search looked for all articles containing the phrase "local food," while excluding the phrases "local food drive" or "local food bank."[3] This yielded an initial data set of 615 articles dated between 1985 and 2015. Within this data set, the first usage of local food as a dedicated term to denote food produced or grown locally took place in an article dating back to 1987 discussing the benefits of "Agri-cities" and their role in land preservation (Steen 1987).

From this point forward, I logged all the articles in the search and then made further exclusions, yielding a focused final data set of 224 articles that was used for basic analysis. I excluded travel reports, event listings, and brief policy announcements, as well as articles that were written about locations outside Ontario or articles that only included local food as a descriptor or as part of a compound modifier (i.e., local food courts). The basic analysis consisted of cataloging articles by name, date, author, type, location, a brief summary, whether a definition of local was offered, whether immigrants were mentioned and how (positively, negatively, or neutrally), reasons provided for supporting (or not) local food, and key quotations. I used this catalog of 224 articles to quantify the main reasons that are given for supporting local food. From this data set, I selected for close reading and in-depth analysis 30 articles from 2003 to 2015. These articles were selected based on richness (prioritizing interviews and reports over lists or recipes) and variety. Despite there being dedicated food columnists, I selected multiple authors to ensure that one journalist's voice was not prioritized too heavily. Throughout this chapter, my arguments rely on the findings of the in-depth discourse analysis of the articles, with some context drawn from the basic analysis of the larger data set of 224 articles.

Understanding that newswriting has a very specific structure, I focused my analysis not on syntax or grammatical structure but on the ideas and content presented (Fairclough 2003). I followed Potter and Wetherell's (1987) understanding of discourse analysis: a careful reading and rereading in search of patterns of variability and consistency, followed by a second phase of analysis that focuses on determining the functions and effects of the texts analyzed. My analysis also follows Fairclough's idea that "text analysis is an essential part of discourse analysis, but discourse analysis is not merely the linguistic analysis of texts" (Fairclough 2003, 3). To this end, textual analysis was employed to the degree that it was useful for analyzing phrasing, but the larger focus remained on unveiling the ways in which the stories and subjects presented in the articles promote specific understandings of history, identity, and space in Ontario through local food.

The Changing Use of Local Food

Before delving into the in-depth analysis, it is worth acknowledging that the use of the term "local food" has changed over time. Currently, local food is generally understood as food produced within a specific geographical area in close proximity to the consumer, often with underlying sociopolitical motivations. In the original dataset, older articles did not use local food in this sense. Most articles in the 1980s used the term local food either to describe food consumed locally, such as travel articles describing local cuisine, or as a compound modifier referring to the place's location (i.e., a local food market, a local market that sells food). Slowly, however, articles began to mention local food as a political term, denoting food intentionally consumed *because* it was produced locally. After the first Agri-city article in 1987, an article in 1989 mentions the Slow Food movement growing in Rome, followed by another a year later celebrating World Food Day and discussing local food as an alternative to globally produced industrial foods. Despite these early mentions, it is not until about a decade later that this usage of local food is frequently featured.

The *Toronto Star* articles that mentioned local food production's political challenge to the global industrial food system go back about three decades, but more than 75% of them dated from 2008 or later. The articles did not give strict definitions or geographic boundaries to the term, instead focusing on discussions of local food's benefits. Articles in the 1990s and early the following decade largely focused on the benefits of freshness and quality, but by 2005 articles had shifted to arguments of the environmental and economic advantages, especially in relation to food miles. In 2008, when coverage of the topic increased dramatically, the "local food movement" is mentioned by name in several articles, reinforcing the fact that local food had gained a certain prominence in the mainstream. These articles still largely focused on the environment and climate change, alongside concerns of food security. From 2009 onward, the importance of supporting farmers grew, underscored by themes of community and tradition, as local food was linked to Canada's food history and settler populations. Finally, articles in the final five years of the data set carried an implicit suggestion that the reader should already be familiar with the many arguments in support of local food. These trends over the years were not strict boundaries, but they were nonetheless visible shifts in the benefits associated with local food. With this brief overview in mind, focus can now be shifted to the deeper narratives associated with local food and Canadian identity.

Multiculturalism: We Are All Immigrants

Historically, Canadian identity and mythology has long been built around a "northernness" that equaled whiteness, admittedly with two variations: one British and the other French (Berger 1966). However, after decades of explicitly racist immigration laws, the late 1960s and 1970s began the open-door policy for immigrants to Canada, which allowed a much higher number of visible minorities (the official term for nonwhite arrivals) to enter the country.[4] This period was initiated by the Liberals, as the new migrants were to supply the growing Canadian economy with a workforce. With the influx of nonwhite or racialized immigrants in this period came the introduction of Pierre Trudeau's national multicultural policy in 1971, after which Canada increasingly understood and sold itself as a nation built on a diversity of cultures and identities. Through this policy, immigrants become citizens, yet according to the government "keep their identities, take pride in their ancestry and have a sense of belonging" (Government of Canada 2017). This idea of a multicultural mosaic is promoted in contrast to (and as more inclusive than) the melting pot or politics of assimilation that the US government supports (Bannerji 2000). It therefore follows that if multiculturalism does play a major role in Toronto's local food discourse, it is possible that the nation's local food imaginaries themselves could be more diverse than those observed by scholars writing about the United States. While there were several references in the articles that explicitly name or support the discourse of multiculturalism, this chapter's following analysis reveals that the recent discourse on local food is less inclusive than it first might appear.

In their discussions of local food, several articles do follow a distinct narrative supporting the popular Canadian discourse of multiculturalism. Interestingly, some articles that discuss historic migration and its relation to farming portray Canada as having been a country of migrants since its inception—always a diverse blend of cultures. Multiculturalism in these articles is written about not as newly invented as a result of recent migrants but rather only newly named. In an excerpt from a 2011 interview, Dorothy Duncan, a Canadian culinary historian, expresses as much: "Yes, because from the outset, except for the First Nations, we have always been a very multicultural country. People from all over the world came here to live, brought their memories with them of what their favourite foods were, and then found this land with its unusual—in many cases—geography and wildly fluctuating climate" (Turnbull 2011).

This quotation reframes multiculturalism as the Canadian origin story, achieving a simultaneous task of placing all Canadians in the same migrant category, just in varying stages of settlement. In this narrative, immigrants of European descent and racialized immigrants are viewed as simply having different cultures. The power and privilege that European settlers were granted throughout Canada's history are not acknowledged in this picture of multiculturalism. While this quotation does acknowledge the existence of First Nations people, this romantic view of settlement ignores the violent legacy of colonialism and the explicit policies enacted to dispossess Indigenous peoples. Of course, this quotation could be seen as a positive acknowledgment that all Canadians other than Indigenous people are guests; however, this is not the implication. The passage reads "from the outset...we have always been a very multicultural country" (Turnbull 2011). The reference to the existence of First Nations people is an aside, an exception; they are not included in the "we" of Canadians. The quotation works to place Indigenous people as an exception to an inherently migrant and multicultural Canadian identity and history.

Aside from this historical framing of migration and multiculturalism, there is also a consistent thread suggesting that new immigrants have always used farming to achieve success. One article on the use of temporary migrant and immigrant labor on Ontario's farms follows this viewpoint, stating, "Ontario has a long history of foreigners and newcomers working the land. Take Ken Forth's farm near Hamilton. He's a fifth-generation Ontario farmer. In the 1920s, his family sponsored Czech workers. Then, they'd pick up Italian immigrants. Then Vietnamese. For all of them, it was a stepping stone to save money for a house and a better job. 'They did the Canadian dream,' says Forth, who now employs 16 workers from the Caribbean" (Porter 2007).

In this article, farming and working the land becomes a rite of passage. While Canada may be a multicultural nation filled with people of diverse backgrounds, there is an expectation that hard work, especially when connected to the land, is part of becoming Canadian. It is the "Canadian dream." This narrative supports the idea that all Canadians were once hardworking immigrants working the soil and that any difficulties are simply part of the journey.

Both these articles celebrate a narrative of multiculturalism that reframes Canadian identity as a fundamentally migrant identity beginning with the first settlers while simultaneously promoting a Canadian dream that promises new immigrants success and integration through hard agricultural labor. A celebration of multiculturalism and immigrant settlement could be seen as overwhelmingly positive, but it is more likely that as "difference is ideologically evoked it is also neutralized" (Bannerji 2000, 96). Although these articles celebrate difference and promote farming as a rite of passage for new immigrants, further analysis of the articles on local food shows this multicultural Canadian dream to be empty of critical discussion of colonial history, structural inequality, or recognition of differences in power and privilege for racialized Canadians.

A New History: Settling Rural Space

As foodie culture has grown throughout the new millennium, so has its influence on the local food discourse. Several articles, ranging from 2008 to 2015, ask the reader to think of the local food movement as a path to a more defined Canadian food history and cuisine. One author writes, "Making local produce available to consumers is actually a matter of recovering lost practices" (Gordon 2008). The suggestion is that Canadians have lost their way and should return to earlier local food traditions. However, these articles highlight European settler culinary traditions, positioning settlers as the original ancestors while ignoring indigenous foodways. One such article suggests Canadian cooks look to early settlers for inspiration when cooking with the native Canadian variety of gooseberries. It references how "in North America, settlers transformed the tart berries into wine, vinegar, preserves and pies" (David 2015). In this statement, settlers become the original innovators who took what the local landscape provided them and crafted unique dishes and beverages, with no mention of the precolonial past.

In the majority of the articles on local food, history begins with farm settlement. Questions of land ownership are not discussed, and in the larger data set of 244 articles, there is almost no mention of First Nations' existence or relationship to the land. In fact, less than five articles mention First Nations, Indigenous, or native people, and of the few that do, they are never the focus of the article itself. Throughout the discourse, there is rarely an elaboration of what came before; it is assumed that there was simply *terra nullius*—nobody's land—and therefore it was free for the taking (Razack 2002). Instead, the articles show farmers as the first Canadians in the national story. In one article (Baute 2009), a farmers market organizer is interviewed about her role. She responds, "It's more the farmers that I feel should be highlighted, really.... They're the ones that are doing the work of growing responsibly, the real *land stewards*" (emphasis my own). In the absence of indigenous bodies, farmers become the rightful caretakers of this unclaimed rural space. This theme of stewardship and responsibility recurs in several articles, positioning farmers in a role extending beyond simply growing food. In another article (Welsh 2009), a farmer is quoted as saying, "We are the stewards of the land.... We have to save it." The "it" is never clarified, although environmental degradation is implied, but what is clear is that farmers are positioned as the saviors. This theme is echoed in several articles that position farmers in the role of "white savior" and land or the environment as needing to be saved. By excluding Indigenous peoples from the discourse, they are essentially invisibilized from rural space and agricultural history, and a new version of settlement is offered in which settler farmers are the first Canadians, who come to own and care for unoccupied land. As a result, local food is presented as a way to support farmers as they fulfill the role of caretaker and preserve Ontario's farmland.

The Classic Farm, Farmer, and Farming Family

In the articles covering local food, there was a clear emphasis on naming the locations of local farms and highlighting individual food producers, painting a clear picture of the people and places involved. The majority of the articles portray local food as something that happens outside the city, on the farm and in rural regions surrounding the urban core. Mentions of urban farming largely introduce the concept as a novelty or innovation, such as the rising trend of urban gardening. One author in a 2013 article describing community shared agriculture (CSA) writes, "Like the many other CSAs around the province, Kawartha CSA brings two disparate communities together: farmers and city dwellers" (Black 2013). In the author's view, farmers and city dwellers are separate entities, once again emphasizing the location of farms in rural space and the connection of urban and rural spaces through produce delivery. There is little acknowledgment of urban farming as being legitimate, and instead the implication is that the *real* farms growing local food exist in rural space.

Aside from prioritizing rural farms, the articles specifically highlight the farmers themselves. In the 224 articles cataloged for basic analysis, the most popular reason given by authors for why someone should choose local food was to support and/or connect with farmers and their farms. While farming is frequently shown to be difficult in terms of profitability, farmers are portrayed as doing the important work of preserving the practices of their families. Most articles paint the picture of family farms, farms run by several generations of farming families, and often hint at the existence of potential future generations of farmers who are introduced to the work side of farming by being raised in this space. One such example features a wife and husband team running an organic farm and emphasizes that the wife, Haley, "grew up grading chicken eggs before school as the child of third-generation farmers" (Porter 2015). This example shows an overlap of domestic life (preschool routines) and the business side of farm space (grading eggs for sale), highlighting the way children are raised into the farming tradition. Families are portrayed not only as managing the farms but also as the hands directly responsible for cultivating and producing the local food feeding Ontario.

In many of these articles emphasizing generational traditions, there are no references to race or ethnicity. One example reads, "As a fourth-generation farmer, John Hambly is determined to leave his children with soil that is a little cleaner, richer, greener" (Welsh 2009). This article focuses on the farmer's ecologically sound growing practices and never references race, culture, or immigration. Hambly, pictured alongside the article, is a white man, but in the article, he is simply a Canadian farmer. Articles that focus on environmental issues of farming often feature these Canadian farmers without reference to their identity. Since these farmers remain unmarked in terms of ethnicity or race, these articles suggest that these multigenerational farming families are the original farming families. This invokes Lipsitz's point that as "the unmarked category against which difference is constructed, whiteness never has to speak its name, never has to acknowledge its rule as an organizing principle in social and cultural relations" (quoted in Yancy 2012, 6–7). In this case, white farming families are the "unmarked marker," the universal from which all else is measured (Frankenberg 1997).

There are a few exceptions where white farmers are featured and their continuing ethnic and cultural identity is included, usually referencing a distinct cultural background (e.g., Mennonite, Jewish). However, the articles' overall construction of Canadian agrarian tradition remains that farmers are usually white, have multigenerational families, occasionally maintain ethnic cultural traditions, and are generally located in rural regions where they take care of the land and feed city dwellers.

The New Immigrant Farmer

In the articles on local food in Ontario, another distinct farming identity emerges in contrast to this image of the white family farmer: the new immigrant farmer. These articles often feature a backstory of the hardworking immigrant, usually racialized, who comes to Canada, turns to farming as their new source of income, innovates, and grows new, exotic produce—usually in urban or periurban locations. In a 2013 article, one author writes, "My favourite 'farmer' is a sophisticated Korean named Yun Joon (Ben) Park. He arrived in Canada in 2000 and soon found himself managing a high-tech mushroom company specializing in exotic Asian varieties such as slim enoki and king oyster, with its meaty white stem and velvety tan cap" (David 2013). Park's arrival in 2000 is included early in the piece, as was the case in most articles featuring these immigrant farmers, emphasizing their relatively recent entrance and the exciting products they bring with them. The author's description here is interesting because Park is identified as being Korean; he is not a Korean immigrant, a Korean-Canadian, but simply a Korean. This form of identification works to distance Park from Canadian identity; he remains as exotic as the mushrooms he grows.

Another key example of this narrative is a 2008 article titled "Diverse Harvest for Budding Farmers." The article discusses an urban incubator farm in Brampton, stating:

> There are common crops at McVean—tomatoes, cucumbers, pumpkins and strawberries—but also callaloo, okra, chiles, bitter gourds and "dragon hot" Indian peppers. Some of these items are rare in Ontario fields, but pair the ethnic diversity of the GTA[5] with the local food movement—and add new Canadians with agricultural ambitions—and they begin to poke out of the soil. This Thanksgiving weekend the first harvest at McVean drew to a close as the budding farmers picked the last of their crops and began to prepare the land for winter.... Baloch grew up on a farm in Pakistan, where his family grew lemons, oranges, bananas, sugar cane and vegetables. Now Baloch lives in Brampton, works a desk job in Mississauga, and has only dreamed about having his own farm. (Baute 2008)

Here the emphasis is on the novelty of the farmers, the interesting foods that they grow, and the urban and periurban locations of the farms. The farmers are described as new and "budding," their crops as rare, and farming as their dream. This focus on the innovative and exciting aspect of new crops is typical of many of the articles on immigrant farmers, as is the focus on periurban locations. The profile of Baloch checks all the boxes: the article refers to his country of origin and the exotic produce his family grew there, the plot in Brampton (a Canadian city) that he now farms, the exotic produce he now grows, and his agricultural dream that is as yet unfulfilled.

While it could appear that any representation of immigrants as farmers is automatically positive, the very act of emphasizing and naming their cultural difference, their newness, and their "exotic" tastes and cuisines implies that there is an unnamed norm, as Lipsitz suggests, from which they differ and to which they are being compared. Articles on white or unmarked farmers in Ontario tend not to linger on ethnicity, instead showing differences between the farmers themselves in location and motivation. In contrast, new immigrant farmers are described as existing firmly in urban or near-urban space, and their motivations to farm are reduced to growing exotic produce and gaining economic success via a generalized agrarian "dream" occupation. Similarly, the emphasis on their recent arrival and their countries of origin suggests that they are not truly integrated—they are not fully Canadian. These immigrants may be new Canadians, may be citizens, and may be farmers, but in the discourse they are "the other" when compared to the unnamed white Canadian farmer. As Mackey puts it, there are those who are "unmarked, unhyphenated and hence normative *Canadian* Canadians who are thus implicitly constructed as the authentic and *real* Canadian people, while all others are hyphenated and marked as *cultural*" (Mackey 2002, emphasis in original). The novelty and exoticness tied to these narratives solidifies the normal naturalness of the unmarked "*Canadian* Canadians" who have been farming for many generations.

There are a few exceptions in the articles that suggest a different story of immigration and farming. One such exception was a 2006 article focusing on the increasing number of immigrants turning to farming. The author, Nicholas Keung, writes, "At a time when European immigrants were still settling into rural Canada to grow wheat and potatoes, Sam Kang Shin-Bong bought a 35-acre farm near Newmarket with the aim of growing oriental vegetables" (Keung 2006). This article places Kang's farm operation on the same timescale as European farm settlement and challenges the idea that racialized farmers are a new phenomenon. This article breaks from the trend and chooses not to juxtapose a hardworking new immigrant farmer identity and a historic unmarked white Canadian one. Instead, the author emphasizes the similar timelines of European and non-European migration and shows racialized farmers as being equally established in Canadian farming history. In doing so, the article stands out as a near singular example of this narrative. Except for a few small asides featured in a few articles such as this one, the overwhelming trend in the discourse is a celebration of the new immigrant farmer as an innovation. Keung's article shows that there are other stories that could complicate this discourse beyond newly arrived immigrants growing exotic produce. However, stories of racialized immigrants do not fit into the historic white Canadian settler imaginary, so they continue to be portrayed as new and exotic, and their early contributions remain erased.

SAWP Workers

In contrast to the articles' dominant portrayal of small family farms as white or as the exoticized vision of new immigrant farmers, in reality Ontario's farms overwhelmingly use seasonal agricultural workers to fill a domestic farm labor shortage. Ontario relies on the largest population of these workers of any province in Canada (Preibisch and Binford 2007). The Seasonal Agricultural Workers Program (SAWP) was started in 1966 as a visa program that brought in temporary agricultural laborers from Jamaica to fill the shortage on Ontario's farms during the growing season. The program has since expanded to include other provinces as recipients and Mexico and several Caribbean countries as new pools of labor. In the few articles that discuss the SAWP, even fewer mention any of the criticisms leveled at the program, including the precariousness of its workers, deplorable living conditions, racism, and abuse (Preibisch and Binford 2007). Rarely are the workers themselves given a chance to speak.

One article that focuses on the program and includes the workers' voices is titled "United Colours of Berrydom" (Porter 2007). This article accurately notes that most local farm labor is done by temporary seasonal workers or new immigrants willing to fill manual labor positions, in this case Sikh workers from Brampton. It specifically notes that these immigrants and SAWP employees fill "the void of native-born workers," who are unreliable and unwilling to do the work. The article opens with a description of the farm: "At a distance, you can tell the migrants from the immigrants on Bert Andrews' farms by their headwear. Fuchsia turbans and mustard-coloured scarves mean Sikhs from Brampton. Dark baseball caps mean Mexicans. Today, the turbans are picking red raspberries on one side of a thick row of maple trees. On the other side, the baseball caps are deep in the black raspberry bushes—30-odd flats deep, by the look of the mounting crates sitting in the shade of a nearby tree."

This description is interesting because it immediately uses objectivation to describe the immigrant and migrant workers. According to van Leeuwen, objectivation is the representation of a "social actor" by a place or thing closely related to them or by an action in which they are involved (van Leeuwen 2008, 46). This practice succeeds in impersonalizing the subjects and distancing them from the category of human. In the case of this article, turbans and baseball caps—objects—are used to represent the workers.

Although objectivation continues throughout the article, the author does later interview these workers and include direct quotations from their conversations, allowing them to occupy a subject position. However, these workers are quoted half as often in the article as other farmers, experts, and researchers. In fact, in a strict comparison of the lengths of quotations attributed to the two farmers and two workers interviewed, there are only 32 words that are spoken by the farmworkers, compared to 158 words spoken by the farmers. The farmworkers and their living situations are consistently described by the author rather than through the farmworkers' own quotations, while farmers are able to articulate their opinions and statements on the labor situation through direct quotations. Other articles that mention immigrant labor or SAWP seasonal workers follow this trend. Rarely are the workers interviewed and in even fewer cases are they quoted; instead, articles feature direct interviews with farmers that briefly mention the existence of foreign workers. Another such example from a 2009 article focused on a hydroponic farmer who acknowledges the use of migrant labor because of difficulties hiring Canadians: "The Mexican migrants are excellent. They come for a purpose—to work and make money. They're focused on the job. We house them very well—I call it Cadillac accommodation—and try to treat them well. It pays off for everyone" (Taylor 2009).

Again, the farmer speaks on behalf of the farmworkers, who are not interviewed, on their living conditions, motivations, and treatment. There is no interrogation of the fact that local Canadians are unwilling to do this work, and the undertone throughout the exchange is that the migrants have gotten a good opportunity to get ahead. The discourse's continued omission of the complicated legal and social realities of SAWP workers and existing labor disputes is surprising because many farms in Ontario rely on this stream of low-cost and reliable labor to remain profitable (Basok 2002). SAWP workers do not have access to the Canadian dream or even citizenship under the current system, yet the articles portray them as willing participants in the local food system.

Conclusion

The *Toronto Star* articles on local food include a celebratory narrative of multiculturalism woven within their discourses on rural space, Canadian history, and farming identities. However, because multiculturalism and difference are not discussed in relation to power and structural and historical imbalances, the Canadian local food discourse shares several similarities with the US white farm imaginary described by Alkon and McCullen. The articles largely portray Canadian farming as beginning with settlers, depict rural space as the uncontested property of generations of unmarked white farm families, celebrate a one-dimensional image of immigrant farmers, and underrepresent the SAWP workers who do the majority of farm labor in Ontario. In doing so, these articles ignore Indigenous histories, foodways, and land claims, create an "other" that reinforces the image of white famers as traditional and stewards, and leave unquestioned the exploitative and restrictive relationship SAWP workers may have with their employers and the possibility of the agricultural Canadian dream. In the articles' description of multicultural, diverse Canada, difference is celebrated but emptied of privilege or power, so white Canadian farmers who have occupied farmland for generations and people of color who have struggled and continue to struggle to own land and establish themselves in the Canadian landscape are all labeled "immigrant" and positioned as experiencing the same struggle on a level playing field. Indeed, according to this narrative, all Canadians are settlers adapting to the Canadian landscape with the same opportunities, only some people are further along than others.

Since the completion of the original analysis in 2015, there has been more writing on the subject of local food. A new search of articles spanning this period (2015–2018) found 49 relevant works out of 97 total articles returned. Four articles in this new data set mention Indigenous, native, or First Nations people, doubling their presence in the discourse compared to the original data set. One of these articles actually focuses on a new hospital program that provides traditional local food to Indigenous patients. While it is encouraging that in the past three years there does seem to be an increasing acknowledgment of the existence of indigenous people, the majority of articles fall in line with the discourses identified in the original search. Most still promote the image of the family farm absent migrant labor and position the farmer as the original environmental steward. Those tracing local food's connection to Canadian history still place its origins within the colonial period of settlement. The discourse may be changing, but it is happening slowly.

Local food narratives need to include a critical understanding of difference, power, and the racial and colonial structures that undergird agriculture and food production. As long as indigenous land ownership and foodways go unrecognized in conversations on local foodsheds and agricultural land preservation, their agricultural practices and foodways will likewise be marginalized and excluded in practice. Similarly, new racialized immigrant farmers may be celebrated in some articles, but what does this mean when the farming industry functions predominantly on low-wage work with low profit margins? Finally, it will take much longer to improve migrant workers' rights if the industry that benefits most from their work overlooks their existence or speaks for them. At a time when farmers are increasingly deserting the profession because of the high work commitment and low financial rewards (Weiler, Levoke, and Young 2016), it is valuable to reconsider these difficult issues embedded locally in our food communities. Recognizing and prioritizing the voices of indigenous people, migrant farmworkers, and racialized immigrants is a crucial step toward understanding their realities and unveiling the power imbalances that are currently obscured in the discourse on local food.

Notes

1. "Alternative food movement" is an umbrella term defined as commitment to ecologically and socially minded production methods. It includes but is not limited to farmers markets and community-based agriculture.

2. This research was funded in part by the SSHRC Insight Grant no. 76166—Unsettling Perspectives and Contested Spaces: Building Equity and Justice in Canadian Food Activism.

3. The exact search phrase was [(pubid(44892) "local food") *not* ("local food bank" *or* "local food banks")].

4. Although "visible minorities" is still the term used for statistical monitoring, racialized individual or group is considered more appropriate terminology.

5. Greater Toronto area refers to the metropolitan area surrounding the city of Toronto.

References

Alkon, Alison Hope, and Christie Grace McCullen. 2011. "Whiteness and Farmers Markets: Performances, Perpetuations...Contestations?" *Antipode* 43(4): 937–959.

Bannerji, Himani. 2000. *The Dark Side of the Nation: Essays on Multiculturalism, Nationalism and Gender*. Toronto: Canadian Scholars' Press.

Basok, Tanya. 2002. *Tortillas and Tomatoes: Transmigrant Mexican Harvesters in Canada*. McGill-Queen's Studies in Ethnic History 37. Montreal: McGill-Queen's University Press.

Baute, Nicole. 2008. "Diverse Harvest for Budding Farmers; Immigrants Learn Basics of Growing Crops Here; Others Benefit with Fresh Callaloo, Okra, Garlic." *Toronto Star*, October 13, 2008.

Baute, Nicole. 2009. "A Skill to Preserve; The Season's Best is on the Table as Newbies Learn the Time-Honed Art of Canning." *Toronto Star*, September 18, 2009.

Berger, Carl. 1966. "The True North Strong and Free." In *Nationalism in Canada*, edited by Peter Russell, 3–26. Toronto: McGraw-Hill.

Black, Debra. 2013. "Nothing Corny about Locally Grown Food: Reporter Buys Minishare from Community Agriculture Group and Has Been Loving Results." *Toronto Star*, August 14, 2013.

Born, Branden, and Mark Purcell. 2006. "Avoiding the Local Trap: Scale and Food Systems in Planning Research." *Journal of Planning Education and Research* 26(2): 195–207.

Campigotto, Rachelle. 2010. "Farmers' Markets and Their Practices Concerning Income, Privilege and Race: A Case Study of the Wychwood Artscape Barns in Toronto." MEd diss., University of Toronto.

David, Cynthia. 2013. "FRESH BITES: King Oyster Mushrooms: Taste the 'Deliciousness' of This Exotic Asian Offering That's High in Taste yet Low on Calories." *Toronto Star*, June 13, 2013.

David, Cynthia. 2015. "Cookie Dough Gooseberry Cobbler: Irresistible, Translucent Berries Have a Tart Appeal and Can Be Used in Desserts or as an Accompaniment to Rich Meats." *Toronto Star*, July 9, 2015.

Fairclough, Norman. 2003. *Analysing Discourse: Textual Analysis for Social Research*. London: Routledge.

Flavelle, Dana. 2009. "Loblaw Thinking Locally; Aims to Grow Amount of Ontario Produce in Its Supermarkets." *Toronto Star*, September 7, 2009.

Flora, Jan L., Mary Emery, Diego Thompson, Claudia M. Prado-Meza, and Cornelia B. Flora. 2012. "New Immigrants in Local Food Systems: Two Iowa Cases." *International Journal of Sociology of Agriculture and Food* 19(1): 119–134.

Frankenberg, Ruth, ed. 1997. *Displacing Whiteness: Essays in Social and Cultural Criticism*. Durham, NC: Duke University Press.

Gibb, Natalie, and Hannah Wittman. 2013. "Parallel Alternatives: Chinese-Canadian Farmers and the Metro Vancouver Local Food Movement." *Local Environment* 18(1): 1–19.

Gordon, Daphne. 2008. "Chef Sings the Praises of Holland Marsh; Jamie Kennedy Delights in Freshness, Variety of Farming Area's Produce." *Toronto Star*, August 20, 2008.

Government of Canada. 2017. *Multiculturalism*. December 12. https://www.canada.ca/en/services/culture/canadian-identity-society/multiculturalism.html.

Guthman, Julie. 2008. "Bringing Good Food to Others: Investigating the Subjects of Alternative Food Practice." *Cultural Geographies* 15(4): 431–447.

Hinrichs, C. Clare. 2003. "The Practice and Politics of Food System Localization." *Journal of Rural Studies* 19(1): 33–45.

Keung, Nicholas. 2006. "Farms a Growing Niche for Newcomers; an Abundant Supply of Farmland Could Offer Huge Potential to Immigrants as Our Ethnic Diversity Creates a Ready Marketplace." *Toronto Star*, July 12, 2006.

Mackey, Eva. 2002. *The House of Difference: Cultural Politics and National Identity in Canada*. Toronto: University of Toronto Press.

McCullen, Christie Grace. 2008. *Why Are all the White Kids Sitting Together in the Farmers Market? Whiteness in the Davis Farmers Market and Alternative Agrifood Movement*. Davis: University of California.

Porter, Catherine. 2007. "United Colours of Berrydom; Our Farm Series Looks at How Migrant and Immigrant Labour Gets Local Produce to Market." *Toronto Star*, June 15, 2007.

Porter, Catherine. 2015. "Ontario Farmers Grow to Appreciate the Greenbelt." *Toronto Star*, February 28, 2015.

Potter, Jonathan, and Margaret Wetherell. 1987. *Discourse and Social Psychology: Beyond Attitudes and Behaviour*. London: Sage.

Preibisch, Kerry, and Leigh Binford. 2007. "Interrogating Racialized Global Labour Supply: An Exploration of the Racial/National Replacement of Foreign Agricultural

Workers in Canada." *Canadian Review of Sociology/Revue canadienne de sociologie* 44(1): 5–36.

Ramírez, Margaret Marietta. 2015. "The Elusive Inclusive: Black Food Geographies and Racialized Food Spaces." *Antipode* 47(3): 748–769.

Razack, Sherene, ed. 2002. *Race, Space, and the Law: Unmapping a White Settler Society*. Ontario: Between the Lines.

Rotz, Sarah. 2017. "'They Took Our Beads, It Was a Fair Trade, Get over It': Settler Colonial Logics, Racial Hierarchies and Material Dominance in Canadian Agriculture." *Geoforum* 82: 158–169.

Liberal Party of Canada. 2010. "Rural Canada Matters: Highlights of the Liberal Plan for Canada's First National Food Policy."

Slocum, Rachel. 2007. "Whiteness, Space and Alternative Food Practice." *Geoforum* 38(3): 520–533.

Steen, David. 1987. "'Agri-cities' Can Save Food Lands, Group Says." *Toronto Star*, February 17, 1987.

Swainson, Gail. 2010. "Ignatieff Touts Homegrown Food Plan, Says Liberal Government Would Spend Millions Helping Farmers Fill Canada with Good Eats." *Toronto Star*, April 27, 2010.

Taylor, Bill. 2009. "Peter Piper Would Be in a Pickle; High-Tech Hydroponic Farm Melds Science, Agriculture as Green as Unripe Pepper Crop." *Toronto Star*, June 8, 2009.

Turnbull, Barbara. 2011. "Eating Our Way through History: Author Examines Origins of Traditional Celebrations, Holy Days." *Toronto Star*, February 5, 2011.

van Leeuwen, Theo. 2008. *Discourse and Practice: New Tools for Critical Discourse Analysis*. Oxford: Oxford University Press.

Wakefield, Sarah, Kaylen R. Fredrickson, and Tim Brown. 2015. "Food Security and Health in Canada: Imaginaries, Exclusions and Possibilities." *The Canadian Geographer/Le Géographe canadien* 59(1): 82–92.

Weiler, Anelyse M., Charles Z. Levoke, and Carolyn Young. 2016. "Cultivating Equitable Ground: Community-Based Participatory Research to Connect Food Movements with Migrant Farmworkers." *Journal of Agriculture, Food Systems, and Community Development* 6(2): 73–87.

Welsh, Moira. 2009. "Compost Can Create More Contented Cows." *Toronto Star*, January 24, 2009.

Yancy, George. 2012. *Look, a White! Philosophical Essays on Whiteness*. Philadelphia: Temple University Press.

11 Planning for Whom? Toward Culturally Inclusive Food Systems in Metro Vancouver

Victoria Ostenso, Colin Dring, and Hannah Wittman

Introduction

Food system planning involves envisioning and implementing structures and processes that influence the food supply chain. Planning is embedded in formal governance structures, people's organizations, social movements, and the private sector, at multiple scales, including municipal, regional, national, and multinational levels. Food policy councils (FPCs), as regional organizations that aim to influence food system planning, began to emerge in North America and across the globe in the 1980s (McCullagh and Santo 2012). They seek to address citizen disempowerment in the food system (Lang 1999) and the absence of food and agriculture in municipal policy and planning (Pothukuchi and Kaufman 1999). They have the potential to contribute to strategic food system planning by building alliances between diverse stakeholders, conducting system-level research and consultation to address a broad range of concerns related to public health, social justice, and ecological integrity (McRae and Donahue 2013). Therefore, many FPCs aim to transform urban areas into sites of "food citizenship," where people can actively partake in shaping the food system (Lang 1999; Harper et al. 2009).

Many FPCs have been successful at bringing together diverse food system stakeholders (consumers, farmers, policymakers, scholars, food industry representatives), forming coalitions to increase the strength of alternative food initiatives (Levkoe and Wakefield 2014). However, many alternative food initiatives themselves, including FPCs, have been criticized for not effectively or equitably engaging members from diverse racial, cultural, and socioeconomic backgrounds (Horst 2017; McCullagh and Santo 2012). Research on

FPCs in North America has also identified shortcomings in advancing racial justice. For example, a study of three FPCs in the Mid-Atlantic found representation of underrepresented groups, including racialized groups, was not being achieved (Boden and Hoover 2018). In that study, group members felt that "the blinding effects of ideological homogeneity" within their mostly white groups made it difficult to put food justice into practice (Boden and Hoover 2018, 10). Similarly, research on food planning in Washington State identified that while the Puget Sound Regional Food Policy Council strives to enhance democratic participatory decision making in the food system, it "has not yet succeeded in attracting, engaging and retaining diverse members," referring to both racial and geographic diversity (Horst 2017, 60). These examples point to the difficulty that many FPCs face in achieving their aim of equitably representing citizen-driven food system change.

On the other hand, some FPCs are taking explicit measures to achieve racial justice. For example, the Los Angeles Food Policy Council (LAFPC) website calls attention to what they call "food apartheid" in the city and uses municipal policy change to address environmental disparities in healthy food access for low-income communities and communities of color (LAFPC 2019). In another example, the Oakland Food Policy Council (OFPC) vision states, "We center racial equity in our radical approach to food justice" (OFPC 2019).

This chapter explores the significance of cultural inclusion in FPCs through the lens of critical race theory and a politics of difference (Young 1990). We examine the context of Metro Vancouver, a region characterized by multiple waves of immigration from Europe and Asia beginning in the late 1800s. Immigration continues to play a strong role in shaping the region, characterized as one of the most racially, culturally, and ethnically diverse regions in Canada. Given this context of immigration that unsettles binaries of white and other, we examine the politics of cultural inclusion in food systems to capture the ways in which racialized settlers are both implicated and marginalized within the context of settler colonialism. Through the politics of difference, Young calls attention to the oppressive consequences that ignoring difference has for individuals who do not fit into the "neutral standard" (Young 1990, 165) created by implicit dominant norms. In Canada, these norms are historically informed by British and French settler colonization and white cultural superiority (Thobani 2007). By making dominant norms explicit and promoting participation from differently

Planning for Whom?

positioned individuals in group decision-making processes, Young (1990) suggests that governance can move toward processes that are more representative and produce more just outcomes.

Critical race theory provides a framework to analyze the influence of race, as a construct that is invented and reinvented by society to preserve power, on the outcomes of people who are racialized differently (Delgado, Stefancic, and Harris 2012). It makes explicit that "White, wealthy, and masculine epistemologies" have dominated Western history, shaped spatial relationships, and ignored or erased the history of marginalized groups (Slocum 2011, 304). The concept of race and processes of racialization have far-reaching impacts on individuals, cultural practices, and the food system (Slocum 2011). In order to plan for a more just food system, the experiences and expertise of individuals from minority groups must be acknowledged and considered.

In Canada, different processes of racialization inform the exclusion of Indigenous peoples and historical and current immigrants of Asian, African, and Latinx descent based on their otherness in relation to the "exalted" characteristics of those European settlers who "belong" to the colonial nation-state (Thobani 2007). Furthermore, the prominence of race-neutral discourse within multicultural policies has the potential to further the inequality faced by racialized groups by treating race as a social factor not warranting explicit attention (Li 2001). Examples of this exclusion in the Canadian food system include; the institutional exploitation of Indigenous people from their food sources and the repression of traditional food knowledge by the Canadian government (Coté 2016), which has contributed to the higher prevalence of food insecurity (Tarasuk, Mitchell, and Dachner 2016); historical exclusion of Chinese Canadian and other racialized farmers from land ownership (Lim 2015); the labor precariousness of migrant food workers today (Otero and Preibisch 2015); and the lack of support for visible minority communities in alternative food initiatives (Gibb and Wittman 2013).

Food Policy and Planning in Metro Vancouver

Considering this framework, this chapter asks: How are leaders and members of FPCs in Metro Vancouver currently recognizing difference within their municipality and supporting participation of diverse racial, ethnic, and cultural groups? Metro Vancouver is the third-largest metropolitan area in Canada, with over 2.4 million people. The region is the second most culturally and ethnically diverse area in Canada, with over one million people (48%) identifying as members of a visible minority group[1] (Statistics Canada 2016). It is also home to 23% of BC's Indigenous population, including 12 local First Nations (Metro Vancouver 2018). Metro Vancouver is distinct from urban settings in the United States, where food justice has been practiced and studied more extensively, because of the narrative and policy approach of "multiculturalism" in Canada (Bottez 2011) and the racial and socioeconomic diversity of immigrant groups, including recent waves of affluent Chinese and other East and South Asian immigrants (Statistics Canada 2016). In this context, immigrants' experience of racialization cannot be understood through black/white and correlated poor/affluent binaries that often frame understandings of who belongs in North America. These binaries do not adequately describe the experience of racialization for diverse racial and ethnic groups in Canada or the United States (Alcoff 2003). This chapter will consider the positionality of immigrants within racial dynamics in this region more broadly by investigating how culture, race, and ethnicity inform outcomes of inclusion in food policy.

This study utilized qualitative methods, including participant observation, interviews with key stakeholders, and document analysis, to understand processes and practices toward cultural inclusion in three Metro Vancouver FPCs. We attended FPC meetings and interviewed FPC members, municipal staff, and nonprofit organization staff and members. We also reviewed the language regarding cultural inclusion in municipal and FPC documents, including Food Strategies, Food Charters, Terms of Reference, Municipal Food Systems Assessments, and the FPC websites.

Two Approaches to Cultural Inclusion

Our engagement with Metro Vancouver FPCs identified two broad approaches to cultural inclusion (color-blind and racial justice approaches) that were often employed by FPC members. A color-blind approach assumes that all people have equal opportunity and capacity to participate in FPCs. In utilizing this approach, some FPCs do not actively recognize implicit group bias or make special efforts to accommodate individual or group differences, or engage in governance processes that utilize specific strategies to assess whether FPC discourses and practices are exclusionary. This approach can fail to acknowledge the influence that structural oppression may have in determining the likelihood that individuals within nondominant racial, cultural, and ethnic groups will participate in FPCs. By not being attentive to difference, this approach can perpetuate social inequality by reinforcing a white spatial imaginary of the local food system (Lipsitz 2011).

In contrast, a racial justice approach makes difference explicit by naming implicit group norms, identifying power relations, and making specific efforts to include individuals from diverse racial, cultural, and ethnic backgrounds to address structural oppressions in the food system (Slocum 2006). This approach acknowledges whiteness as a set of structural privileges that is reinforced culturally through politics and practice (Guthman 2008) and strives to incorporate new strategies to involve members of nondominant racial, cultural, and ethnic groups in decision making.

In the following sections, we provide a brief description of several FPCs in Metro Vancouver, outline their goals for cultural inclusion, and then identify their diverse approaches toward these goals. We find that documentation, meeting discussions, and participants' viewpoints do not fit neatly into color-blind or racial justice categories. In conclusion, we offer insights for those within the alternative food movement to consider as they strive for cultural inclusivity.

Vancouver Food Policy Council

The Vancouver Food Policy Council (VFPC) is a 21-member civic agency whose members are city appointed and represent various sectors of the food system. The VFPC has five working groups that focus on specific food policy areas (food waste, development, children and youth, urban agriculture, and food justice). As an official civic agency, the VFPC submits an annual work plan and report of accomplishments to the city council, holds public meetings, and publishes meeting minutes online. The VFPC is rooted in a framework of environmental sustainability, guided by municipal goals outlined in the Greenest City Action Plan (City of Vancouver 2012). In 2016, the VFPC began to actively incorporate more of a sociocultural lens in its work, with leadership from the Food Justice Working Group (FJWG).

Cultural inclusion is implicated in the VFPC's Terms of Reference (VFPC n.d.), which states that its goal is to act as a "bridge between citizens and civic officials" for topics regarding the food system. The Food Charter (City of Vancouver 2007) and Food Strategy (City of Vancouver 2013) identify as goals to celebrate the diverse food cultures in Vancouver, ensure that residents have access to culturally appropriate food, and enable participation in food system activities that reflect the city's ethnocultural diversity. One example of a way that FPCs in general can further a color-blind approach can be found in the definition of what resources are important features of the local food system. In this instance, the Vancouver Food Strategy defines a list of food assets, or "resources, facilities, services, or spaces that are available to Vancouver residents and are used to support the local food system" (City of Vancouver 2013, 23). The list features elements associated with urban agriculture (community gardens, urban farms) and local food consumption (food networks, farmers markets, street food vendors) (City of Vancouver 2013, 24). Many of these spaces are very recent additions to the Vancouver food landscape, emerging from the environmental and sustainable food movements in the 1990s, and are primarily white spaces (Gibb and Wittman 2013; Seto 2011).

Some participants identified this list of spaces as symptomatic of a white spatial imaginary of Vancouver. One Chinese Canadian participant, for example, reflected on how the VFPC's list of food assets includes "feel good" places in the landscape of the food system, such as community gardens, but not the places that are feeding people, such as the produce wholesale district. Another Chinese Canadian participant described the current list as "whitewashed" because the food places that are valued by other cultural groups are not included, providing an example of a café that is an important gathering space for elderly Chinese people in the Downtown Eastside but is excluded from the list. By overlooking cultural food assets, the VFPC employed a color-blind approach to formulating and implementing their vision for the food system.

Another color-blind approach expressed in policy and programming efforts was the prioritization of nutrition-based perspectives over cultural perspectives. Access to healthy and culturally appropriate food is a core goal of all FPCs in this study. Interviewees indicated, however, that the healthiness of food from a conventional nutritional science perspective often took precedence over cultural diets and food practices. For example, this white participant explained how she experiences the tension between healthy foods and cultural foods in her work as a dietitian: "Speaking about cultural food is definitely on everybody's radar. So it has come out in terms of when nutritious food can overcome cultural foods, such as brown rice versus white rice. White rice is much more cultural, however, brown rice prevents high blood sugars, chronic disease, and heart disease.... And to be frank about this one, I'm on the [side] of choosing the brown rice because that's my nutrition space."

In this nutrition-forward standpoint, white rice is not considered healthy despite its being a staple in many food cultures. Without representation of other cultural ways of understanding and relating to food, the nutrition perspective based on a construct of "health" in food and the "neutral" rationality of a dominant version of science biased toward Western culture will continue to guide FPCs.

Of the FPCs in this study, the VFPC is most deeply embedded in a municipal governance structure. Participants identified an environment of colorblind professionalism that is reinforced by membership selection processes, attitudes toward community engagement, and meeting structure. One white participant described the leadership positions held by FPC members, which allow them to speak for their own and other communit(ies):

> A lot of the reason why the council... [is] diverse, or from different facets of the food system, is that a lot of us are on the pulse of what's happening in Vancouver, in BC, at the national level in the food system and food policy. Therefore, we're on the pulse of things that may not be affecting everyone but they may be affecting one particular cultural group in Vancouver. So I mean [a Chinese Canadian cultural leader] was very present but if they would not have been there... to point out the loss of Chinese cultural heritage in [Downtown Vancouver] and the food assets there, I think one of us would have known and brought this to the table.

In this excerpt, this participant indicates that because the council is comprised of individuals with experience in many areas of the food system (production, distribution, retail, access, and waste), they are "on the pulse" of the food system and can speak on behalf of the communities that they represent (or aim to represent). Furthermore, this participant asserts that the presence of a Chinese Canadian cultural leader in the group was not essential to the FPC addressing the priorities of that cultural group.

Participants also used professionalism to redirect the work required to further a more culturally inclusive agenda to a subgroup of diversity experts and cultural community organizations. In this case, the VFPC delegated the responsibility of cultural inclusion to a group of members, the Food Justice

Working Group (FJWG), which is comprised of activists of different ethnocultural and racial backgrounds who pursue antioppression and antiracism interventions in their professional and personal lives. However, FJWG members recounted that advancement toward racial justice is constrained by a lack of shared understanding, interest, and action from the FPC as a whole. The following excerpts, the first two from FJWG members and the latter two from other VFPC members, reflect the divergent viewpoints on the action necessary to put cultural inclusion goals into practice:

> [The first step is] being able to name that race and culture is not being addressed and that [race] has historically and continues to segregate and reinforce a Western base of food practice.—Chinese Canadian, male, FJWG

> It gets back to... what it means to be white and privileged and speak out about racism.... And how do we work on inclusion together and start to pick up on some nuance? Like the things that I can say or do that my colleagues who come from communities of colour, who don't carry white privilege, or have been exhausted by saying over and over, can't? I think that's the next level where we recognize that power and privilege exists in our movement but for those of us that have power and privilege, what do we do with that?—White female, FJWG

> You know, it would be great to have a cheat sheet for people for which [cultural inclusion] is not really [their] expertise. I'm well versed in the social aspect of food systems but that's not what I do on a day to day basis. It's not my background. This being said, I recognize the value in having a little simple cheat sheet like, "Here are five ways you can think about [inclusion] a bit more."—White female

> I think that the failure is that we don't have enough resources to advance the diversity and inclusion and representation as best we can... but if [cultural inclusion] had its own set of activities then we could really advance on diversity and inclusion through various community groups.—White male

In the first two excerpts, FJWG members assert that current members, who hold power and privilege within the movement, are responsible for being able to talk about race, understand how it operates within the food system, and critically engage in their work and learn how to use their power to be more culturally inclusive. In the second set of excerpts, FPC members acknowledge the importance of cultural inclusion but do not see themselves in a leadership role in the ongoing work required to advance that priority. In the third excerpt, the FPC member suggests that cultural inclusivity can be achieved through following guidelines in a simple template. This positions cultural inclusion as something that can be quickly taken care of through a "cheat sheet" rather than something that involves personal and group reflexivity to recognize and unlearn oppressive tendencies and challenge power imbalances. In the fourth excerpt, the FPC member reflects on how with more resources they could advance diversity and inclusion by using cultural community groups to do their inclusion activities. In this way, both latter participants are outsourcing the work of cultural inclusion.

A standard meeting format is one of the structures that facilitates the partitioning of responsibility among FPC members. Subgroups are allocated most of the meeting time to summarize their project progress and report on the following steps. As a result, some participants felt that there was inadequate time for the FJWG to engage the rest of the group in nuanced conversations about privilege and develop a shared sense of responsibility in cultural inclusion work.

Despite the divergent approaches to cultural inclusion held by group members, the VFPC has made strides to offset color-blind professionalism within the council and challenge power hierarchies through racial justice approaches. For example, members referred to efforts to build relationships with and recognize the food system work of organizations led by visible minority community members, such as sponsoring an event organized by Indigenous leaders to protect wild salmon habitat in British Columbia, endorsing the work of the Hua Foundation on the loss of cultural food assets in Chinatown (see Ho and Chen 2017), and supporting a study of the cultural food retail environment in the city of Vancouver (see But and Bencio 2017).

In summary, within this food policy space, some members are actively working toward cultural inclusion with racial justice approaches, while other members simultaneously employ color-blind perspectives. These color-blind perspectives have been shown to deflect conversations that may address power relations and contribute to an internal tension between color-blindness and racial justice.

Richmond Food Security Action Team

Richmond is the most ethnically diverse municipality in this study, with 76% of the population identifying as a visible minority from over 150 ethnic origins (Statistics Canada 2016). The first FPC formed in 2002, when the Richmond Poverty Response Committee created the Richmond Food Security Action Team (RFSAT), a group of representatives from local government, the health authority, and nonprofit organizations, chaired by the Richmond Food Security Society (RFSS).

The RFSAT's 2014 Terms of Reference contain multiple goals for representation, such as "to collaborate with community partners and individuals," "to provide shared leadership...with community members," and to "foster relationships between diverse stakeholders." The Richmond Food Charter (2016) development process (2014–2016), which was spearheaded by RFSAT members, made specific efforts to include cultural groups beyond those who were already involved (e.g., foodies, environmentalists, and food philanthropists/charitable food agencies or health agencies). Leaders described their motivation to develop a Food Charter as a starting point to involve a greater diversity of voices in food system planning. An FPC member describes the effort: "We ended up doing 26 different focus groups...and we tried to get engagement in terms of a wide variety of people from the community. So I remember doing groups with Chinese people, South Asian people, with Somali people.... [Our city] is just a very diverse community."

This engagement strategy offered an opportunity for cultural groups to provide input on a local food policy document. However, RFSAT members also discussed limitations of this engagement, including the inability of focus group participants to inform agenda setting or provide feedback.

RFSAT disbanded in 2016, and since that time the main voice for food policy work in Richmond has been expressed through Richmond Food Security Society and has primarily involved advocacy for community gardens and farmland protection. At the time of this study, RFSS was in conversation with the city of Richmond, Vancouver Coastal Health, and other nongovernmental organizations (NGOs) to form a municipal advisory committee for food security issues. This shifting governance structure of Richmond's local food policy efforts is not uncommon for FPCs and is an example of the ongoing, and sometimes uneven or sporadic, relationship building that occurs between grassroots leadership and grant or government support.

Since RFSAT disbanded, cultural community engagement in food policy and programming has been less explicit. While RFSS's organizational goals are to "identify and understand the diverse audiences that we serve and adapt our programs to reflect these demographics" (RFSS 2017), interviews indicated a color-blind approach to cultural inclusion work in their programming efforts. For example, one white RFSS staff member explained,

> None of our programs are exclusive, they are open to everybody. We don't even ask what people's nationality is, or their gender, we don't really care. We want everyone who is interested in our work to be involved.... We just haven't had capacity to be that targeted in our program offerings and I am not sure it would be the right approach for us. We certainly will make every effort to not exclude anybody for financial reasons, for cultural reasons.

This participant aims to include everyone through a color-blind approach that assumes equality of access and that communications and outreach will reach the diverse populations within the municipality. This participant dismisses a targeted approach both because of a lack of capacity and because they are not convinced it would make a difference.

Another white former RFSAT member felt that measuring cultural inclusion by asking participants about their race, culture, or ethnicity was inappropriate from a privacy standpoint. This participant was not in favor of tracking the ethnic background of participants because questions of identity are "sensitive" and could be "very problematic." They attributed this to not wanting to threaten participant comfort by asking questions of identity and to legal privacy concerns with the organization storing sensitive personal information. While these data storage concerns are relevant, this participant's insistence on not measuring inclusion may inhibit policy and programming analysis aimed at achieving culturally inclusive outcomes that would benefit from such data.

According to participants, changes in leadership capacity and a lack of explicit efforts to maintain relationships with these cultural community members because of color-blind ideologies mean a racial justice approach has not been sustained. As one white female city planner and former RFSAT member explained, food security and diversity work often relies on unpaid labor and gets put off to the side:

> Because our work in food security is ad hoc, or off to the side a little bit, it's the same in relation to diversity. So it's not to say that we aren't doing anything but what we end up doing might be just, sort of ad hoc and off the side of our desk.

Because the core group of food system planning leaders in Richmond were white professionals, it seemed easier for them to put cultural inclusion "off to the side" and continue their work. This speaks to the importance of capacity, knowledge, and representation of policy leaders to challenge white spatial imaginaries within food system work.

North Shore Table Matters

Table Matters is a network of North Shore residents guided by a 14-member steering committee that includes representatives from three municipalities, the school district, nonprofit organizations, businesses, and other community members. The Table Matters network "supports food policy and community development projects that build sustainable food systems and make healthy food accessible for everyone living on the North Shore" (Table Matters n.d.). Their work is guided by the North Shore Food Charter (Table Matters 2013) and coordinated by a paid staff person. Their work reflects an environmental emphasis; for example, coordinating a carbon footprint diet challenge and food waste reduction challenge. Table Matters advocates for and develops food security policy and supports community members who wish to present local food system issues to the city council.

The three municipalities represented by North Shore Table Matters—West Vancouver, district of North Vancouver, and city of North Vancouver—are located on the unceded territory of three First Nations (Musqueam, Squamish, and Tsleil-Waututh). The Squamish and Tsleil-Waututh reserves are located in multiple places across the community. There is also a high level of ethnocultural diversity within the region: 36.2% of community residents are recent immigrants, including Filipino, Indian, and Chinese (Statistics Canada 2016). North Shore Table Matters's guiding principles, outlined by the North Shore Community Food Charter, refer to the importance of cultural inclusion in regard to a celebration of diversity: "Food Culture & Education: Our community becomes proficient in food literacy and celebrates all food cultures." A white female member described the Food Charter development process as "grassroots" and "very low-barrier" (e.g., in the evenings, free food) but went on to reflect on the cultural inclusion of the process: "I am trying to think about cultural diversity.... You know I think I don't think that there's been a lot of direct outreach to specific communities beyond the First Nations." This member recognized the contradiction between the group's goal to "celebrate all food cultures" and their historical failure to include cultural groups and was keen to be more inclusive moving forward: "I think that what we need to do first is go and talk to [people from various cultural groups] and be like, 'What does [the food system] mean to you? What would you like us to be for you?'"

Table Matters members also discussed past efforts that the group has made to support and consult with organizations that work more directly with immigrants and First Nations. During one meeting, the group discussed organizing a reconciliation workshop for the steering committee and a desire to have First Nations members represented on the FPC.

However, there was also tension between a desire by some members to "efficiently move ahead" on group objectives and established timelines and taking the time to build relationships with cultural and Indigenous groups. In this example, a white female member reflects on how the ideal timeline for creating a Food Action Plan conflicted with the amount of time it takes to build relationships with a local First Nations community:

> In speaking with one of the people who I am in contact with at the Tsleil-Waututh Nation, she has talked to me about how much time this is going to take and she's really trying to facilitate the connections with the right people. And I said to her, you know, this is going to be really challenging because this [action plan development] process isn't going to slow down. It's going to keep on going.

This tension between continuing to move ahead on objectives with current core supporters and pausing to transition processes and adjust timelines to be more inclusive of cultural groups demonstrates how conflicting group priorities can impact outcomes of inclusion.

Members of this FPC recognized that their past efforts haven't been very culturally inclusive, and they expressed interest in change. At the time of this study, their cultural inclusion efforts focused on engagement with members of the three First Nations within the region. Members expressed a desire to recognize settler colonial positionalities in an effort to meaningfully incorporate First Nations perspectives into FPC work. These efforts to listen to traditional knowledge holders indicate an initial application of racial justice approaches in their work. As members build competence through this reflexive work, they may begin to prioritize relationships over timelines and understand the context for cultural inclusion in their community as more expansive than the current First Nations/settler binary in order to answer the question posed by the member earlier, "What would you like us to be for you?," in a way that considers other racialized immigrant groups within their community as well.

Conclusion

This project assessed how FPCs in Metro Vancouver put cultural inclusivity goals into practice by identifying two approaches operating simultaneously within these organizations: a color-blind approach and a racial justice approach. We suggest that color-blind approaches limit FPCs' ability to achieve cultural inclusivity by defining the food system through a dominant white perspective (e.g., food assets and nutritious food), claiming to be inclusive of "everyone" while reinforcing this perspective. The approach is categorized by minimal efforts to include structurally marginalized groups and conducting processes in spaces and formats that privilege professional (read: primarily white) voices. The color-blind approach paradoxically contributes to the maintenance of a white/other binary and a racial imaginary that cannot be separated from the category of "immigrant." In contrast, racial justice approaches, recognizing difference, sought to develop alliances among FPC members, develop social learning, and challenge unequal power relations to reciprocally engage with cultural groups.

Many FPC members were quick to refer to time, resource, personal, and structural constraints that kept them from achieving their group's goals for cultural inclusion. However, this research has shown that FPCs cannot achieve cultural inclusion simply by overcoming these barriers. In this context, members who are forging a path toward cultural inclusion through racial justice approaches must be persistent and resilient to challenge norms of exclusivity and demonstrate the importance of considering member positionality, valuing cultural difference, and moving toward representation that is more equal. In order for cultural diversity to shift from being "off to the side" to being central to FPC work, cultural community members need to be valued as creators and contributors, not just receivers, of food system change efforts.

Members of FPCs focusing on racial justice approaches in Metro Vancouver suggest that to achieve a more socially just food system, current FPC members must be open to acknowledging what they do not know, learning from and alongside members of cultural communities, and restructuring FPC processes to be more inclusive. Conversations with FPC members resulted in the identification of several principles for FPCs to advance culturally inclusive outcomes:

- Transparency
  - Outline cultural inclusion goals, including how to achieve and assess.
  - Make modes of participation (formal membership) and decision making public.
- Reflexivity
  - Understand how positionality shapes priorities (Lyson 2014).
  - Actively attend to the roles of race, power, and structural oppression and their relation with cultural inclusion (Slocum and Cadieux 2015).
- Social/Emotional Practice
  - Embed a compassionate, healing-centered approach to planning practice (Lyles, White, and Lavelle 2017).
  - Engage with dissent, ensuring counternarratives and alternative viewpoints are recognized and included (Clark et al. 2017).
- Accessibility
  - Acknowledge and cultivate different institutional arenas as nodes of a broader food policy network.
  - Actively seek out and build relationships with cultural knowledge holders, leaders, and issues.

Our study has highlighted areas where food justice practice is both supported and hindered by FPCs' approaches to inclusivity. We echo calls by others (Cadieux and Slocum 2017) for those advancing food justice, including FPCs, to be explicit about how their practices constitute racial justice and social change for equity. Further participatory and action research is needed to identify emergent and iterative approaches employed by FPCs to achieve visions of inclusivity and racial justice.

The achievement of equitable and sustainable food systems necessitates the participation of racial, cultural, and ethnic minority groups to challenge the dominance of white spatial imaginaries. In a culturally pluralistic society such as Canada, approaches to participation must pay attention to difference, including the unique rights of and the injustices faced by racial, cultural, and ethnic minority groups, in order to accommodate the range of voices that have a stake in food system change.

Acknowledgments

We acknowledge that this research has taken place on the unceded territories of the Coast Salish peoples in what is predominantly referred to as Metro Vancouver, Canada. Additionally, we gratefully recognize and applaud the municipal and nonprofit staff, and food policy council volunteers, for courageously sharing their insights and perspectives on this subject.

Note

1. Visible minority is the term that has been employed by the government of Canada to refer to "persons, other than Aboriginal peoples, who are non-Caucasian in race or non-white in colour" and includes Chinese, South Asian, Black, Filipino, Latin American, Southeast Asian, Arab, West Asian, Japanese, Korean, other visible minorities and multiple visible minorities (Statistics Canada, 2015). It includes some people with mixed European and non-European origin. Notably, the term visible minority does not include Aboriginal groups. This term has been critiqued for subtly implying race without actually naming it (Li, 2001).

References

Alcoff, L. 2003. "Latino/as, Asian Americans, and the Black-White Binary." *Journal of Ethics* 7(1): 5–27.

But, I., and C. Bencio. 2017. *The Role of Small Grocers in Neighbourhood Food Access: A Study of Residents' Perceptions in Renfrew-Collingwood*. Report commissioned by the City of Vancouver in partnership with the University of British Columbia. https://sustain.ubc.ca/sites/sustain.ubc.ca/files/GCS/2017_GCS/Final_Reports/The%20Role%20of%20Green%20Grocers%20in%20Neighbourhood%20Food%20Access_But_2017%20GCS.pdf.

Boden, S., and B. M. Hoover. 2018. "Food Policy Councils in the Mid-Atlantic: Working toward Justice." *Journal of Agriculture, Food Systems, and Community Development* 8(1): 39–51.

Bottez, M. 2011. "Critical Multiculturalism in Canada and USA." In *Towards Critical Multiculturalism: Dialogues between/among Canadian Diasporas = Vers un multiculturalisme critique : dialogues entre les diasporas canadiennes*, edited by E. Bujnowska, M. Gabryś, and T. Sikora, 27–43, Katowice, Poland: PARA.

Cadieux, K. V., and R. Slocum. 2015. "What Does It Mean to Do Food Justice?" *Journal of Political Ecology* 22(1): 1–26.

City of Vancouver. 2007. *Vancouver Food Charter*. http://vancouver.ca/files/cov/Van_Food_Charter.pdf.

City of Vancouver. 2012. *Greenest City: 2020 Action Plan*. https://vancouver.ca/files/cov/Greenest-city-action-plan.pdf.

City of Vancouver. 2013. *What Feeds Us: Vancouver Food Strategy*. http://vancouver.ca/files/cov/vancouver-food-strategy-final.PDF.

Clark, J. K., J. Freedgood, A. Irish, K. Hodgson, and S. Raja. 2017. "Fail to Include, Plan to Exclude: Reflections on Local Governments' Readiness for Building Equitable Community Food Systems." *Built Environment* 43(3): 315–327.

Coté, C. (2016). "Indigenizing" Food Sovereignty: Revitalizing Indigenous Food Practices and Ecological Knowledges in Canada and the United States. *Humanities*, 5(16), 1–14.

Delgado, R., J. Stefancic, and A. Harris. 2012. "Introduction." In *Critical Race Theory*, edited by R. Delgado and J. Stefancic, 1–18. New York: NYU Press.

Gibb, Natalie, and Hannah Wittman. 2013. "Parallel Alternatives: Chinese-Canadian Farmers and the Metro Vancouver Local Food Movement." *Local Environment* 18(1): 1–19.

Guthman, Julie. 2008. "'If They Only Knew': Color Blindness and Universalism in California Alternative Food Institutions." *Professional Geographer* 60(3): 387–397.

Harper, A., A. Shattuck, E. Holt-Giménez, A. Alkon, and F. Lambrick. 2009. *Food Policy Councils: Lessons Learned*. www.farmlandinfo.org/sites/default/files/Food_Policy_Councils_1.pdf.

Ho, A., and A. Chen. 2017. *Vancouver Chinatown Food Security Report*, August. http://www.huafoundation.org/work/food-security/choi-project/Vancouver-Chinatown-Food-Security-Report.html.

Horst, M. 2017. "Food Justice and Municipal Government in the USA." *Planning Theory and Practice* 18(1): 51–70.

Lang, T. 1999. "Food Policy for the 21st Century: Can It Be Both Radical and Reasonable?" In *For Hunger Proof Cities: Sustainable Urban Food Systems*, edited by M. Koc, R. McRae, L. Mougeot, and J. Welsh, 216–224. Ottawa: IDRC.

Levkoe, C. Z., and S. Wakefield. 2014. "Understanding Contemporary Networks of Environmental and Social Change: Complex Assemblages within Canada's 'food movement.'" *Environmental Politics* 23(2): 302–320.

Li, P. 2001. "The Racial Subtext in Canada's Immigration Discourse." *International Migration and Integration* 2: 77–97.

Lim, S. R. 2015. "Feeding the 'Greenest City': Historicizing 'Local', Labour, and the Postcolonial Politics of Eating." *Canadian Journal of Urban Research* 24(1): 78–100.

Lipsitz, G. 2011. *How Racism Takes Place*. Philadelphia: Temple University Press.

Los Angeles Food Policy Council (LAFPC). 2019. *Food Equity and Access*. https://www.goodfoodla.org/food-equity-and-access/.

Lyles, W., S. S. White, and B. D. Lavelle. 2018. "The Prospect of Compassionate Planning." *Journal of Planning Literature* 33(3): 246–266.

Lyson, H. C. 2014. "Social Structural Location and Vocabularies of Participation: Fostering a Collective Identity in Urban Agriculture Activism." *Rural Sociology* 79 (3): 310–335.

McCullagh, M., and R. Santo. 2012. *Food Policy for All: Inclusion of Diverse Community Residents on Food Policy Councils*. https://www.jhsph.edu/research/centers-and-institutes/johns-hopkins-center-for-a-livable-future/_pdf/projects/FPN/how_to_guide/getting_started/Food-Policy-All-4-8.pdf.

McRae, R., and K. Donahue. 2013. *Municipal Food Policy Entrepreneurs: A Preliminary Analysis of How Canadian Cities and Regional Districts Are Involved in Food System Change*. Toronto Food Policy Council, Vancouver Food Policy Council, Canadian Agriculture Policy Institute, Toronto.

Metro Vancouver. 2018. *Metro Vancouver's Profile of First Nations with Interests in the Region 2018*. http://www.metrovancouver.org/services/first-nation-relations/AboriginalPublications/ProfileOfFirstNations.pdf.

Oakland Food Policy Council (OFPC). 2019. *Our Work*. http://oaklandfood.org/our-work.

Otero, G., and K. Preibisch. 2015. *Citizenship and Precarious Labor in Canadian Agriculture*, November 18, 2015. Vancouver: Canadian Centre for Policy Alternatives. https://www.policyalternatives.ca/publications/reports/citizenship-and-precarious-labour-canadian-agriculture.

Pothukuchi, K., and J. L. Kaufman. 1999. "Placing the Food System on the Urban Agenda: The Role of Municipal Institutions in Food Systems Planning." *Agriculture and Human Values* 16(2): 213–224.

Richmond Food Security Society (RFSS). 2017. *About Us*. https://www.richmondfoodsecurity.org/about-us-2/.

Richmond Food Charter Working Group. 2016. *Richmond Food Charter*. https://www.richmondfoodsecurity.org/richmond-food-charter/.

Seto, D. 2011. "Diversity and Engagement in Alternative Food Practice: Community Gardens in Vancouver, B.C." Master's thesis, University of British Columbia.

Slocum, R. 2006. "Anti-racist Practice and the Work of Community Food Organizations." *Antipode* 38(2): 327–349.

Slocum, R. 2011. "Race in the Study of Food." *Progress in Human Geography* 35 (3): 303–327.

Slocum, R., and K. V. Cadieux. 2015. "Notes on the Practice of Food Justice in the U.S.: Understanding and Confronting Trauma and Inequity." *Journal of Political Ecology* 22(1): 27–52.

Statistics Canada. 2015. *Visible Minority of Person.* http://www23.statcan.gc.ca/imdb/p3Var.pl?Function=DEC&Id=45152.

Statistics Canada. 2016. *Census Profile, 2016 Census: Vancouver [Census metropolitan area], British Columbia and British Columbia [Province].*

Table Matters. n.d. *The North Shore Table Matters Network.* http://www.tablematters.ca/about-table-matters/.

Table Matters. 2013. *North Shore Community Food Charter.* http://www.tablematters.ca/about-table-matters/the-north-shore-community-food-charter/.

Tarasuk, V., A. Mitchell, and N. Dachner. 2016. *Household Food Insecurity in Canada, 2014.* Toronto: Research to Identify Policy Options to Reduce Food Insecurity (PROOF). http://proof.utoronto.ca.

Thobani, S. 2007. *Exalted Subjects: Studies in the Making of Race and Nation in Canada.* Toronto: University of Toronto Press.

Vancouver Food Policy Council. n.d. *Mandate and Terms of Reference.* http://www.vancouverfoodpolicycouncil.ca/about/working-groups-2/.

Young, I. M. 1990. *Justice and the Politics of Difference.* Princeton, NJ: Princeton University Press.

12 "Here, We Are All Equal": Narratives of Food and Immigration from the *Nuevo* American South

Catarina Passidomo and Sara Wood

Introduction

> I don't want to speak ill of Americans—but often, they are very simple. Or rather, you eat something [American], and it has no taste. And I imagine that since Mexican food has flavor, I think that they try to identify a bit with us. I think that right now, Americans try your food because they want to involve themselves with you. Sometimes, there are bad, discriminatory people, but there are many, many good people. I think that they let us know that, that they're with us, yes? That they love the food, that has flavor—our food. Because that's what many people have told me. "It has flavor!" It's that this has something special. I think that's also why they come back and come back. Because they like how we treat them, and the flavor that we have. I'm trying to buy [this building]. And to make sure that it's not "fancy." It's not an elegant place. But it's somewhere where we're going to receive you with—humility. That none of us are mightier, none of us are lower here. Here, we are all equal.—Laura Patricia Ramírez

The preceding excerpt is from an oral history interview with Laura Patricia Ramírez conducted by Gustavo Arellano for the Southern Foodways Alliance (SFA) in Lexington, Kentucky, in 2015. Laura was born in Guadalajara, Jalisco, Mexico, in 1968. She and her husband arrived in Paris, Kentucky, in 1985 so he could work in the horse racing industry. When they arrived, Laura estimated that there were maybe 30 or 40 Latinx living in Bluegrass Country. Now, she runs Tortillería y Taquería Ramírez in a section of Lexington nicknamed "Mexington" for the large number of Latinx immigrants living in the area.[1] Laura worked as a housecleaner before she and her husband opened their first restaurant, in 1997. At one point, they operated three restaurants and a nightclub, but they scaled back to the tortillería in 2000. Laura's story is part of a collection of oral histories collected by collaborators for the Southern Foodways Alliance (SFA) called "Bluegrass and Birria" (https://www.southernfoodways.org/oral-history/bluegrass-and-birria/). The full collection features stories of restaurant and *heladeria* (ice cream shop) owners, chefs and cooks, and other food entrepreneurs across Kentucky. In the introduction to the oral history collection, Gustavo Arellano writes, "The majority of Mexican restaurants [in Kentucky] still offer combo plates of cheesy enchiladas and sizzling fajitas. But now *taqueros* sling chicharrones (fried pork skin with the meat still attached) and both *lengua* and *cabeza* (beef tongue and head). Restaurateurs prepare regional dishes like *birria* (goat stew), *barbacoa* (slow-roasted lamb or mutton), and *chilaquiles* (fried tortilla strips bathed in salsa and beans)." Indeed, these are just some of the tastes of the *Nuevo* American South.

We begin this chapter with Laura because her perspective provides a lens through which to consider immigrant food stories throughout the "New"/ *Nuevo* American South. This chapter features the voices and stories of Latina immigrant food entrepreneurs across the region, but we want to demonstrate that these women do not represent isolated or ethnically homogenous communities; rather, they are more accurately indicative of a diverse and dynamic globalizing South. We dedicate the bulk of this chapter to these women's stories, told in their own words to SFA oral historians. First, though, we offer context for understanding these unique voices. While we do not contend that these stories are representative of the vast and diverse experiences of Latina food entrepreneurs in the South, we find that their perspectives offer insight into broader trends while also illuminating the uniqueness of each woman's experience. Also, drawing on Garcia, DuPuis, and Mitchell (2017), we contend that these women's stories collectively work to "disrupt the comforting notion that recipes and foodways...have traveled with us, unchanged, over many miles and generations" (Garcia, DuPuis, and Mitchell 2017, 3). Indeed, the stories we share demonstrate hybridity, adaptability, and change at the same time that they represent lasting connections to alternative homelands.

Context: Creating the "Nuevo" American South

Until quite recently, the US South received relatively fewer immigrants than did the industrial areas of the North and along the coasts. While other parts of the United States experienced rapid rates of industrialization and urbanization during the twentieth century, the southern economy continued to be dominated by agriculture, primarily relying on the labor of African Americans. These conditions, combined with postbellum labor practices, including hostility to unions, made the South historically inhospitable to immigrants and perpetuated a "binary racial configuration" in the region (Winders and Smith 2012, 224). For this and other reasons, "for most of the twentieth century, observers and social scientists have seen the American South as locked in cultural isolation, first from the presumed mainstream life in the wider United States, and, even more, from the wider modern world beyond U.S. borders" (Peacock, Watson, and Matthews 2005, 1).

While that perception may persist, the modern reality is quite different: driven primarily by immigration from Latin America, the "Nuevo" South (Guerrero 2017) is the region of the United States experiencing the largest increase in its foreign-born population, from less than one million in 1960 to 13 million in 2013. Between 2000 and 2013, the South's foreign-born population increased by 55% (Pew Research Center 2015). These demographic shifts in the South pose challenges to the black-white binary that became prevalent during the nineteenth and twentieth centuries (Watson 2005, 280–281). The contemporary "transnational South" is a reality, forcing all to take new stock of the question of southern identity and the meaning of globalization in a formerly isolated region (Watson 2005, 278).

While the making of a "global" US South is a relatively new phenomenon, the dynamics that drive immigration to the region have deep historical roots. As Bankston (2007, 25) demonstrates, immigration early in the South's history, though much more limited in scope, was a product of the same social and economic forces that have fostered more recent movement to southern states. Specifically, the push and pull factors driving flows of people into the region (economic or political devastation in sending countries, economic opportunities in the United States, family reunification, and geographical access) have remained the same, even as their intensity has increased, and sending countries have changed over the course of the last century. Prior to the American Civil War (1861–1865), most immigrants coming to the region came through the port cities of New Orleans and Baltimore and were primarily of Irish, German, or French descent. Small numbers of Chinese laborers also settled in the region, but the vast majority of immigrants to the United States were attracted to industrializing urban areas in the Northeast, Midwest, and West, or to commercial agricultural work in the Southwest.

New waves of immigration to the US South have been dominated by immigrants from Latin America and Mexico, with the majority of those immigrants settling in what Bankston (2007) refers to as the "access states" of Florida and Texas—by sea or over land, these are obviously the two states with greatest proximity to sending locations throughout Mexico and Latin America. Because of the relatively long-standing presence of Latinx immigrants in Florida and Texas, Winders and Smith (2012) omit these two states from their thorough study of more recent waves of Latinx settlement throughout the US South, restricting it to what the authors refer to as "nontraditional destinations." The states with the fastest-growing immigrant populations are what Bankston calls "opportunity states": Georgia, North Carolina, Maryland, and Virginia. Among these, the agricultural and construction industries have attracted large pools of Mexican and Central American laborers to Georgia and North Carolina, while white- and blue-collar workers from South and Southeast Asia have been attracted to the service and professional industries of Maryland and Virginia. Even "limited migration" states such as Louisiana and Mississippi have seen increases in first- and second-generation Latinx migration as opportunities arise in the construction and restaurant industries. The ten states with the highest per capita growth in "Hispanic" population, as measured by the US Census Bureau, are almost all in the South. South Carolina saw the highest growth rate, followed by Alabama, Tennessee, Kentucky, Arkansas, North Carolina, Maryland, Mississippi, South Dakota, Delaware, Georgia, and Virginia (Lopez 2011).

This new migration reflects historic trends linking "events south of the border and movement across the border" (Kochhar, Suro, and Tafoya 2005, 42). Since the 1980s, economic restructuring, new immigration policies, and deteriorating conditions in several Latin American countries have drawn Latin American immigrants to the US South in ever-increasing numbers, challenging long-standing ideas and expectations about both "the South" and "Latino experiences" in the United States (Odem and Lacy 2009). As Winders and Smith (2012) argue, studies of the "Latinization" of new destinations within the traditional South can offer insights into the complexities of Latinx experiences across the United States. In particular, the authors argue, southern Latinx migration can refine our understanding of Latinx transnational practices in new southern locations, of racialization and how racism operates against Latinx in southern locales, and of neoliberal globalization and practices of flexible labor experienced by Latinx workers at southern worksites (Winders and Smith 2012, 223).

Each of these themes resonates in the oral history excerpts that follow. Our narrators, who migrated from Mexico to Tennessee, North Carolina, and Kentucky, and from El Salvador to Virginia, to varying degrees embody "transnational" identities: the "multiple ties and interactions linking people or institutions across the borders of nation-states" (Vertovec 1999, 447). They engage in what Cravey (2003, 604) has referred to as "translocal ways of life," enabling them to maintain a connection to their place of birth while also cultivating connections to their "new" home community in the United States. Food—ingredients, recipes, preparations, and the sharing of meals with family, friends, and customers—is central to translocality for these women.

Food and food businesses are also essential to the somatic and cultural survival of each of these women within the United States. We are inspired by Abarca's (2007) analysis of Mexican women food entrepreneurs, who use their businesses to build familial wealth and strengthen community ties. Through a series of *"charlas culinarias"* (culinary chats) with women who own small food stands in El Paso, Texas, and Cuidad Juárez, Chihuahua, Mexico, Abarca highlights the deep philosophical resonance of these businesses for much beyond capital gain. Though each of the stories we share in this chapter are distinctive, we find they share an emphasis on using food businesses to build familial and community "wealth" beyond the monetary.

These oral histories are part of a vast archive of stories collected by oral history collaborators for the Southern Foodways Alliance. Sara Wood, the second author of this chapter, was the SFA's lead oral historian from 2014 until 2017 and worked directly with a number of the collaborators featured here. These and all SFA oral history interviews are archived, and transcripts are available on the SFA website, along with the following description of the SFA's oral history program: "Our documentary work gives voice to the complex expressions of people and place through food, exploring race, class, gender, religion, labor, and other cultural issues. By collecting these stories, we honor the men and women whose hard work enriches the landscape of Southern food and culture" (http://www.southernfoodways.org/oral-history/). Photographs and audio clips accompany most interviews; we encourage readers of this chapter to explore the oral history archive to encounter more stories connecting immigrants—individuals SFA director John T. Edge (2017) refers to as "active Southerners"—and their foodways. The three stories that we feature here, by Karla Ruiz, Zhenia Martinez, and Argentina Ortega, are just a taste of the continuing transformation of the US South. At the beginning of each excerpt, we list the narrator's name, business, location, oral history project title (and corresponding URL), interviewer name, and date of interview. We also provide background information on each narrator. After each woman's narrative, we offer a brief analysis to situate her story within the context of the "Nuevo" American South.

Karla Ruiz

> Karla's Catering & Prepared Foods, Mesa Komal Kitchen, Nashville, Tennessee Project: Nashville's Nolensville Road (http://www.southernfoodways.org/oral-history/nashvilles-nolensville-road/)
> Interviewer: Jennifer Justus
> Date: February 11, 2016

Karla Ruiz came to Nashville from Mexico City in 2000 to visit extended family members. She was offered a job waiting tables at a Mexican restaurant and never left. She eventually found work in the back of the house at Belle Meade Plantation with one of Nashville's most celebrated chefs, Martha Stamps, who helped her hone the lessons she had picked up—but had never written down—at her grandmother's elbow back in Mexico. Karla learned to blend Mexican techniques with southern American ingredients, like empanadas bulging with southern peaches. After several years working with American chefs, Ruiz took the leap to start her own company, Karla's Catering, in 2005. She prepares her dishes in Mesa Komal, the shared commissary kitchen of the Casa Azafran community center. Working at the community kitchen brings Ruiz close to other chefs and food entrepreneurs from a variety of countries—each of them aiming to share where they come from through their food. She explains:

> I come from Mexico City and also a small town in Michoacán, [where] my grandmom used to live, and I came to visit friends. And the first time I came to Nashville I decided that's going to be my home forever.
>
> I learned to cook in Mexico but I don't know that I was good at it, or I don't even know that I enjoy it back then, because I do it as an everyday thing to do. I find out here under Martha Stamps that I can do different recipes, or I can create

recipes, and I think when I was here working for her is when I find out that I can do more than just Mexican or home cooking. I can create and identify some of the flavors and then create it again.

In Mexico, I remember watching my grandmom, watching her making Dulce de Leche. I remember every step that she showed me. What I think it makes me [able to] get here into big business is because I know the real flavors of food. I know the good flavors, or a Mexican dish, [and I would] learn that from my grandmom.

When I was in Belle Meade Plantation the dishes that I remember to present—and I was a little scared—that was empanadas, because they were very Latin-American dish that not many people in Belle Meade will know what is that or how they eat it. So, I present it with local squash blossom and cheese, and that was a hit. That's what I did that they love it, as well as I love Southern food, I love Southern culture, and I learn from [Martha] the love of the culture, the importance of cook to keep the family together. When I was working in a Mexican restaurant, I was thinking to myself, "Wow. Why do they think this is a Mexican food?" because [none] of that was really a Mexican food. I don't even think the rice we prepared that way. [*Laughs*] So, and we don't eat chips and salsa. So when I work for Martha Stamps I learn about American culture in general. I fall in love [with] the culture more than I was before, and I learned that they also have their dishes, and I think we all have the same kind of [dishes]. Like, I compare the cheese grits with tamales. It's the same ingredient, just prepared different way.

I was very nervous. I came with my son; he was six years old by then. I was nervous because I came from a home that was super, super protective. I was a single mother but my parents still treated me like I'm still their baby, and my son was their only grandson. We lived with them so it was very protective. Coming here [to the United States] and deciding to stay was very scary because now my son goes to school in a bus, which I was freaking out. He suffers because he doesn't understand anything what the teacher says and it was heartbreaking to go to school every day crying and he doesn't know what to do and ask [you] yourself every day, "Am I doing the right thing? Should I go back?" It was hard. People bullied him because he spoke Spanish and looks Latin. There wasn't a lot of Spanish people in that area, so it was difficult. It was heartbreaking also and every day I asked myself, "Should I go back?" and I just keep trying hard and staying, and now today I know that I did the best decision. I stay. My son now finish college and he's a very successful teacher, and I'm so proud of him. But back then it was hard to keep thinking: go back or stay here?

It was not long. The business started getting busier and busier. I remember that I thought, "This is my year. I'm going to make it [just with] the catering." I know that this is my life. I want to be in the kitchen every day the rest of my life.

Karla's story emphasizes the adaptability and resilience required of immigrant food entrepreneurs. To be successful, Karla had to learn to

prepare foods that blended ingredients and techniques she learned from her grandmother in Mexico with traditional southern foods more familiar to her clientele in Nashville. This strategy of adapting immigrant foodways to the palate of the host culture is well documented among scholars of both food and immigration. In fact, as Donna Gabaccia argues, the blending of disparate cuisines is precisely what created "American" food, if there is such a thing. She argues, "the American penchant to experiment with foods, to combine and mix the foods of many cultural traditions into blended gumbos or stews, and to create 'smorgasbords' is scarcely new but is rather a recurring theme in our history as eaters" (Gabaccia 1998, 3). The penchant for experimentation and combination is particularly prevalent in the New American South, where tamales, banh mi, yaka mein, bubble tea, and Karla's sweet potato empanadas increasingly have a place in the southern food canon alongside fried chicken, collards, cornbread, and sweet tea (Edge 2017).

In Karla's story, we also see the questioning and doubt that so often accompany the decision of voluntary migrants to leave their country of origin. Upon arrival in Nashville, Karla confronted challenges as well as opportunities. Karla had to weigh her young son's experiences of discrimination against the opportunities she perceived would be available to him if she stayed in Nashville. At the time that Karla settled in Nashville, anti-immigrant sentiment was still relatively rare in comparison to other parts of the country (Odem and Lacey 2009, xvi). However, as the concentration of immigrants, particularly from Mexico and other Latin American countries, increased, so did native perceptions that immigrants were a threat to both jobs and "American" culture. This broader trend is evident in Karla's concern that her son would face discrimination growing up in the American South. In the intervening years since this interview was conducted, anti-immigrant (particularly anti-Mexican) sentiment has reached new heights, in many cases reinforced and exacerbated by impassioned and unsubstantiated political posturing.

Zhenia Martinez

Las Delicias Bakery, Charlotte, North Carolina
Project: Charlotte's Central Avenue Corridor (http://www.southernfoodways.org/oral-history/central-avenue-corridor/)

"Here, We Are All Equal"

Interviewer: Tom Hanchett
Date: April 24, 2017

Zhenia Martinez co-owns Las Delicias, a Latina *panadería* (bakery) on Charlotte's Central Avenue. Her parents, Margarita and Aquiles Martinez, came from Mexico to Columbia, South Carolina, in 1985, when Zhenia was 11, gaining citizenship under President Reagan's 1986 amnesty legislation. A government bureaucrat in Mexico, Aquiles went to work in a restaurant, then moved into construction, where an accident disabled him. He stocked a van with Mexican groceries and drove to small towns to sell to newly arrived Latinx. Seeing the need for a Mexican bakery in Charlotte, he and Margarita apprenticed to a baker back in Mexico, then in 1997 opened Las Delicias. When Aquiles retired in 2011, Zhenia took up the work with her partner, Colombian-born Manolo Betancur. She says:

> My dad studied economics. He was from a family that liked knowledge and sought knowledge. When we lived in Mexico, we had at least one wall that was completely covered from floor to ceiling with books. He didn't have your typical life. It felt like my dad was a nomad. He just liked going to different places, because we had lived in Mexico City and from there we moved to southern Mexico, and then to Puebla. And I think on one of the parades that we had for some celebration in Mexico—and my mom tells me the story—she said she saw so many kids that she thought, "My god, how are my kids going to find a job here?" We had an aunt [Licha Carrillo] living in the U.S., and that's when they decided that they wanted to move us to the U.S. They came in 1985. I started seventh grade.
>
> When they first came, they worked as cooks. I don't think they ever saw it as bad. I think because it was an honest way to make a living and feed your family. Her clientele was mostly American. I don't remember seeing any Hispanic clientele. My parents left working for my aunt to find something to have. While working on construction [sites], my dad had already been through these smaller towns throughout North Carolina and South Carolina. So he bought Mexican products, put them in a van, and he started going to small farming towns to sell his products. And one of the things that he bought was pan dulce, and that's when he started to see how people were really looking forward to that, and I think that's when he got the idea that he wanted to open up a bakery. They went to Chihuahua [Mexico] to a small bakery and they basically said, "Would you teach us?" They learned everything that they could about pan dulce. It's so artful. I wish anybody that underrates pan dulce would go spend a day trying to make the stuff and make it come out as beautiful as it does, because everything has to be shaped by hand.
>
> It was 1997. I'm twenty-one. I was on my own, and I was going to school and working and going part-time at Cracker Barrel. [Laughs] Of all the places, I know.

I didn't quite grasp all of that was going on back then, so it just seemed kind of odd and out of place for me. But there were a few times that people came in there, in all honesty, said, "I don't want her to be my waitress," for the exact same reasons that you would think, "because she's colored." We actually had people, regular customers, that didn't accept me as a waitress, but they accepted an African American. I asked somebody else, "Why is it that she can wait on them, but I can't?" I mean, not that I minded, because if I was going to be mistreated, then I'd rather not do it. But they said, "Well, it's because it's what they're used to." It's just little things like that, and it happened more than once where we would get customers that were set in their ways, to put it nicely.

I remember their first spring that [my parents] opened [the bakery]. We literally sat at the door waiting on customers, because there was nobody that would come in. It didn't look alive at all. I had come [back] in the early nineties [to] Central Avenue. I think it was a hard decision, because I know I had seen the toll that it took on my parents. But at the same time, I couldn't see the bakery going to anybody else. And to this day I can't. I think given all the years that we lived in South Carolina the community wasn't as big. So now when they moved here and opened up the bakery, I think it felt like the community grew, and it grew around them. Over the years bakers from other Latin American cultures would come and leave a recipe that we would get new products. So what started out as a Mexican bakery, we're now a Latin American bakery. And I think it's the same in the environment in the community we're growing, that it's no longer centralized. I think what's happening in Charlotte is that as we're growing so intertwined we're seeing a lot more influx of other cultures and sharing with them, not tailoring to just one customer but to everyone.

Zhenia's oral history is part of a collection of stories from Charlotte's Central Avenue Corridor. As the introduction to the collection on the SFA website explains, Central Avenue reveals Charlotte, North Carolina's, shifting demographics, "from working class textile mill employees in the twentieth century, to new immigrants in the 1990s" (https://www.southernfoodways.org/oral-history/central-avenue-corridor/). By 2000, Charlotte was the country's second-largest banking center, behind New York City, spurring employment opportunities for both white- and blue-collar workers. By 2005, the Brookings Institution ranked Charlotte as the second-fastest-growing Latinx metro area in the United States (Hanchett 2013, 174).

In a study of Charlotte's Central Avenue Corridor, Tom Hanchett uses the analogy of the "salad bowl suburb" to describe the city's ethnically diverse corridor, where "many ingredients come together to create a new dish" (Hanchett 2013, 169). He describes a "jumble of little shopping plazas" where one can "walk to a Vietnamese grocery and two Vietnamese restaurants, a

Mexican grocery and taquería, a Salvadoran deli and two Salvadoran eateries, a Somali restaurant and grocery, an Ethiopian bar-restaurant-nightclub, and a Lebanese grocery-restaurant" (166). This is the dynamic context in which Zhenia's family's *panadería* continues to thrive, although the business was one of the first "ethnic" establishments along Central Avenue. Zhenia's family's experience of incorporating recipes from throughout Latin America in what started out as a Mexican bakery is reflected in Hanchett's observation that "people who have grown up in separate countries with separate cultures are now coming together to form a new 'Latino' community" (176). This is perhaps no more evident than in the hybridized Latinx foodways that flourish throughout the South.

Zhenia's account of the racism she experienced at Cracker Barrel and the preference among white customers for black (rather than Latinx) waitstaff also illuminates the complexities of the southern context. In *Latinos Facing Racism: Discrimination, Resistance, and Endurance*, Joe Feagin and Jose Cobas argue that the dominant white power structure shapes the fate of racialization for Latinx—that is, whether and which Latinx will be perceived as white, black, or some other racial category (Feagin and Cobas 2013). Despite persistent antiblack racism in the South, the relatively "new" presence of Latinx in the region complicates historic binary racial formations and perhaps presents new threats to whites' perceived racial hegemony. Certainly the perception of Latinx presence as a "threat"—to jobs, to "American culture," to the demographic dominance of whiteness, to notions of legality and justice—has only increased in recent decades (Plaza 2009, 22). In the South, this tension may be complicated by cultural and industry rhetoric that pits African Americans against Latinx despite their common oppressor (Stuesse 2009).

Argentina Ortega

> La Sabrosita Bakery, Richmond, Virginia
> Project: Women at Work in Richmond (http://www.southernfoodways.org/oral-history/women-at-work-in-rva/)
> Interviewer: Sara Wood
> Date: December 10, 2012

At 19, Argentina Ortega left Sensuntepeque, El Salvador, and moved to the United States. Shuttling back and forth between El Salvador, Southern

California, and Houston, she took baking classes and began earning a living working in a bakery. In 2005, Argentina settled in Richmond to be with her three sons: Mario, Eduardo, and Jorge Dawson. With a small business loan, she purchased La Sabrosita Bakery. At the time, the bakery had a poor reputation, with only a handful of clients. Gradually, she built an incredible reputation and customer base. In 2009, her sons' construction business slowed to a halt, so the four became partners and opened a larger space on Midlothian Turnpike, where the business sits today. Her customers hail from all corners of the world, and hundreds of deliveries are made each week, stretching beyond Virginia into Washington, D.C., Maryland, and North Carolina. As she tells it:

> In El Salvador the days I feel they are longer. You have time for everything. You get bored because you have too much time. The University of El Salvador; it was a government university. There was a lot of women in the university but the career that I picked [business administration] there were just men. We were about five women. That university was a dream, beautiful, they took good care of it. But then the guerillas came and they started hitting the university. They tried to convince students about their ideas and some were okay and some were not. And I said, "I don't think I would like that." That's why I asked my family to send me to California. I was sad but I wanted to come. I tell my sons that at that time coming here was like going to the moon.
>
> I wanted to start first a bakery in my country but never did it. And then when I came to California I worked at my husband's bakery. But [the marriage] didn't last too long, so I was in California having a bakery for four or five years and then I came [to Richmond, Virginia] and my son asked me, "Do you still want to have a bakery?" And I said, "Yes I want it," and luckily somebody that we knew was selling the bakery and I bought it. It was really tiny. Most of the recipes I have are from El Salvador. When I came [to Richmond] it was 2002 and there were a lot of Hispanics. Here in Richmond, I don't think we missed any of the food from our country. We looked for other foods just to have a variation but any food that we want to have is here because I see it here in the bakery. We have customers from South America, Central America, Puerto Rico and Middle East and Africa.
>
> But the thing works like this. You come and then you're here in the states if you work. If you are honest you have a better life, a lot better than in our country and we like it and so why not share that with our family? And they send for a cousin, for family, for friends. People like it because here if you work, you have a home, you have your dress, you have your food, and that's your country. I think the 99-percent of the Hispanics moved here to survive, to have a house, to have food, to have dress, to have some money to help their families over there because there is a poverty in my country that you cannot imagine. I worked every day

around the house and when I was working I was talking to the Virgin, to God and saying "Just let me have my home, be independent; help me. I'm your daughter. Help me to achieve that." I just want to have money to pay my expenses to have an honest life and be happy and have my family. [*Laughs*] I never was thinking of having success or being a business woman. I just wanted to survive and have an independent life.

Argentina's story is just one in a collection of SFA oral histories from Richmond that feature "women at work." The collection shares stories of women entrepreneurs: cooks, oyster shuckers, farmers, restaurateurs, and bakery owners like Argentina. As the introduction to the collection describes the women, "They are gracious, bossy, patient, fierce, and kind. They elude the spotlight. They are busy, and they have to get back to work" (https://www.southernfoodways.org/oral-history/women-at-work-in-rva/).

Argentina emphasizes her familiarity with doing "men's work," dating back to her time in business school in El Salvador. Like many women in the food industry, Argentina had to balance inward and outward expectations of her as a mother and woman with her ambition to be independent. Women chefs, in particular, have faced considerable obstacles as the professionalization of cooking has emphasized masculinity and diminished the presence and role of women (Harris and Guiffre 2015). For Argentina, those challenges are compounded by having to adjust to a new home that felt, initially, like "going to the moon." Despite those challenges, and in partnership with her sons, Argentina was able to establish a tremendously successful business, using food to unite disparate communities in Richmond, whose Latinx population grew by 165% between 1990 and 2000 (Schleef and Cavalcanti 2009).

Argentina also comments on the wide availability of foods from El Salvador as well as other Latin American countries. While the "typical" (stereotypical) immigrant food story often involves someone lost among a sea of hot dogs and longing for an arepa, Argentina reminds us that many immigrants find both familiar and new foods in the United States. Argentina appears eager to try them all.

Conclusion

In this chapter's epigraph, interviewee Laura Patricia Ramírez speculates, "Americans try your food because they want to involve themselves with you." She sees food as a way to break down barriers, real and perceived, between herself and native-born Americans. This perspective, mirrored in such recent publications as Ying and Redzepi's (2018) edited volume *You and I Eat the Same*, may come across as naive or overly optimistic, and it is certainly the case that discrimination against and animus toward immigrants in general and Latinx in particular have heightened in the years since the SFA interviewed Laura in 2015. As other chapters in this volume demonstrate, the rhetoric and policies of the Trump administration have exacerbated anti-immigrant sentiment and threatened the safety and livelihoods of hundreds of thousands of Latinx individuals living in the United States. It would be foolhardy to claim that food is or has the potential to be a panacea for widespread structural and interpersonal racism. It is also tempting, yet overly simplistic, to make general claims about the lived experiences of diverse groups of people.

We share the stories of Laura, Karla, Zhenia, and Argentina because each offers unique perspectives on the immigration-food nexus and demonstrates the power of narrative. They do not tell a single or unified story, but they do offer some themes to consider. Through these stories, we see the hybridization of diverse Latin American cultures and cuisines with one another and with more "traditional" southern foods and cultures. We see resilience, but also fear and doubt, in the face of discrimination. We see the unique challenges that women food entrepreneurs face, and strategies for balancing multiple responsibilities and identities. Perhaps most importantly, at a time when immigrants' stories have been overshadowed by rhetoric and policies that dehumanize them on the one hand and well-meaning activists who would speak *for* them on the other, these narratives invite us to *listen*. Within food system scholarship and activism, narratives like these offer a way of restructuring who is telling the story. Through oral history, we can listen to these stories from real people, not statistics—from women who use food to draw connections between disparate communities and cultures, one bite at a time.

Note

1. Dr. Steven Alvarez coined the term "Mexington, KY" while teaching a course called "Taco Literacy" at the University of Kentucky in Lexington. The course, which has an accompanying blog (https://tacoliteracy.com/), was the subject of a

Southern Foodways Alliance *Gravy* podcast episode (https://www.southernfoodways.org/gravy/bluegrass-tacos/).

References

Abarca, Meredith E. 2007. "Charlas Culinarias: Mexican Women Speak from Their Public Kitchens." *Food and Foodways* 15(3–4): 183–212.

Arellano, Gustavo. 2015. "Introduction: Bluegrass and Birria." Southern Foodways Alliance. https://www.southernfoodways.org/oral-history/bluegrass-and-burria/

Bankston, Carl. 2007. "New People in the New South: An Overview of Southern Immigration." *Southern Cultures* 13(4): 24–44.

Cravey, A. 2003. "Toque Una Rachera, Por Favor." *Antipode* 35(3): 603–621.

Edge, John T. 2017. *The Potlikker Papers: A Food History of the Modern South*. New York: Penguin Press.

Feagin, Joe R., and Jose A. Cobas. 2013. *Latinos Facing Racism: Discrimination, Resistance, and Endurance*. Boulder, CO: Routledge.

Gabaccia, Donna R. 1998. *We Are What We Eat: Ethnic Food and the Making of Americans*. Cambridge, MA: Harvard University Press.

Garcia, Matt, E. Melanie DuPuis, and Don Mitchell. 2017. "Food across Borders: An Introduction." In *Food across Borders*, edited by Matt Garcia, E. Melanie DuPuis, and Don Mitchell, 1–23. New Brunswick, NJ: Rutgers University Press.

Guerrero, Perla. 2017. *Nuevo South Latinas/os, Asians, and the Remaking of Place*. Austin: University of Texas Press.

Hanchett, Tom. 2013. "A Salad Bowl City: The Food Geography of Charlotte, North Carolina." In *The Larder: Food Studies Methods from the American South*, edited by John T. Edge, Elizabeth Engelhardt, and Ted Ownby, 166–184. Athens, GA: University of Georgia Press.

Harris, Deborah A., and Patti Giuffre. 2015. *Taking the Heat: Women Chefs and Gender Inequality in the Professional Kitchen*. New Brunswick, NJ: Rutgers University Press.

Kochhar, Rakesh, Roberto Suro, and Sonya Tafoya. 2005. *The New Latino South: The Context and Consequences of Rapid Population Growth*. Pew Research Center. http://www.pewhispanic.org/2005/07/26/the-new-Latino-south/.

Lopez, Mark Hugo. 2011. *Pew Hispanic—Census 2010—Reaching Latinos Online*. Slide 7. https://www.slideshare.net/bixal/lopez-reaching-Latinos-online-phcapril-2011.

Odem, Mary E., and Elaine Lacy, eds. 2009. *Latino Immigrants and the Transformation of the U.S. South*. Athens: University of Georgia Press.

Peacock, James L., Harry L. Watson, and Carrie R. Matthews, eds. 2005. *The American South in a Global World*. Chapel Hill: University of North Carolina Press.

Pew Research Center. 2005. "Modern Immigration Wave Brings 59 Million to U.S., Driving Population Growth and Change Through 2065." September 28, 2005. http://www.pewhispanic.org/2015/09/28/chapter-5-u-s-foreign-born-population-trends/.

Plaza, Rosio Cordova. 2009. "New Scenarios of Migration: Social Vulnerability of Undocumented Veracruzanos in the Southern United States." In *Latino Immigrants and the Transformation of the U.S. South*, edited by Mary E. Odem and Elaine Lacy, 18–33. Athens: University of Georgia Press.

Schleef, Debra J., and H. B. Cavalcanti. 2009. *Latinos in Dixie: Class and Assimilation in Richmond, Virginia*. Albany: State University of New York Press.

Stuesse, Angela. 2009. "Race, Migration, and Labor Control: Neoliberal Challenges to Organizing Mississippi's Poultry Workers." In *Latino Immigrants and the Transformation of the U.S. South*, edited by Mary E. Odem and Elaine Lacy, 91–111. Athens: University of Georgia Press.

Vertovec, S. 1999. "Conceiving and Researching Transnationalism." *Ethnic and Racial Studies* 22(2): 447–462.

Watson, Harry L. 2005. "Southern History, Southern Future: Some Reflections and a Cautious Forecast." In *The American South in a Global World*, edited by James L. Peacock, Harry L. Watson, and Carrie R. Matthews, 277–288. Chapel Hill: University of North Carolina Press.

Winders, Jamie, and Barbara Ellen Smith. 2012. "Excepting/Accepting the South: New Geographies of Latino Migration, New Directions in Latino Studies." *Latino Studies* 10(1–2): 220–245.

Ying, Chris, and René Redzepi, eds. 2018. *You and I Eat the Same: On the Countless Ways Food and Cooking Connect Us to One Another*. New York: Artisan.

13 Boiled Chicken and Pizza: The Making of Transnational Hmong American Foodways

Alison Hope Alkon and Kat Vang

The Vang Family Thanksgiving

In the Sacramento suburbs, all the houses look alike, but the weekend before Thanksgiving, one house feels different. Over 20 parked cars are clustered around the home and sprawled across the quiet streets. Murmurs of conversation and the gleeful shouts of children are audible from outside. Because the Thanksgiving holiday conflicts with Sacramento's Hmong New Year's celebration, and because some individuals were spending the official holiday with in-laws, the four generations of this Hmong American family have decided to celebrate Thanksgiving early. Though they have been in the United States for only a few decades, like many American families, they enjoy the chance to gather and share a holiday meal.

Over 100 people fill the two-story home. In one room, men huddle around a large table, engaged in a lively conversation about sports and popular culture in a mix of Hmong and English. One young father holds his newborn daughter wrapped in a fuzzy pink blanket. Children are everywhere. The younger ones weave around the men, playing with toy cars and animal figurines, while older children line the stairs. A quiet group of teenagers are sprawled on the floor beneath a bright chandelier, each with an electronic device in hand. In an adjoining living room, the family matriarch, a petite, white-haired woman, sits on the couch. Her legal age is 97, but she claims to be 120. Either is possible, as the Hmong village from which she immigrated, deep within the jungles of Laos, did not issue birth certificates, and her birthday had to be approximated when she came to the United States in 1981. Although thin and aged, she wears a large smile as she animatedly talks and laughs with her daughters and granddaughters, many of whom are swaddling newborns or watching toddlers nearby. Over the past century, this matriarch has seen her culture transform from self-reliant villages of subsistence farmers in rural Southeast Asia to a contemporary Hmong American community emphasizing higher education and dependence on modern amenities.

This gathering is more than a Thanksgiving dinner: it is an intimate family celebration and culinary affair. Two long kitchen tables are piled high with food. Three trays of papaya salad, a spicy mixture of shredded papaya, sauces, and chilies, sit next to a honey-glazed ham adorned with pineapple slices. A large bowl of mashed potatoes and bread stuffing neighbors a tray of beef and bamboo mixed into vermicelli noodles. Small plastic lunch bags packed with handfuls of balled rice sit next to a platter of Chick-Fil-A fried chicken. Trays of steamed vegetables and carved turkey crowd a large bowl of fruit salad. Deviled eggs, chicken potpies, and Hmong pepper dipping sauce are piled onto paper plates along with fried fish, a creamy salad, and store-bought dumplings.

This mélange of typical Thanksgiving dishes alongside rice, pepper dipping sauce, and vermicelli noodles represents the different influences and cultural fusions that inform the making of Hmong American foodways. Embedded within the rich and diverse display of dishes is a cuisine born from the interaction of Hmong and American traditions, adapted by generations of Hmong cooks to please the palates of their families. This chapter investigates Hmong American foodways, meaning not only what kinds of foods this community grows, cooks, and eats but also the integral role of food in the making of culture, identity, and meaning. Because food is both an important marker of one's sense of self (Robinson 2014; Douglas 1996) and a means through which individuals claim membership in a community (Mannur 2007), it can reveal much about the processes through which refugees and their descendants forge hybridized identities in their new homelands, even as they experience new forms of racialized oppression. Cultural geographers often refer to these sorts of identity formation processes as translocal, emphasizing that they are agency oriented and simultaneously grounded in multiple locales (Brickell and Datta 2011). In their study of Hmong translocal placemaking practices, Michael Rios and Joshua Watkins (2015) write that "the use of visual materials and other material objects enable[s] the symbolic and affective bridging between locations as well as a heightened sense of home even when the possibility of return is nonexistent." In our study, food serves not only as a material object but also as a process through which translocal identities are forged.

The little existing research conducted on Hmong foodways comes largely from nutritionists and public health researchers, who often cast their eating practices as an example of "dietary acculturation," the process by which a migrating group adopts the foods of their new environment (Satia-Abouta 2003). This previous research was motivated by concern for Hmong Americans' physical health, as evidenced by attention to rates of diet-related diseases and obesity, and often concluded by prescribing that Hmong Americans return to a traditional diet,[1] mainly rice and vegetables, with smaller portions of meat (Wilcox and Kong 2014). Problematically, this suggestion regards Hmong culture as static and unchanging, ignoring the role that migrant food practices can play in creating new notions of identity in their new homelands (Baker 2004; Hondagneu-Sotelo 2014; Garcia, DuPuis, and Mitchell 2017).

Following a review of the literature and our research methods, we present data from a survey of 125 Hmong Americans that examines their food practices and the meanings they associate with them. In contrast to the public health research, our work reveals that Hmong Americans often enjoy Hmong cuisine, eat it regularly, and see it as integral to their senses of culture, identity, and family. However, they also eat a wide variety of dishes available in the United States, and sometimes even adapt Hmong dishes to incorporate ingredients and techniques from both American and other Southeast Asian cuisines, such as Thai and Vietnamese food. Through our emphasis on food as culture, rather than mere sustenance and nutrition, we have come to understand dietary practices as a means through which Hmong Americans create translocal, hybridized identities that are both distinctly Hmong and distinctly American.

History of the Hmong People, Migration, and Resettlement

The Hmong people are a transnational ethnic group who have lived in Southern China, Thailand, and Laos. The details of their origin story are contested by Hmong scholars and historians such as Mai Na M. Lee and Gary Lee, who employs postcolonial analysis to challenge the starring role too often given to colonial powers and attempts to reconstruct the Hmong origin story on their own terms (Lee 1997, Lee 2008).

Historically, the Hmong people lived in small villages, where they farmed rice, grew vegetables, and kept livestock (Ross 2013). In many ways, the Hmong people were autonomous and self-reliant, and these qualities helped preserve the Hmong culture amid social and political climates that marginalized and discriminated against them. In eighteenth-century China, for example, the Hmong did not share the same rights and privileges as the Han Chinese despite paying taxes and performing corvée labor.[2] Instead, the Hmong were barred from attending school, forbidden from visiting Han towns, and subjected to discriminatory laws with harsher punishments (Lee 1997). Categorized throughout dominant Chinese history as the Miao/Meo people, the Hmong were not recognized as a distinct and sovereign culture and were not historicized in an accurate and specific manner.

The Hmong began migrating to the United States in the 1970s, in the aftermath of the Secret War in Laos. For 25 years, the Hmong were recruited by the CIA to support the Royal Lao government in suppressing the Pathet Lao and North Vietnamese Communist forces (Ross 2013). When the Pathet Lao came into power in 1975, the Hmong were forced to flee, first finding safety in Thai refugee camps. Since then, Hmong refugees have resettled in France, the United States, Australia, Canada, Germany, French Guiana, and Argentina (Morrison 2008). Currently, there are over 4.5 million Hmong people across the globe, with nearly 300,000 in the United States (Lee 1997; Moua and Vang 2015). The two largest clusters of Hmong Americans are found in St. Paul, Minnesota, and Fresno, California, and many Hmong initially resettled elsewhere have moved to these areas, signaling the importance of translocal connections within transnational migration (Rios and Watkins 2015; Brickell and Datta 2011).

Hmong People in the Academic Literature

A small but growing body of scholarly literature attends to Hmong American experiences. In their historical discourse analysis, Kao Nou L. Moua and Pa Der Vang (2015) write that the scant publications that attended to Hmong experiences in the 1980s emphasized their lack of English skills and tribal culture as preventing assimilation. Such studies focused on unemployment (Yang and North 1998) or problems with acculturation (Sherman 1988). Summarizing these early studies, Ross writes that, "Consequently, many stereotypes of Hmong people persist: they are the least prepared of all refugee groups to succeed in modern society; they are resistant to change...and they are unable to assimilate fully into American culture" (Ross 2013, 3). Decolonial scholar Eve Tuck describes this sort of research as "damage-centered" and seeking to "rationalize a group's brokenness" (Tuck 2009, 413). In the 1990s, analyses shifted to a "social problems" approach in which researchers sought to improve rather than merely condemn the Hmong community. This research is characterized by a focus on acculturation, generational differences, and social and economic difficulties but continues to emphasize community deficits rather than assets. This perspective on Hmong immigrants was popularized by Anne Faidman in her well-intentioned but problematic 1997 best seller *The Spirit Catches You and You Fall Down*. While she is sympathetic to the Lee family at the center of the account, they are nonetheless presented as unable or unwilling to assimilate into or navigate the American medical system.

Presently, these themes are joined by more critical and intersectional analyses of Hmong American identities and experiences as, in the words of Hmong scholar Pao Lor (2012), "neither static nor stable." Elaborating on this, Jacob Hickman argues that Hmong American identity "is much more than a unilineal shift from more-or-less Hmong to more-or-less American, rather the experience is nuanced and often fraught with multiple sets of meaning and ethics" (Hickman 2011, 248). These more nuanced approaches to Hmong culture and identity are the hallmark of Hmong studies. Hmong communities are a part of the constructed racial category of Asian and commonly experience what Kandice Chuh calls "Asiatic racialization," meaning that hegemonic strains of US nationalism have "repeatedly denied or 'nullified' potential citizenship, by creating 'Asians' as different from 'Americans'" (Chuh 2003, 15). However, Asian American studies has been predominantly focused on East Asian communities, rendering Hmong experiences invisible (Schein and Thoj 2009). Thus, Hmong studies remains in conversation with scholarship in Asian American studies and often employs similar approaches, including critical race theory (Wilcox 2012; DePouw 2012; Vue, Schein, and Vang 2016), poststructuralism (Vang and Nibbs 2016), and intersectionality (Boulden 2009), but remains a distinct field focused on Hmong experiences.

This chapter brings the Hmong studies approach to bear on questions of food, health, and agriculture, which have previously been studied mainly from the "social problems" perspective. With regard to farming, scholars have documented the ways that university extension programs (applied programs through which land grant universities engage in research to assist farmers) attempt but often fail to serve Hmong farmers. For example, Goldberg (2008) describes how, despite extension agents' efforts, Hmong growers in Sacramento did not know how to access extension services, did not trust extension agents, and did not believe these services fit their needs (see also Ostrom, Cha, and Flores 2010). This mistrust may be rooted in Hmong experiences with the US government, which has applied labor and safety laws written for industrial agriculture to Hmong microfarms, in some instances fining them far more than their annual incomes for failing to cover extended family members (Minkoff-Zern et al. 2011; Sowerwine, Getz, and Peluso 2009). Other scholars have depicted Hmong agriculture as a solution to social problems. Some have analyzed the importance of Hmong farming and gardening traditions for their communities' food security and culture (DeMaster 2005; Ross 2012) and for preserving agricultural biodiversity and ethnobotany (Corlett, Dean, and Grivetti 2003). Others have highlighted the important roles Hmong farmers can play in creating more racially diverse alternative food systems and opportunities for friendly interracial interactions through food (Morales 2011; Slocum 2008; Alkon and Vang 2016).

Relatedly, research on Hmong health generally aims to document growing rates of obesity and diet-related illnesses, explaining this in terms of "dietary acculturation" and the consumption of fast food. Lisa Franzen and Chery Smith, for example, argue that "environmental changes and increased acculturation have negatively impacted the weight and health of Hmong adults" (Franzen and Smith 2008, 173). Rather than blaming individual eaters, scholars have critiqued federal nutrition guidelines for a lack of sensitivity to ethnocultural norms (Trapp 2010; Nibbs 2010) and have argued for what Stang et al. (2007) call "culturally tailored" nutrition interventions (Franzen and Smith 2008; Goto et al. 2010; Vue, Wolff, and Goto 2011). In response, organizations such as the American Diabetes Association have released literature aimed at orienting health professionals to Hmong American food practices (Ikeda 1999).

While these studies reflect important advances in public health research toward cultural appropriateness, they remain rooted in a damage-centered approach legitimating professional interventions designed to "bring good food to others" (Guthman 2008). Hui Niu Wilcox and Panyia Kong (2014) write that "implicit in this narrative and the rhetoric of intervention is the notion that Hmong American food practice is not only different, but also deficient, and that Hmong Americans must have external 'help' to fix their diet." This public health approach is built on an understanding of Hmong culture as premodern and static that fails to recognize the complexities through which refugee communities and their descendants engage with their new homes, including the crafting of new translocal cultural foodways. In contrast, our cultural approach is more similar to recent studies of other immigrant groups that are focused on creativity, hybridity, and meaning making (Garcia, DuPuis, and Mitchell 2017; Peña et al. 2017; Chapman and Beagan 2015). This approach has moved beyond academia; for example, a 2018 NPR story highlighted the fusion of "ethnic" and traditional American dishes on the Thanksgiving tables of many immigrant groups in a way that is very similar to Kat's family's experience (Gharib 2018), while the crafting of hybridized identities is the primary narrative animating celebrity chef Marcus Samuelson's new program *No Passport Required*, in which he visits, profiles, and eats alongside immigrant communities (Alkon and Grosglik in preparation).

In sum, early scholarship on Hmong refugees and American-born Hmong criticized the community for not assimilating enough, while recent public health research criticizes them for becoming too assimilated. Neither recognizes the nuanced processes through which culture and identity are constructed and reconstructed through lived experiences. Furthermore, Wilcox and Kong (2014) argue that much of the research on Hmong American food and health attends only to populations with very low incomes but explains health problems solely in terms of culture. Instead of emphasizing acculturation, they argue that researchers should interrogate the interrelated effects of racism, sexism, and economic pressures.

While poverty does exist, contemporary Hmong communities are economically mixed (Zhou 2007). Our survey of Hmong Americans begins to disentangle the effects of culture and poverty by investigating the foodways of a group that is broadly middle class, college educated, and upwardly mobile. While previous scholarship too often historicizes Hmong people, we did not expect that Hmong Americans who have lived most or all of their lives in the United States would eat only Hmong foods. Our survey respondents maintain a strong preference for Hmong foods, the presence of culinary knowledge, and strong associations between Hmong foods, family, and culture. Understanding these foodways helps to highlight the creativity, self-determination, and agency of Hmong communities as they craft new and evolving cultural identities that include Hmong, Asian, and American influences.

Research Approach

This chapter represents our first collaborative study of Hmong food and farming. As a Hmong American undergraduate studying sustainable agriculture, Kat is motivated by a desire to understand the cultural knowledge embedded within Hmong food pathways as well as to explore the ways food engages self-determination, knowledge production, and identity formation in communities of color. Alison is a white, Jewish professor who studies and writes about food justice. Her work is guided by a deep interest in the ways that communities of color create and engage in local food systems and the senses of meaning, place, and community that they derive from them. She became interested in Hmong food and farmers through a previous project on the predominantly Southeast Asian Stockton Farmers Market.

Research on Hmong communities depicts them as difficult to study, even for community members, a dynamic that Kat has experienced during past research. However, we have both found that Hmong Americans born or raised in the United States are far more open to researchers than earlier generations. For this reason, as well as the gaps in the literature noted earlier, we designed a survey to better understand the food practices of Hmong Americans. We distributed this survey via our social networks and relevant organizations through Survey Monkey, receiving 125 responses in just a few days.

Our sample was predominantly female and between 18 and 35 years of age; 70% were American born, while the remainder immigrated as children. Social class is often complex to determine, and this was especially true in our case. Most respondents placed themselves in the highest income categories, with approximately 20% earning between $75,000 and $100,000 and an additional 30% earning $100,000 or more. However, Hmong Americans tend toward nonnuclear family arrangements, making it possible that this reflects more earners rather than higher individual incomes. Approximately 40% of our sample identified at least three adults living in the home, though it is unclear how many are contributing to their household

Boiled Chicken and Pizza

income. With regard to education, 42% of the respondents, their siblings, or their spouses had graduated from college, and an additional 32% had earned an advanced degree. Taken together, we believe these variables characterize our sample as middle class, educated, and upwardly mobile. This is quite a departure from the previous public health research, which tends to examine people with very low incomes, even while framing its findings exclusively in terms of cultural difference.

Our survey consisted of a mix of closed-ended and open-ended questions. Beyond demographics, we were broadly interested in what people cooked and ate, with whom, and what these foods meant to them. We coded many of the open-ended questions into categories, allowing us to generate a statistical overview of responses while also including examples in respondents' own words. Several open-ended questions regarded respondents' preferred or most commonly eaten foods. For these questions, Kat used her own cultural knowledge to code the responses as indigenous Hmong dishes, Hmong American dishes, Hmong adaptations of other Southeast Asian dishes, and American food. In our analysis, we contextualize many of the foods listed by our respondents with material from public culture, including Sami Scripter and Sheng Yang's (2009) *Cooking from the Heart: The Hmong Kitchen in America*, which we believe is the first commercially published Hmong cookbook, as well as several blogs and videos devoted to Hmong cuisine. Indeed, we were both surprised to learn how much new media dedicated to Hmong cultural foods exist. This showcases enthusiasm for Hmong cuisine among young people, pushing back against the common historicization of Hmong culture as premodern. Blogs and videos especially demonstrate a sense of cultural fluidity through which young people express interest in traditional and evolving foodways.

Eating Hmong Foods

A large majority (71%) of our respondents eat Hmong food at least a few times per week, 92% prefer Hmong food to American food, and 85% believe Hmong food is healthier than American food. All but one of our respondents reported knowing how to cook Hmong food, with a roughly even split between those who knew how to cook a few dishes versus many dishes. This is clearly a community with a strong knowledge of and love for cultural foods. Specific dishes mentioned included *kopia* (Hmong chicken soup), boiled chicken, pork ribs boiled with cabbage and ginger, and Hmong sausage, a coarsely ground pork sausage flavored with ginger, garlic, chilies, and lime juice that is commonly eaten in Pho soup or over steamed rice.

Cooking from the Heart offers an overview of Hmong cuisine as "simple, earthy, fiery and fresh" (Scripter and Yang 2009). The authors describe a traditional Hmong diet as white rice eaten with plentiful vegetables, small amounts of meat, and various fresh herbs and spices such as cilantro, lemongrass, hot pepper, and ginger. Condiments like fish sauce, oyster sauce, soy sauce, sriracha, and hoisin are common as well, and Hmong pepper dipping sauce is both so essential and varies so widely that the book offers 11 different preparations for it. Traditionally, food is steamed or boiled, and simple soups and stews are common. More recently, Scripter and Yang note, Hmong cuisine has been in dialog with other Asian cuisines. Cooks have developed Hmong takes on common dishes such as papaya salad and egg rolls and embraced techniques like stir-frying, deep frying, and broiling. As Hmong people have encountered new places, first in other parts of Southeast Asia and then around the world, their cuisine has mingled with other foodways.

However, several survey respondents pushed us to question the category of "Hmong foods" itself by expressing complexity as to what Hmong foods are. Several directly asked the question, "What is Hmong food?" Another wrote, quite lyrically, "I feel that anything I cook is Hmong food because I am Hmong. I always add my own spices (whether grown in my garden or bought from a store). My techniques are Hmong and my cooking tools are Hmong. Even if it's lasagna, it's Hmong because I made it." These responses depict Hmong food as dynamic and inventive, shifting over time with the experiences and tastes of cooks and eaters. They also show the creative work involved in the process of translocal identity formation, where places become linked through material objects such as food and everyday practices such as gardening and cooking.

And yet we can speak of "Hmong food" because the same dishes came up over and over again. The most common were boiled chicken and chicken soup, which Scripter and Yang refer to as Hmong "signature dishes," with significance in healing and religious rituals. Many of our respondents described these as among their most favorite comfort foods. For example, one American-born college student offered the following account: "I love freshly killed chickens with herbs. It just reminds me of my childhood because my parents raised chickens so we ate it quite often, like almost every day. Also, when I first went away to college and would go back home to visit my family, my mom would pack a boiled chicken for me to take back home to college."

In addition to foods specific to Hmong cuisine, many respondents enjoy dishes influenced by other Asian cuisines. Several respondents mentioned eating rice porridge—a dish common to many parts of Asia—as a comfort food when ill. Other respondents mentioned *nava* or *nab vam*, which they especially associated with Hmong New Year. The cooking blog Hmongfood.info describes these as a Vietnamese three-color sweet drink or dessert. Interestingly, a food associated with one of the most important Hmong events of the year is a Vietnamese dessert, despite the fact that sweets were not traditionally a part of Hmong cuisine. Through indigenous Hmong foods as well as foods influenced by other Asian cuisines, those people we studied create and re-create foodways that both symbolize and constitute their translocal Hmong American identities.

Indeed, 80% of survey respondents believe it is important to eat Hmong food, and their most commonly described reasons revolved around the themes of culture and identity. Several mentioned these words specifically, but others described Hmong food as "a part of me," "deeply rooted," or conveying "a sense of home." One woman, for example, wrote that Hmong food "shows who I am and where I came from. Food is culture, identity and history." For several respondents, and in contrast to the academic literature that emphasizes intergenerational conflict, food is linked to a strong sense of family. In the words of another woman, "It's home cooked food that I crave. And it's good to know parts of me and where I'm from. I think it somehow preserves our culture and provides a connection between our older Hmong generation and younger generation." As these quotations exemplify, many Hmong Americans view food as a way to craft identities that incorporate and respect their cultural heritage and families. In addition, a smaller number of respondents believe it is important to eat Hmong foods because of health benefits (14%) or because it is tasty (6%).

Our survey respondents maintained this love for Hmong food despite the fact that just under half of them (49.5%) reported experiencing racism or microaggressions when eating Hmong food in public. For example, one young woman describes bringing "simple Hmong food such as plain steamed rice that has been kneaded with boiled egg and a pinch of salt" to her elementary school cafeteria: "We'd eat this using just our hands in the school cafeteria and white kids would bully us for eating yellow poop." Sometimes, these microaggressions pressured individuals to change their eating habits. For example, another young woman responded that instances such as this "made me never want to pack lunch to school, unless it was American food," while another young woman from a more affluent home wrote, "Of course! [She has experienced this]. Hahaha. It made me not want to eat the food in public. But I still eat it often."

Others, however, were not affected or were even emboldened by these remarks. One highly educated woman in her late twenties recalls being called a savage during high school for eating purple *ncuav* (Hmong style mochi) with her hands but claims she "didn't think too much of it during that time, and it did not affect [her] desire to eat Hmong foods." A highly educated, affluent man in his early thirties described a similar experience among adults: "Caucasian co-workers once complained that our food smelled and made a comment saying we should not bring Hmong food to work or warm up our food and eat in the same break room. No, it did not affect me one bit. I brought in crazier food with stronger smell and spices the next day."

Interestingly, several respondents replied that they did not experience these sorts of microaggressions as children because they received federally funded school lunch. Poverty seems to have sheltered at least some of those we surveyed from a common racist experience.

Regardless of the perception of Hmong foods by others, it is beloved by the Hmong Americans we surveyed; 92% of our respondents preferred Hmong food to American food, mainly for its flavor. Respondents describe Hmong cuisine as tastier and more satisfying, with several referring specifically to spice and a love for pepper dipping sauce. Others prefer Hmong food because it is healthier, easily customizable to personal tastes, and easier to cook.

Growing Hmong Food

This preference for Hmong food helps to support and is supported by a strong interest in food cultivation, confirming Lee's (2005) observation that "agriculture is closely interrelated with other aspects of Hmong society." While none of our survey respondents are commercial farmers, 79%

described family involvement in growing food. Older generations were often the primary growers, but 42% of respondents participate in some way. These families cultivate a wide variety of vegetables and herbs, including those common to Hmong cuisine. Examples include green mustards, cilantro, *taub* (Asian pumpkins), cucumbers, and bitter melon. Several respondents mentioned ingredients for specific dishes such as "medicine herbs for chicken soup." Only 21% of our survey respondents' families sold food, and they did so mainly at farmers markets (12%) and Asian grocery stores (8%).

Previous research has largely painted growing food as of interest mainly to older generations of Hmong (DeMaster 2003; Brady 2011; Rose 2013). However, most of our respondents (72%) actively grow or plan to grow food with their own children. Their reasons for doing so were rarely culturally specific but instead focused on the food itself as healthier, fresher, cheaper, and more accessible. Several even offered responses common to supporters of the broader national movement for local, sustainable, and/or organic food, such as, "I believe in knowing where my food comes from and what's in it." This is similar to Alkon and Vang's findings from interviews and surveys of Hmong farmers and customers at predominantly Southeast Asian farmers markets. The cultural nature of food was often unstated amid emphasis on price and quality (Alkon and Vang 2016).

Despite these practices of food cultivation, the bulk of our respondents purchased their foods from Asian Markets (94%) or mainstream supermarkets like Safeway or Wal-Mart (84%). In addition, nearly half of those we studied made use of farmers markets, farms, and their own gardens. From her research in Rhode Island 15 years ago, Kathryn DeMaster found that "obtaining preferred Hmong foods would be difficult, if not impossible, were they not grown by Hmong agriculturalists persisting in their knowledge and practices" (DeMaster 2005, 116). In contrast, those we studied show that they can find appropriate food from a wide variety of sources, but many choose to cultivate food or purchase directly from those who grow it.

But Not Only Hmong Food

Our survey respondents expressed preferences for Hmong food but did not eat it exclusively. We asked respondents to write down everything they had eaten the previous day; 58% mentioned at least one traditionally Hmong or Hmong American dish. Only 5% of respondents, however, ate these foods exclusively, with the vast majority eating some Hmong food, some food from other Asian cuisines, and some American food throughout the day. For example, one young, college-educated woman listed "squirrel stew, papaya salad, boiled chicken, waffle, sweet rice in banana wrap, mini pizza from Food Co." Several of these dishes are included in *Cooking from the Heart*. The book offers a recipe for *Nqaij Nas Hau Xyaw Txuj Lom*, or squirrel stew with eggplant, in which the squirrel meat is soaked in dark beer to remove some of the wild flavor and then boiled with garlic, lemongrass, chilies, lime leaves, galangal, and ginger before adding the eggplants, basil, Sichuan pepper, and fresh cilantro. The cookbook explains that, traditionally, "Hmong men enjoy hunting wild squirrels. In a Hmong home, no meat is ever wasted, so Hmong women still make this old-fashioned stew." The authors also include a recipe for *Txhuv Nplaum Qhwv Nplooj Tsawb*, or sticky rice in banana leaves, a dish common to many Asian cuisines. In this version, the rice is steamed with coconut milk and sugar and then stuffed into a banana leaf, folded up, and steamed again.

In addition to these typical Hmong dishes, the respondent also listed papaya salad, which Kat classified as a Hmong American dish. In contrast to the sweeter versions served at Thai restaurants, the Hmong version is generally spicier and more sour. While this is not a traditional Hmong food, it has been embraced by Hmong cooks and is often sold at New Year celebrations. And of course, waffles and pizza are commonly considered to be American food. This respondent demonstrates the ways that Hmong Americans weave together different ingredients and dishes to create a way of eating that is sometimes culturally specific and other times generically American, much as their larger identities are both Hmong and American in a way that is neither straightforward nor linear.

Conclusion

In contrast to the public health literature problematizing dietary acculturation, our survey reveals that Hmong Americans continue to grow, cook, and eat Hmong food. However, they also enjoy American food and food from other cuisines. These foodways are a part of their hybridized, translocal identities; they are both Hmong and American and move between their heritage and new homelands in ways that defy the linearity common to metanarratives of assimilation but are consistent with contemporary scholarly works on immigrant foodways.

The literature on Hmong food and health prescribes a return to traditional foods without a fine-grained understanding of current Hmong Americans' food practices. These researchers are working from a damage-centered perspective (Tuck 2009) with very small samples of mainly high-poverty individuals, and it is possible that their analyses conflate class with culture. Given that the overwhelming majority of the Hmong Americans we surveyed reported regularly consuming Hmong food, it is also possible that the public health experts are advocating that Hmong Americans should eat Hmong food exclusively. This is not a reasonable expectation for any ethnic group and historicizes Hmong Americans as premodern, ignoring their agency in processes of contemporary racial identity formation.

Cuisines evolve over time, especially when carried by immigrants and refugees around the world. Our respondents, as well as public media such as cookbooks, cooking blogs, and videos, demonstrate that Hmong foodways as they are maintained in the United States have come to encompass variations on other Southeast Asian dishes and techniques. In this sense, Hmong foodways parallel the stance of Hmong studies as an academic discipline. They are Asian and intersect with a variety of Asian cultures, but maintain a uniqueness that cannot be collapsed into the Asian immigrant experience. This suggests an approach, not only to the study of foodways but also to culture more generally, that emphasizes the ongoing construction and maintenance of translocal identities.

This study represents the very beginning of an ongoing research project aimed at understanding Hmong American food and agricultural practices. While we have documented the consumption patterns of American-born and American-raised Hmong individuals, we have only begun to explore issues of meaning and identity in a rich and substantial way. We have not yet heard the stories that Hmong Americans tell about various dishes, nor have we witnessed their roles in cultural traditions. We have begun to ask questions about gender and cooking practices but have not been able to understand how they reproduce or shift patriarchal norms. Future research can provide a more detailed understanding of food practices and their cultural meanings, and can attend to the role of agriculture in Hmong senses of self, community, and place.

As relatively recent immigrants with a rich cultural history, Hmong Americans are an important yet understudied group who are currently weaving their ethnic traditions into the fabric of American identity. In the tradition of ethnic studies more broadly, the Hmong studies literature aims to document and analyze this process, highlighting both the challenges this community faces as a racialized minority and their agency in creating new ways of being. The literature on Hmong food and health, however, has not incorporated this approach, despite the fact that it is increasingly common to studies of immigrant foodways. This chapter seeks to build bridges between these fields of scholarship, aiming for a deeply nuanced, culturally informed approach to health. Moreover, it demonstrates the ways that food can serve as a powerful lens through which to examine immigrant experience, documenting structural barriers while highlighting the agency through which communities draw on their histories to create new cultural identities.

Notes

1. We are cautious about terming this diet as traditional, as it implies something static and unchanging in precisely the way we are arguing against. However, alternative terms, such as precolonial, do not map well onto the Hmong experience, and "traditional" is the word that many contemporary Hmong Americans use to refer to these foodways (Scripter and Yang 2009).

2. Corvée labor is a feudal system of unpaid, often forced, labor exacted in lieu of taxes (Lee 1997).

References

Alkon and Grosglik. In preparation. "Eating (With) the Other: Race in American Food Television."

Alkon, Alison Hope, and Dena Vang. 2016. "The Stockton Farmers' Market." *Food, Culture and Society* 19(2): 389–411.

Baker, L.E.. 2004. Tending Cultural Landscapes and Food Citizenship in Toronto's Community Gardens. Geographical Review 94(3): 305–325.

Boulden, W. T. 2009. "Gay Hmong: A Multifaceted Clash of Cultures." *Journal of Gay and Lesbian Social Services* 21(2–3): 134–150.

Brickell, Katherine, and Ayona Datta. 2011. *Translocal Geographies*. Burlington, VT: Ashgate.

Chapman, Gwen E., and Brenda L. Beagan. 2015. "Food Practices and Transnational Identities." *Local Environment* 16(3): 367–386.

Chuh, Kandice. 2003. *Imagine Otherwise: On Asian Americanist Critique*. Durham, NC: Duke University Press.

Corlett, Jan L., Ellen A. Dean, and Louis E. Grivetti. 2003. Hmong Gardens: Botanical Diversity in an Urban Setting. *Economic Botany* 57(3): 365–379.

DeMaster, Kathryn Patrice. 2005. "Sowing the Mustard Green Seed: Hmong Gardeners Cultivate Knowledge, Land, and Culture." In *Discourses and Silences: Indigenous Peoples, Risks and Resistances*, edited by G. Cant, A. Goodall, and J. Inns. Canterbury, New Zealand: Canterbury University Press.

DePouw, C. 2012. "When Culture Implies Deficit: Placing Race at the Center of Hmong American Education." *Race Ethnicity and Education* 15(2): 223–239.

Douglas, Mary. 1996. *Purity and Danger*. New York: Routledge.

Faidman, Anne. 1997. *The Spirit Catches You and You Fall Down*. New York: Farrar, Strauss and Giroux.

Franzen, Lisa, and Chery Smith. 2008. "Acculturation and Environmental Change Impacts Dietary Habits among Adult Hmong." *Appetite* 52:173–183.

Garcia, Matthew, E. Melanie DuPuis, and Don Mitchell, eds. 2017. *Food across Borders*. New Brunswick, NJ: Rutgers University Press.

Gharib, Malaka. 2018. "Turkey and Tamales: People of Color Share Their Multicultural Thanksgivings." *The Salt*, National Public Radio. https://www.npr.org/sections/thesalt/2018/11/20/669354317/turkey-and-tamales-people-of-color-share-their-multicultural-thanksgivings.

Goldberg, Anna. 2008. "Understanding Hmong Growers' Livelihood Strategies in Sacramento." MS thesis, University of California, Davis.

Goto, Keiko, Wa Mee Vue, Tong Xiong, and Cindy Wolff. 2010. "Divergent Perspectives on Food, Culture and Health among Hmong Mothers with Middle-School Children." *Food, Culture and Society* 13(2): 182–198.

Guthman, Julie. 2008. "'If They Only Knew': Color Blindness and Universalism in Alternative Food Systems." *Professional Geographer* 60(3): 387–397.

Hickman, Jacob. 2011. "Morality and Personhood in the Hmong Diaspora: A Person-Centered Ethnography of Migration and Resettlement." PhD dissertation, University of Chicago.

Hondagneu-Sotelo, Pierrette. 2014. *Paradise Transplanted: Migration and the Making of California Gardens*. Berkeley: University of California Press.

Ikeda, J. 1999. *Hmong American Food Practices, Customs and Holidays*. Chicago: American Dietetic Association and American Diabetes Association.

Lee, Gary Yia. 2005. "The Shaping of Traditions: Agriculture and Hmong Society." *Hmong Studies Journal* 6:1–33.

Lee, Gary Yia. 2008. "Diaspora and the Predicament of Origins: Interrogating Hmong Postcolonial History and Identity." *Hmong Studies Journal* 8:1–25.

Lee, Mai Na M. 1997. "The Thousand-Year Myth: Construction and Characterization of Hmong." *Hmong Studies Journal* 2(1): 1–23.

Lor, Pao. 2012. "Hmong American Professional Identities: An Overview of Generational Changes since the 1970s." In *Hmong and American: From Refugees to Citizens*, 147–160. St. Paul: Minnesota Historical Society Press.

Mannur, Anita. 2007. "Culinary Nostalgia: Authenticity, Nationalism and Diaspora." *MELUS*. 32(4): 11–31.

Minkoff-Zern, Laura-Anne, Nancy Peluso, Jennifer Sowerwine, and Christy Getz. 2011. "Race and Regulation: Asian Immigrants in California Agriculture." In *Cultivating Food Justice: Race, Class, and Sustainability*, edited by Alison Hope Alkon and Julian Agyeman, 65–85. Cambridge, MA: MIT Press.

Morales, Alfonso. 2011. "Growing Food *and* Justice." In *Cultivating Food Justice: Race, Class, and Sustainability*, edited by Alison Hope Alkon and Julian Agyeman, 149–176. Cambridge, MA: MIT Press.

Morrison, Gayle. 2008. *Sky Is Falling: An Oral History of the CIA's Evacuation of the Hmong from Laos*. Jefferson, NC: McFarland.

Moua, Kao Nou L., and Pa Der Vang. 2015. "Constructing 'Hmong American Youth.'" *Child and Youth Services* 36:16–29.

Nibbs, Faith. 2010. "A Hmong Birth and Authoritative Knowledge: A Case Study of Choice, Control, and the Reproductive Consequences of Refugee Status in American Childbirth." *Hmong Studies Journal* 11:1–14.

Ostrom, Marcia, Bee Cha, and Malaquias Flores. 2010. "Creating Access to Land Grant Resources for Multicultural and Disadvantaged Farmers." *Journal of Agriculture, Food Systems and Community Development* 1(1): 89–106.

Peña, Devon, Luz Calvo, Pancho McFarland, and Gabriel R. Valle. 2017. *Mexican-Origin Foods, Foodways and Social Movements: A Decolonial Reader*. Little Rock: University of Arkansas Press.

Rios, Michael, and Joshua Watkins. 2015. "Beyond 'Place': Translocal Placemaking of the Hmong Diaspora." *Journal of Planning Education and Research* 35(2): 209–219.

Robinson, Eric. 2014. "What Everyone Else Is Eating: A Systematic Review and Meta-Analysis of the Effect of Informational Eating Norms on Eating Behavior." *Journal of the Academy of Nutrition and Dietetics.* 114(3): 414-439.

Ross, Hannah. 2013. "The Hmong Farmers of Providence RI." MA thesis, Brown University.

Satia-Abouta, Jesse. 2003. "Dietary Acculturation: Definition, Process, Assessment, and Implications." *International Journal of Human Ecology* 4(1): 74–93.

Schein, Louisa, and Va-Megn Thoj. 2009. "Gran Torino's Boys and Men with Guns: Hmong Perspectives." *Hmong Studies Journal* 10: 1–11.

Scripter, Sami, and Sheng Yang. 2009. *Cooking from the Heart: The Hmong Kitchen in America.* Minneapolis: University of Minnesota Press.

Sherman, Spencer. 1988. "The Hmong in America: Laotian Refugees in the 'Land of the Giants.'" *National Geographic Magazine*, October, 586–610.

Slocum, Rachel. 2008. "Thinking Race through Corporal Feminist Theory." *Social and Cultural Geography* 9(8): 849–870.

Sowerwine, J., C. Getz, and N. Peluso. 2009. "The Myth of the Protected Worker: Southeast Asian Farmers in California Agriculture." *Agriculture and Human Values* 28(4): 579–595.

Stang, J., A. Kong, M. Story, E. Eisenberg, and D. Neumark-Sztainer. 2007. "Food and Weight-Related Patterns and Behaviors of Hmong Adolescents." *Journal of the American Dietetic Association* 107(6): 936–941.

Trapp, Micah. 2010. "What's on the Table: Nutrition Programming for Refugees in the United States." *North American Association for the Practice of Anthropology Bulletin, Special Theme Issue: Anthropological Perspectives on Migration and Health* 34: 161–175.

Tuck, Eve. 2009. "Suspending Damage: A Letter to Communities." *Harvard Educational Review* 79(3): 409–423.

Vang, Chia Youyee, and Faith Nibbs. 2016. *Claiming Place: On the Agency of Hmong Women.* Minneapolis: University of Minnesota Press.

Vue, Pao Lee, Louisa Schein, and Bee Vang. 2016. "Comparative Racialization and Unequal Justice in the Era of Black Lives Matter: The Dylan Yang Case." *Hmong Studies Journal* 17: 1–21.

Vue, W., C. Wolff, and K. Goto. 2011. "Hmong Food Helps Us Remember Who We Are: Perspectives of Food Culture and Health among Hmong Women with Young Children." *Journal of Nutrition Education and Behavior* 43(3): 199–204.

Wilcox, Hui Niu. 2012. "The Mediated Figure of Hmong Farmer, Hmong Studies, and Asian American Critique." *Hmong Studies Journal* 13(1): 1–27.

Wilcox, Hui Niu, and Panyia Kong. 2014. "How to Eat Right in America." *Food, Culture and Society* 17(1): 81–102.

Yang, D., and D. North. 1998. *Profile of the Highland Lao Communities in the United States*. Washington, DC: US Department of Health and Human Services, Family Support Administration, Office of Refugee Resettlement.

Zhou, M. 2007. "Divergent Origins and Destinies: Children of Asian Immigrants." In *Narrowing the Achievement Gap: Strategies for Educating Latino, Black, and Asian Students*, edited by S. Paik and H. Walberg, 109–130. New York: Springer.

14 Recipes for Immigrant Lives: Crossing, Cooking, Cultivating, and Culture at a Shared-Use Commercial Kitchen

Situational Strangers

Ingredientes: Introduction

> Empanadillas de Angeles Reescrito Auténtico
> 4 lb carne molida de res...

This is the story of recipes loosely belonging—in that recipes are sociocultural narratives—to an immigrant from Guanajuato, Mexico. This is the story of the ways in which these recipes have evolved and taken on new cooks, ingredients, and customers as they have traveled through kitchens and meals over time and space, affirming that food "defies unitary categorization" (Deb 2014). This is therefore also a single "foodways" thread woven within the collective human story that speaks of recipes from all cultures carried in memories, on folded and stained pieces of paper, in pockets and bags like identity papers only meaningful to the beholders, only fully *real* once cooked and eaten. This is a true story of recipes loosely belonging to an immigrant from Mexico, while also being an archetypical story, a thread of a much larger cable-laid story, given that truth is relational and that those who are not socially connected, whose experiences and recipes are not documented, cannot access the solidity of such a claim.

This is the hybrid story of our attempts to tease out some of these "entangled" (Barad 2007) threads, many of which have resulted from intimate "*charlas culinarias* (culinary chats)" (Abarca 2004) with immigrant friends, family members, colleagues, and eaters, who like ingredients are combined to make a single dish. Some of these immigrants cook these recipes directly from their thoughts, while others work from given instructions or from written words, others revise methods and ingredients into new forms and tastes, while still others are its eaters. Some tell their stories in Spanish, which we have translated, others in the adopted language of these lands,

English, so there are nuanced layers of meaning being lost and constructed. Our *charlas culinarias* have taken place at CLiCK, Inc. (Commercially Licensed Co-operative Kitchen), a shared-use kitchen in eastern Connecticut, and in the surrounding community over the past few months and the past few years, creating an overlapping and layered sense of time, just as recipes themselves can be recent but are never really new, given that their roots are always in the past.

> Reescrito continua … 1 taza de sofrito casero
> 30 tortillas de harina blanca

Recipes may be from unsullied memories or written down, giving the impression of being fixed. Nevertheless, ingredients must change with the moment, with the geography of their production site, just as the people who cook them and eat them must be and are also changed. In this manner, recipes once familiar become new and strange even to their owners, just as the people, who in their homelands were once known, through the act of immigration become part of a vague group type identity making them strangers to others and even to themselves. As Simmel states, though, immigrants are not strangers in "the usual sense of the term, as the wanderer who comes today and goes tomorrow, but rather as the man [or woman] who comes today and stays tomorrow—the potential wanderer, so to speak, who, although he has gone no further, has not quite got over the freedom of coming and going" (Simmel 2004, 139). For Simmel, the epitome of such a stranger is the trader, the ones who are "near and far at the same time" and who are "not really perceived as individuals but as strangers of a certain type" (Simmel 2004, 141). Such "strangers of a certain type" are also the ones whose food is strange "of a certain type." In fact, all new immigrants are strangers, and if you, "as one who belongs to a given place," do not recognize their strangeness by their language, clothing, or rituals, then you may well identify them by their strange food. Our shared-use kitchen is such a place, where sights and smells, recipes and products, are created by immigrants and nonimmigrants alike, blending the strange with the familiar, challenging the binary concepts of local and global food.

> Reescrito continua …
> Comenzar sofriendo el sofrito … una vez mezclas todos los ingredientes comienza a formar las empanadillas.

In not being perceived as "known," strangers are not privileged names in the same manner as those who belong to a given place. In being nameless or renamed in the language of the ones who belong, their strangeness is continually reinforced, even potentially to themselves. When asked the everyday questions, "Who are you?," "Where are you from?," "Where do you live?," and "Where do you work?," they may not be able or willing to give "truthful" answers. For the stranger, these answers are *political*; how to answer them and to whom can mean the difference between security and threat, anonymous residency and possible exposed deportation. Those immigrants in our story are known to us and have their "real" names used, some do not, and some we don't know except in passing. But, like all of us, if we become strangers we may still be able—if we can eat the food and savor the tastes from the places whence we once belonged—to sigh and utter to ourselves, "Well, at least our food *knows* our names" (Klindienst 2006).

Reescrito continua...
Recuerda incluir un pisca de tu propia cultura.

To recognize this loss of names, we, too, have given up our "real" names—becoming situational strangers, as opposed to being reified ones. As researchers, we can choose to be in the normal role of the stranger as "the potential wanderer"; however, once our subjects of investigation have been interviewed and our observations made, we can return to where we belong, along with our data and our reclaimed names, while those whose stories we seek to illuminate often remain obscured and unnamed. In our case as authors, although we do not share the same race or ethnicity—one of us is a recent immigrant from Puerto Rico and the other is a first-generation American whose parents came from England—in our roles as researchers, we are not strangers; rather, we are the entrusted storytellers, the documenters of truth. In contrast, those who are strangers are not trusted by those who belong, even if and when they do speak their truths. For this reason, we also seek to play with concepts of what is fully true and false, what is real or imaginary. Rather than flashing our social science cards with claims to be documenting "empirical truths," as only those who are privileged can claim, we are intentionally engaging in "critical ethnographic" (Madison 2012) methods, evoking a "methodology of the oppressed" that seeks to thwart systems of power based on constructions of hierarchy and separation (Sandoval 2000). Those who are methodologically fastidious may see such methods as suspect, but they are nonetheless, as Sandoval asserts, grounded in "'love', understood as a technology for social transformation" (Sandoval, 2000, 2). In this manner, we, too, are moving into the borderlands (*la frontera*), as conceptualized by Gloria Anzaldúa (1987), calling on her complexity of identity as *la mestiza*. Developing "a tolerance for ambiguity...characterized by movement away from set patterns and goals and toward a more whole perspective" (Anzaldúa 1987, 101), Anzaldúa conceptualized, "the work of mestiza consciousness is to break down the subject-object duality" so that "duality is transcended" (102). Likewise, we are inspired to transcend the imagined duality between truth and falsehood within our social science research and between us as the named ones and those who are the strangers.

Thereby, in the spirit of "inclusion," included here are pieces of stories from immigrants we have not met but who have spoken of their stories and had them documented by others, as well as pieces of stories that have been said to us by some stranger, somewhere, at some time, in some manner, and therefore the words we share are "original," but not all are necessarily "authentic" (Abarca 2004). We take this distinction from Abarca's work, where she identifies the problematic politics of attempting to claim a recipe (or anything else) as being "authentic," having cultural legitimacy, as opposed to being "original," as in "the production belonging to that person" (Abarca, 2004, 19). Likewise, we seek to focus on our stories not as being "authentic" but rather as being "original," in that we seek to recognize that "culture is always changing, [and]...as active agents we are always defining new cultural practices" (20). In short, we are weaving together stories from our *"charlas culinarias"* as well as from secondary sources and immigration archives, thereby creating "a meeting of multiple sides in an encounter with and among others" (Madison 2012, 10). By blurring the artificial borders between ourselves, those immigrants who are part of these stories, and imaginary others, we are evoking the alchemy of cooking. As Hauck-Lawson identifies, "food is a potent mode of communication," a mode that she termed a "food voice" (Hauck-Lawson 1992, 6). In this spirit, we seek to create a "food chorus," playing the role more of cooks than of researchers, for not only are we changing "the recipes," making them "original" (hence subjective) as opposed to being "authentic" (hence objective), but we are asking readers to trust that our methods will enhance the emergent taste of our results, even as we keep some ingredients secret.

In following the threads of recipes from Mexico, we honor their literary tradition of magical realism that teases reality and invites (im)possibility. Borges gave the definition of magical realism when he said, "I imagine a labyrinth of labyrinths, one sinuous spreading labyrinth that would encompass the past and the future and in some way involve the stars" (Martin 1989, 3). In our case, such a labyrinth has been spun from food, much as it is in Esquivel's 1989 novel *Like Water for Chocolate*. In taking our cue from that novel, we must utter that not only is food *magical* but so is everything else in the universe, if magic is taken to mean a "reality" other than the fractured one we perceive. As feminist physicist Barad recognizes, echoing the sentiments of Anzaldúa, it is "impossible to differentiate in any absolute sense between creation and renewal, beginning and returning, continuity and discontinuity, here and there, past and future" (Barad 2007, ix). The same can be said about the recipes for immigrant lives and how we have chosen to weave their stories and enhance them with our own flavoring, thereby connecting strangers collectively known and unknown.

Reescrito continua...
Para empanadillas saludables hornea las empanadillas en vez de freírlas.
¡Buen provecho!

Otros Ingredientes: The Authenticity of Place

In our small deindustrialized town in Connecticut, ask anyone in the Mexican community who makes "authentic" Mexican food, and they will tell you: Maria! "Maria... she makes the real thing. When I eat her food I feel like I am back in my home," says Jose, a local Mexican stranger who happens to be one of our neighbors. Shown here, authenticity in food is not only political in terms of who gets to claim it, how it is claimed, and under what conditions (Abarca 2004); it also requires that the eater possess the cultural and even regional palate to recognize and affirm that authenticity. For the rest of us who claim to like Mexican food, the taste remains merely delicious, as opposed to being "the real thing." For Maria, a 44-year-old who is short and plump, authenticity means "made by hand" (*tortillas de maiz* becoming *flautas de pollo* or *quesadilla*) and the fact that "all my ingredients come from Mexico." To verify this, she goes off to get a large bag of white onions and some containers of spices, all of which have been imported. Speaking in Spanish, she continues, "Sometimes I change the ingredients depending on what I have but I try to use only things that come from Mexico...like the maize." The connection between corn and Mexican food goes back to 3500 BC, with the "holy trinity" being maize, beans, and squash, intricately linked with Mexican national identity (Pilcher 2005). To get her imported ingredients, Maria orders them directly from food distributors. "You know there are food smugglers," she says, raising her eyebrow to indicate that she has shared something secretive. "They cross the border and bring back items that we can't get here," she says, producing a large piece of cactus from a bag on the floor. "It's ingredients like these that are from Mexico that make my food authentic. My recipes connect back to my country," she adds, evoking in our minds an image of intricate food webs that run across physical spaces and back through time.

Martin, who opened a Mexican restaurant, shares another example of a stranger's linkage between national identity and the authenticity of his food. When asked by researcher DePue to whom his restaurant catered, he replied, "The restaurant is for pretty much American people. It's Mexican food, but it's like—" (DePue 2008, 15), and then his sentence ended there. When asked if the food is "Americanized Mexican food, or...the authentic stuff," he affirmed, "No, I think it's the real Mexican. It's stuff we make" (15). In this manner, we see that what is made, how it is made, who it is made for, and who ultimately eats it is highly complex in terms of how national and cultural labels are used in the face of claims to authenticity. Obviously, for cooks, as well as writers of cookbooks, claims to authenticity can augment their status, especially if they themselves are not "originally" from the culture. Such claims, however, can also be expressions of colonialism and cultural appropriation, as literary critic Goldman (1992) explores. While we do not dismiss the inequalities embedded within "cultural appropriation," we also seek to celebrate cultural *entanglements* and the ways in which claims to "authenticity" often require individualized cultural identities that "strangers" are not always privileged to have. As with Maria, her food, and those connected to her, we seek to recognize them all as separate and "original" individuals while also seeing them all as cultural abstractions, as "authentic" archetypes that, like food itself, defy singular categorization.

What is not abstract, though, is the place on which we now focus our attention: the physical shared-use commercial kitchen where the recipe and the individuals we are discussing intersect, as well as the other kitchen users, many of whom are also immigrants coming from other countries and cultures and as a result have also become abstracted, entangled, and yet enhanced. Serving over 20 different small food businesses, our shared-use kitchen acts as the incubator wherein our small local food entrepreneurs can benefit from the cross-pollination of ideas, knowledge, and social networks.

It took seven years of working on the idea of a shared-use commercial kitchen to actually make it a reality (Godfrey and Freake 2016; Godfrey 2017), and during that whole time, Maria would often appear at board meetings and say, "*¿Cuando viene la cocina?*" (When is the kitchen coming?). The answer would always be "*Pronto, pronto*" (soon, soon), even though the answer did not become true for seven years. Maria's interest in CLiCK was because for 15 years she cooked food in her house to feed over 100 Mexican wholesale nursery workers. As a result, her living room had vats of rice, beans, and *tortillas* in places where most people have furniture. The walls of her apartment were covered in steam and grease, and her children were teased at school because they "always smelled like food." People would stop in all the time to buy food at her house, and she could never relax. Fast-forwarding to her relocation in the shared-use kitchen, what took her 10 hours to cook in her house now takes only 5 hours at CLiCK because, she says, "It's a real business kitchen... and I have more employees and support." The shared-use kitchen for her meant not only that she could reclaim her house and furniture, and that her children would no longer "smell," but that her food business could become legal and therefore expand its markets.

And so it has.

Otros Ingredientes: Crossing, Cooking, and Starting a Food Business

"Many people here in the US think that all of us [Mexicans] come to this country looking for money, but it is not always the case," says Maria, dropping her *empanadillas* into the hot bubbling commercial fryer at CLiCK's shared-use kitchen. The kitchen is full of steam, food smells, and ingredients, as well as the two other women who work for her to get the *empanadillas* ready for lunchtime delivery at the wholesale nursery where about 500 mostly Mexican immigrants work, including Jose.

"I was born and raised in Guanajuato, Mexico," Maria continues in Spanish, "I had a normal and happy life as a child and adolescent. As a young woman, I had a small but successful flower shop in my community...money was not a problem." As she says this, she emphasizes her point by taking up the fryer basket and placing it on the counter. "One day I received a phone call from a Mexican friend in the US to let me know my father was very sick. So I decided to take a few weeks off from my business and 'cruzar como mojada' [cross the border] to the US to get my dad and bring him back home, Mexico. That was the moment when my journey began."

Of Mexican immigrants currently in the United States, half of them are here illegally (Gonzales-Barrera and Krogstad 2017). Maria first came here illegally, but, as she says, she came not for money but for family reasons. This is also the case for many Mexican immigrants (and others around the world) in that their journeys over the border, into the borderlands/*la frontera*, are not "choices" in the full sense of the word but rather are the result of family crises or the devaluation of farm products within their own countries, resulting most specifically from the North American Free Trade Agreement (National Farm Workers Ministry 2017). Maria did not intend to stay in the United States, but, she continued,

> I had my business back in Mexico but then I broke my leg and then I met my husband here.... I ended up living in a house with 14 other immigrants in a three bedroom apartment here in CT. I was not working so I cooked, making lunches [*comida Mexicana*] for everyone. I would do it for free but then when they got paid they each started to pay me $20 a week. I never decided to make a business...it just happened. I never really learned to cook.... I just somehow knew how to do it. I'd look at different Mexican foods, I'd taste them and then I'd make them myself. My recipes come from me, from my memory...you could say the ingredients speak to me...my recipes come from...*mi cultura* [my culture].

In starting an underground food business in 2000, Maria joined what in the United States amounts to a $2 trillion economy (Godfrey 2017). This underground food economy is of course illegal in the United States, but in the developing world, it is standard fare and plays a significant role in creating income for women (Chen 2000). As for how Maria knew how to cook and her notion that the ingredients "speak to her," we are reminded of the description of Nacha, the cook in *Like Water for Chocolate*, who "didn't know how to read or write, [but] when it came to cooking she knew everything there was to know" (Esquivel 1989, 6). Maria does know how to read and write in Spanish, but she does not cook from written recipes and rather looks to her ingredients to tell her what they want to be in that moment.

As well as listening to her ingredients, Maria cooks with her heart. She stated, "I put a lot of passion and emotion into my cooking. I also still make flower arrangements like in my first business in Mexico.... I do what I do with love." Hearing Maria describe her feelings in relation to her recipes reminded us of Lorde's work on the uses of the erotic and her desire to reclaim it as "the sensual—those physical, emotional, and psychic expressions of what is deepest and strongest and richest within each of us, being shared: the passions of love, in its deepest meanings" (Lorde 1984, 56). Linking Lorde's definition of the erotic with food and with Maria's statement that she cooks and arranges flowers with love gave us insight into her secret ingredients. Maria notes how even though she can tell you a recipe for all the foods she cooks, nevertheless, she says, "Every time I cook I change things just a little." In other words, the food is always "original" (Abarca 2004). Through the alchemy of cooking, which brings disparate ingredients together, there are always the unknown factors such as "energy." The cook's energy, the "energy" (freshness) of the ingredients, the "energy" of the heat source, and the "energy" of the kitchen all combine to make each cooking experience unique. Then of course there is the eating, and the recognition that once we begin the act of eating, the separation (which Barad would argue is an illusion anyway) between ourselves as the eaters and that which we eat dissolves.

"Do you want food?," Maria asks only one of us in English, knowing that the one who speaks Spanish is vegan. "Si, gracias, siempre" (yes, thanks, always) is the attempted Spanish reply, and she rushes off to the walk-in cooler, only to return with large white Styrofoam boxes filled with food. "You want the red or green salsa?," she asks, holding up the little plastic containers. "*Ambos*, both, of course," is the reply.

Otros Ingredientes: Cooks, Gardening, and Community

"Due to my mental conditions I had not worked for over 13 years. I suffer from depression, social phobia and extreme anxiety. I do not like leaving my apartment. It is the only place I feel comfortable," Lula shares in Spanish. Lula is a slight 50-year-old woman from Puerto Rico who, like Maria, moved to the mainland because of family matters. Puerto Ricans coming to the mainland is *not* illegal, but it does dramatically change their surroundings (especially if they come north to a temperate climate like Connecticut) while challenging their identities. Lula came to escape problems with her husband, a story she shares while doing dishes at CLiCK. She explains, "He cheated on me with another woman.... *No me molesta decirlo* (it doesn't bother me saying it anymore). But back then in my mind was sadness and in my heart was pain. I was leaving a whole life behind.... I was depressed and suffering a crisis of anxiety all the time. It was a bad time." Later, during a conversation at her home, Lula continued to share her story: "My mental condition created problems and arguments among my family as they help me with everything.... Then last summer my daughter invited me to help her in a community garden at CLiCK. I started to work at the community garden as a volunteer at CLiCK."

As she shared about CLiCK, Lula became much more animated: "Working as a volunteer at the CLiCK community garden changed my life. Volunteering at the community garden became an everyday activity I started to add to my routine. Working with the plants is so relaxing for me. My mind is occupied with positive thoughts about the plants, weeding, watering, and harvesting. I feel in peace mentally and active physically. I just wanted to be alone at the garden taking care of the plants. I was always outside in the garden, until one day my daughter asked me to help in the kitchen at CLiCK."

Lula's experience is again not unusual; gardens and gardening have been shown to be healing, if one is *choosing* to garden and has some autonomy over what is done and how it is done as opposed to being a "farmworker." Klindienst, in her 2006 book *The Earth Knows My Name: Food Culture and Sustainability in the Gardens of Ethnic Americans*, spoke to immigrants around the country about their chosen gardens and the power of gardening. She states, "Many of the gardeners spoke to me about the spiritual power of the act of gardening. The land is said to 'speak,' and the gardener learns" (Klindienst 2006, xxiii), much as the ingredients speak to Maria, in terms of her knowing what to make. Additionally, Klindienst shares examples of other Puerto Ricans, who like Lula have found solace in gardening. In one such example, Hilda, who works for Nuestras Raíces (Our Roots), a grassroots urban agriculture organization in Holyoke, Massachusetts, observed that, "Community gardening is powerful.... It preserves the earth, it preserves good health nutrition, it preserves so many things—friends, families, cultures, values, traditions" (Klindienst 2006, 201). Then, in comparing gardening in Puerto Rico to that in the United States, Hilda made the astute statement that, "In Puerto Rico having a garden is about growing your own food.... Here it's not only about food... it's a way of screaming out, 'I want to keep my culture. I want to give this tradition to my children and leave them with this gift, this pride'" (Klindienst 2006, 205). Hilda's expression of pride in sharing her culture with her children is one that Maria shares in relation to her food. Maria's pride extends not just to sharing her food with her family or other Mexicans but also with North Americans. Maria explains, "I feel pride because there are many *gringos* who have learned about traditional, authentic Mexican food from my cooking.... They get trained by my food to identify traditional authentic Mexican food as opposed to fake Mexican." This theme of authentic versus "fake" is very important for Maria and Martin and for many other immigrants who, like Hilda, want to scream "I want to keep my culture."

The issue then becomes how to preserve the balance between "keeping one's culture," avoiding being culturally appropriated, while at the same time benefiting from the richness of cultural diversity, in particular when it comes to food. For example, when Lula met Maria, a cross-cultural bond was created both through a shared common language (Spanish) and the language of food. Lula explains how she went from the CLiCK garden to the kitchen: "I was afraid to go in the kitchen because I knew there are more people working and they seem to be English speaking only. I was as I thought that only white people worked at CLiCK. I always feel uncomfortable around white people because I cannot understand them. I do not want them to think that I am rude, but I cannot communicate so I just look away."

Lula's fear of "white people" because she cannot speak English is again not uncommon, given that only about 38% of foreign-born or first-generation Latinos speak English (Taylor et al. 2012). However, what she found once she did go into the CLiCK kitchen were other Spanish speakers. She explains, "I was wrong. In CLiCK I met Maria, who is the owner of a Mexican catering business. I also met three other Mexican ladies and one Guatemalan lady, all who work at CLiCK and like me don't speak English. I was amazed!" Therefore, slowly Lula began working—something she had not done since coming to the United States. She continued,

> At the CLiCK kitchen the more I learn, the more I want to help. I have even learned how to make Mexican food. It's different but similar to our food... they eat more *maize* and *chilies* than we do... but I am getting used to it. I have even shared with Maria something about our food. Every weekday I wake up early to go

to CLiCK to work with Maria. I am not afraid to leave my house and I have made many friends. I am not looking away when people talk English to me. Everybody at CLiCK is also trying to learn Spanish words in order to communicate with me. Working at CLiCK has changed my life. I feel part of the community.

Just as Lula has learned from working with Maria, Maria in turn has learned from working with her. Lula has taught her to challenge stereotypes of Puerto Ricans and how to make her Mexican food just a little bit Puerto Rican. As she explains in another conversation at her house, sitting on her sofa, where once there would have been large vats of food, "Many of us Mexicans...we think Puerto Rican women are lazy but Lula breaks this for me.... She is such a hard worker. And you too," she says laughing, as she puts her hand out to the one among us who is Puerto Rican. In our Connecticut town and in many other places around the country, conflicts and tensions between Mexicans and Puerto Ricans are not uncommon, as they compete for jobs, housing, and open spaces while also interfacing very differently with issues of legality, taxes, and public assistance (Feuer 2003). But happily at CLiCK, cooking together has created new "original" cross-cultural communities that are like the "borderlands/*la frontera*" (Anzaldúa 1987) in their intersections, even as national identities are proudly expressed through "authentic" recipes (as claimed) and their manifested food items.

Additionally, Lula's daughter, who is a vegan (and who is also one of us), has been learning Maria's recipes from her and from her mother to use them in her cooking classes held at CLiCK. Employed by our state university's agriculture extension program as the nutrition outreach educator to our county, she teaches "healthy eating" to local Latino families, including children. To do this, she changes ingredients to traditional Latino dishes to make them, as she says, "healthy" and even "vegan." "It's really hard being a Latina and a vegan," she tells the other one of us. "When I first came here from Puerto Rico, I ate healthy but I didn't eat apples or kale—I'd never seen kale before—but I am learning too and then teaching the kids that we can put apple and kale in quesadillas and maybe even some soy cheese...but then they aren't really Mexican anymore." She continues, "I think it helps the kids that I too am Latina and that I too am learning. If I were white the kids and their parents might feel like I was judging them and their culture...but we know that in our home countries we eat so much fruits and vegetables but here eating healthy is crazy expensive. It's not like it's our culture that's the problem—I see it as a money problem."

This last point speaks to a prevailing perception held by many whites within "alternative food institutions," that people of color don't value "healthy foods" but would "if only they knew" (Guthman 2008). Such dominant views reinforce racism and classism, as opposed to recognizing the ways in which race and class privileges and oppressions, combined with relationships to physical places and cultural spaces, shape food access, food choices, and food identities. Returning to how she has been changing Maria's recipes, she adds, "When I use soy cheese with the kids I like to tell them that we are making Mexi-*gan* quesadillas. Get it?" We do and we laugh, but we are not sure Maria would approve, as they may now be "original" but certainly not "authentic!" "What matters," she adds, "is that the kids eat them. You know I really want the kids who are born here to get the same healthy food exposure I had in Puerto Rico or that they would have if they had money...it's important to me."

In being a first-generation immigrant, like Maria, Jose, and Lula, Lula's daughter brings her cultural knowledge of healthy eating, which in her new role as a stranger is often assumed by the nonstrangers—especially within the white "alternative food" community—not to exist. This is why she is passionate about teaching healthy eating. Her focus is not on cultivating the *desire* for better, healthier foods—that already exists. Nor is her inclusion of ingredients like apples and kale about making recipes fit within the white "alternative food" health label. Rather, her focus is on working knowledgably within the confines of the food region and economy that shape many immigrant lives. In her educator role, she provides other immigrants with knowledge of what to *do* with such food items as kale and even apples while honoring cultural tastes and food expectations as well as valuing a tradition of health. She concluded our *charlas culinarias* by noting, "People try to blame us immigrants for eating bad food as if that is what we like, as if that is our culture. It's not, it's yours." We nod together, affirming our agreement.

Otros Ingredientes: Cultivating, Consuming, and Conclusion

"I hate to look at your asparagus...it brings back memories," said Jose, who is in his late fifties and is wearing a Mexican cowboy hat and black work boots. He had wandered over into our garden to borrow our wheelbarrow and saw our emerging spring crop of asparagus. Jose, who speaks near fluent English, can talk endlessly, so he begins to tell us about when he first came to the United States about 30 years ago. "When I first crossed over from Mexico,...you know I grew up on a Lemon Farm in Vera Cruz...I used to work in the asparagus fields out in Eugene, Oregon. You have to water asparagus all the time...and picking it...*hay Dios mío* (oh, my God)...it kills your back."

Jose continued, "I have been farming for over 30 years and now I finally get my own garden, even though when I get home or on the weekends I am so tired I can't do as much as I would like to." Jose drives a truck for the wholesale nursery where Maria sells her food, and during the summer months he works 10–12 hours a day. He says, "Some days I don't have time to eat but when I do, I always buy Maria's food...you know she makes it like real Mexican food." In fact, Maria hears such compliments all the time. She had one stranger say to her, "Ah, your food reminds me of the *enchiladas* of *mi Madre* (my mother)." She has even had her customers in tears: "I have even seen people crying because they have not eaten our traditional food in years," she says. Depending on the season, Maria can sell hundreds of meals in a day, and her customers never get bored because, as she likes to say, although she always cooks Mexican, it is "*comida!*" (whatever we got!), and in that way it is both *authentic* and *original*. Jose agrees, saying, "Her food is...you know...you never get tired of it. It's like home."

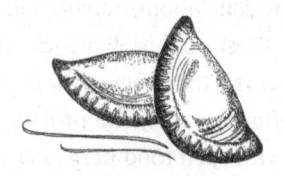

How much of our relationships to our recipes, to our lives, are held within the dreams we have as to what we will make or do, could make or could do, if we had the time, money, ingredients, skill, or opportunity? Recipes for immigrant lives are made not just from food items or from actual experiences but also from an intersection of dreams, many of which we create while we are doing those tasks that form the material stability of our everyday existences.

Maria may stand in a hot kitchen six hours a day during the week, but her dream is to "train people to take over her role," and she would like to have "a buffet style restaurant for Mexican food...¿quién sabe? (Who knows?)...Maybe it will happen."

Lula may work with Maria cooking and clearing, but her dream is to continue to feel better and to do more enjoyable things with her family using her money from working for Maria. "I went to Orlando, FL, with my daughters and grandchildren for a vacation. It made me so happy...my life has really changed," and with that, she smiled.

Jose may drive a truck up and down the New England highways, but his dream is to return to Vera Cruz in order to run his family farm. "I was born under a lemon tree," he likes to share, "I hope one day I can die under one as well."

In our dreams, the threads of our particular parts of the collective human tapestry get woven in the colors and patterns we most desire, regardless of what our actual lives may be like. In our dreams of the lives we wish to live, we are never the strangers, the others; we are always home.

Our strangers who find community together through CLiCK are, as Simmel theorizes, both "near and far," not only to us but also to each other, and even to themselves. In such a space, the distinctions between the "authenticity" of one culture and the "originality" of its ongoing time- and place-based interpretation through intersections with other cultures and our own "original selves" is both the challenge and the richness of the borderlands. However, the question of how to preserve the balance so that the dominant nonstranger culture does not demean, co-opt, or erase the other(s) is an ongoing struggle in our increasingly hegemonic and globalized world. Yet in CLiCK's shared-use commercial kitchen, the universal human threads that bind all us strangers and nonstrangers together within these stories are formed not just through the geographies of intimacy in the kitchens, gardens, neighborhoods, and streets where we have had *charlas culinarias* but also through the recipes and the culturally diverse foods whose textures, smells, and tastes we have shared and will continue to.

After all, you the reader will not meet our strangers or even the strangers and nonstrangers who are us. And even though you most likely will not eat Maria's food, you can nevertheless still find some food somewhere that has the essence of "authenticity"—Mexican or otherwise—with a pinch of creative "originality" in it. It is hoped that some of its ingredients have been grown by you or by strangers you know in the specific place where you are, or created by the same or other specific strangers in that place in order to now make it new, yet ever linked to all our collective pasts. Thereby, in that intentionally cross-pollinating moment, you, too, can become part of the millions of other disparate threads that nevertheless weave together the global recipes of *"mi cultura"* (my culture), *"tu cultura"* (your culture), and *"nosotros culturas"* (our cultures)—and hence all our foodways.

Empanadillas de Angeles Reescrito Original

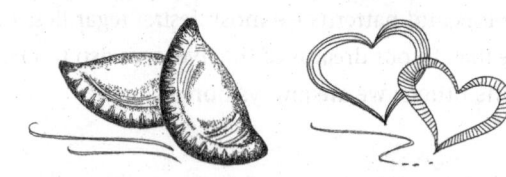

Ingredientes:

Reescrito continua...
Mucho Amor...
Mas Pasión...
Tanto Energía...
Sueños colectivos...
Suficiente de mi cultura... y un poco de tu cultura

Mezcla despacio con conciencia....
¡Buen provecho!

References

Abarca, M. E. 2004. "Authentic or Not, It's Original." *Food and Foodways* 12:1–25. https://www.tandfonline.com/doi/abs/10.1080/07409710490467589.

Anzaldúa, G. 1987. *Borderlands / La Frontera: The New Mestiza*. San Francisco: Aunt Lute Books.

Barad, K. 2007. *Meeting the Universe Halfway: Quantum Physics and the Entanglement of Matter and Meaning*. Charlotte, NC: Duke University Press.

Chen, M. A. 2000. *Women in the Informal Sector: A Global Picture, the Global Movement*. Kennedy School of Government Coordinator, WIEGO. http://www.cpahq.org/cpahq/cpadocs/module6mc.pdf.

Deb, P. 2014. "The Journey of Food from 'When Mr. Pirzada Came to Dine' to 'Mrs. Sen's' in Lahiri's *Interpreter of Maladies*." *South Asian Diaspora* 6(1): 121–135.

DePue, M. 2008. Interview with Martin Mauricio, July 20, 2008. Abraham Lincoln Presidential Library, AIS-V-L-2008–042. https://www.illinois.gov/alplm/library/collections/oralhistory/ImmigrantStories/Documents/MauricioMartin/Mauricio_Mar_4FNL.pdf.

Esquivel, L. 1989. *Like Water for Chocolate: A Novel in Monthly Installments*. New York: Doubleday.

Feuer, A. 2003. "Little but Language in Common; Mexicans and Puerto Ricans Quarrel in East Harlem." *New York Times*, September 6, 2003. http://www.nytimes.com/2003/09/06/nyregion/little-but-language-in-common-mexicans-and-puerto-ricans-quarrel-in-east-harlem.html.

Godfrey, P. 2017. "Reflexive Food-Truck Justice: A Case Study in CLiCK, Inc., a Nonprofit Shared-Use Commercial Kitchen." In *Food Trucks, Cultural Identity, and Social Justice: From Loncheras to Lobsta Love*, edited by J. Agyeman, C. Matthews, and H. Sobel, 149–165. Cambridge, MA: MIT Press.

Godfrey, P., and H. Freake. 2016. "Feeding Community: A Case Study of a Shared-Use Commercial Kitchen in Eastern Connecticut." In *Feeding Cities: Improving Local Food Access, Sustainability, and Resilience*, edited by C. Bosso, 113–128. London: Routledge.

Goldman. A. 1992. "'I Yam What I Yam': Cooking, Culture and Colonialism." In *De/colonizing the Subject*, edited by S. Smith and J. Watson, 169–195. Minneapolis: University of Minnesota Press.

Gonzales-Barrera, A and J. M. Krogstad. 2017. "What We Know about Illegal Immigration from Mexico." Pew Research. http://www.pewresearch.org/fact-tank/2017/03/02/what-we-know-about-illegal-immigration-from-mexico/.

Guthman, J. 2008. "'If Only They Knew': Color Blindness and Universalism in Californian Alternative Food Institutions." *Professional Geographer* 6(3): 387–397.

Hauck-Lawson, A. 1992. "Hearing the Food Voice: An Epiphany for a Researcher." *Digest* 12(1–2): 6–7. http://digest.champlain.edu/archive.html.

Klindienst, P. 2006. *The Earth Knows My Name: Food, Culture, and Sustainability in the Gardens of Ethnic Americans*. Boston: Beacon Press.

Lorde, A. 1984. *Sister Outsider: Essays and Speeches*. Trumansburg, NY: Crossing Press

Madison, D. S. 2012. *Critical Ethnography: Method, Ethics and Performance*. London: Sage Books.

Martin, G. 1989. *Journeys through the Labyrinth*. New York: Verso.

National Farm Workers Ministry. 2017. http://nfwm.org/education-center/farm-worker-issues/farm-workers-immigration/.

Pilcher, J. 2005. "Industrial Tortillas and Folkloric Pepsi: The Nutritional Consequences of Hybrid Cuisines in Mexico." In *The Cultural Politics of Food and Eating: A Reader*, edited by James L. Watson and Melissa L. Caldwell, 235–250. Boston: Blackwell.

Sandoval, C. 2000. *Methodology of the Oppressed*. Minneapolis: University of Minnesota Press.

Simmel, G. 2004. "The Stranger." In *Social Theory: The Multicultural and Classic Readings*, edited by Charles Lemert, 180–184. New York: Westview Press.

Taylor, P., M. H. Lopez, J. Martinez, and G. Velasco. 2012. Language Use Among Latinos. Pew Research Center. https://www.pewresearch.org/hispanic/2012/04/04/iv-language-use-among-latinos/.

Concluding Thoughts

Julian Agyeman and Sydney Giacalone

From our first musings on the need for this edited volume, our central motivation was clear. The intersection of food, policy, and immigration was brought into particularly sharp focus after the November 2016 US presidential election, and a book linking these issues seemed to us to provide something that scholarship on food studies was lacking. Individual scholars had produced compelling and valuable research on individual topics from migrant laborers to ethnic corner stores, but these emerging inquiries into what we have characterized in this volume as the immigrant-food nexus felt like separate pieces to a broader, as yet unassembled picture.

While the chapters in this volume range substantially, they come together to link macro and micro scales: from large-scale policy conversations on immigration and food to intimate immigrant foodway narratives. To talk about immigration, we must talk about food. To talk about food, we must talk about immigration. Moreover, to talk about food, we must *listen to immigrants*. This emerging picture of the multiscalar immigrant-food nexus is pushing us, our authors, and our wider fields of study and activism to raise the need for deeper, more nuanced understandings within food scholarship and the alternative food movement. In the following sections, we discuss several of the common threads running through these chapters, threads still unraveling as we begin to pull at them, revealing knots and twists to be productively developed further.

Food System "Alternatives" Are Already Here, and Immigrants Are Leading Them

Recent attention to food as a political and social topic has consistently focused on envisioning "alternatives" to our current food system. Within the largely white, majority female, middle- and upper-class body of citizens most able to assert themselves visibly as "members" of the alternative food movement, there seems to be a general assumption that going against the conventional food system is a new concept, that the growing networks of "alternative" food production and consumption are novel forms of food system participation only recently imagined, attempted, and achieved.

Our volume shows this to be an overly simplistic assumption: there are and always have been a host of diverse forms of "alternative" participation within our food system, and many of the people envisioning and sustaining them are immigrants. Schmid (chapter 8) details the innovative collective strategies Mexican American women are carrying out to remain competitive within the global capitalist food economy. Passidomo and Wood (chapter 12), Alkon and Vang (chapter 13), Huang (chapter 5), and the authors of chapter 14[1] each demonstrate how immigrants are transforming the stubbornly static concepts of "local" and "authentic" foods to reveal more productive, real meanings and relationships through translocal cuisines in intercultural spaces. These authors, Minkoff-Zern and Sloat (chapter 7), and Khojasteh (chapter 4) break the mold of addressing immigrants within the food system as solely field or restaurant workers earning the minimum or a below-minimum wage, instead showing immigrants as creative farmers and business owners successfully and imaginatively (re)molding industry norms on their own terms.

These narratives represent a clarion call to food policy leaders and actors within the alternative food movement to pay more attention to those already productively and innovatively acting out alternatives through their daily practices. It is of paramount importance to recognize the gendered positionalities of many of these key actors; immigrant women take center stage in so many of these narratives. Likewise, transnational positionality is key: immigrants are using their transnational metaknowledge to conceptualize what "alternatives" within our capitalist food economy actually look like. These immigrants' forms of food system participation are in many ways more true "alternatives" than the emerging prescriptions of much of the "alternative food movement" in that they actually provide truly alternative modes of exchange and relation between people and food while also disrupting traditional power hierarchies in food leadership along gender and citizenship lines.

Concluding Thoughts

Food as Performance, Joy, and Emotion

A second thread running through this volume concerns the complexity of the relationship between emotion and food. Our chapters resist the monolithic portrayal of immigrant experience with food as one of scarcity while also resisting the tokenistic portrayal of immigrant foodways as existing in a bubble of cultural celebration. Rather, we see the way food plays a vital role in the emotional experience of immigrants navigating moments of togetherness, separation, joy, fear, and determination. Ortiz Valdez (chapter 9) shows women and men on New York dairy farms experiencing joy and belonging through food at the very time they are experiencing increasing food insecurity and fear for their physical well-being due to increasing stares from locals and a heavier ICE presence. The authors of chapter 14 detail how, while dealing with depression and anxiety, Lula found comfort in the intercultural bonds she made while cooking in the CLiCK kitchen. Curtis (chapter 1) reveals how small community acts of care toward sequestered, undocumented farmworkers can be radical forms of resistance to an increasingly militaristic state. Within conditions structured to be inhospitable, even unlivable, "these embattled communities are seedbeds for the regeneration of democratic practices, energies, values and visions. And it is perhaps in such locations that the social webs critical to farmworker demands for justice are being woven" (Curtis, chapter 1).

Immigrants use food as a way to perform their transnational identities and ongoing relationships with faraway people and places. With the increasing trauma felt by immigrant families in the United States today—family separation and indefinite detention at the border continue as we write this—these performances through food reveal themselves as a way immigrants are practicing radical self-care. To truly take on immigrant food nexus within food system studies and reform, activists and policymakers must directly recognize mental health as an area of paramount importance in conversations about what food means in the daily lived experiences of immigrants.

Culturally "Appropriate" Food Is Complex

Recent food scholarship holds up access to culturally appropriate foods as one of the highest ideals in food system thinking. While this volume likewise celebrates culturally meaningful foods and the benefits they provide to immigrant communities, many of these chapters push us to complicate the simplistic idealization of the concept.

Within scholarship and policy initiatives on access to culturally appropriate foods, there is often evidence of a white spatial imaginary:[2] "neutral" (read: white) foods are taken as the most common and accessible foods, while "other" foods (read: foods associated with people of color) are taken as the exception, the exotic or authentic but not the usual or commonplace.

Several of our chapters disrupt what may be better described as a spatioculinary imaginary. In Passidomo and Wood's chapter, Argentina, the owner of La Sabrosita Bakery, states, "Here in Richmond, I don't think we missed any of the food from our country. We looked for other foods just to have a variation but any food that we want to have is here because I see it here in the bakery." In a historically southern city where it may seem logical to assume the Latinx community would have difficulty accessing Latinx foods, Argentina paints a picture of abundance of access and intercultural relationships formed through her El Salvadoran recipes. Likewise, Alkon and Vang, Khojasteh, and the authors of chapter 14 demonstrate scenes and relationships saturated with culturally meaningful foods that are not exotic or a rare find but rather the normal everyday way food is experienced and performed.

These chapters also ask us to question exactly what culturally appropriate food is, who defines it, and where one may find it. Common narratives within the alternative food movement and more critical scholarly and political initiatives include the assumption that these foods are made from "traditional" ingredients, found in small ethnic corner stores or grown at home, and not the "standard" ingredients popular in either big box stores or popular alternative points of sale, such as urban farmers markets. These common narratives prove to be a limiting and often false way to understand what constitutes cultural foodways in many immigrants' lives. Ortiz Valdez shows that the point of sale that makes chorizo, queso fresco, and corn husks accessible to women on rural New York dairy farms is not some ethnic market or local vendor but instead the Wal-Mart international aisle. The authors of chapter 14 show hybridity in the classification of "traditional" foods within everyday food economies: kale, apples, and vegan cheese become "Mexi-gan" quesadillas. Several authors show how food can take on cultural meaning by way of the techniques and labor put into making it: women follow Guatemalan steps of plucking a chicken in New York (Ortiz Valdez), and a lasagna becomes Hmong through the hands that make it (Alkon and Vang). In this way, hybridity and adaptation both fuel and deepen the rich meanings of immigrant foodways just as importantly as tradition and connection to past homes and lives.

Finally, these chapters reveal that immigrants' access to culturally appropriate foodways is embedded within complicated power hierarchies. Joassart-Marcelli and Bosco (chapter 3) detail how ethnic food is taking on both meanings of community belonging and meanings of foodie status and tokenized authenticity within gentrification. The analyses of Linton (chapter 10) and Minkoff-Zern and Sloat provide historical contexts to how US and Canadian narratives of immigrant farmers and the foods they cultivate—even within "inclusive" government programs—have always racialized, exoticized, and labeled immigrants and their foodways as "other," and continue to do so. Ortiz Valdez asks us to consider these historical power hierarchies within current anti-immigrant sentiment: how can we acknowledge the mental health benefits farmworkers and their families gain from access to culturally meaningful food, but also acknowledge that the context of this access is increasingly secluding this already marginalized community onto the farm?

Conversations about Immigration, Food, and Change

It is important that we remain wary of the assertion that the current climate of anti-immigrant and xenophobic sentiment is particularly "new" or will certainly inspire progressive action toward change. Minkoff-Zern and Sloat, Curtis, and Linton add to the already significant body of literature on the ways food has always been a racial project, a tool through which colonialism, racism, and xenophobia operate. Neubert (chapter 2) provides a particularly timely analysis of these dynamics at play today, urging us to question how the concepts of agricultural success, environmental protection, and thriving food economies have been and continue to be mobilized as coded concepts embedded with racial hierarchies of labor and power over a community's self-determination.

With this context in mind, what lessons can we take from this volume that may help these conversations lead to real forms of change rather than perpetuating temporary attention and continued structural, physical, and emotional violence toward immigrants?

Research on the Immigrant and the Employer

With the increasing importance of critical scholarship on immigration, this volume points to a research area that deserves particular attention: immigrants and their employers. Several of our authors point to how recent anti-immigrant sentiment and policy has thrown these relationships into new territory. Ortiz Valdez presents a growing reality on many US farms; anti-immigrant policies and ICE violence force immigrant laborers to stay further confined on the farm and force their employers to implement new ways of keeping their laborers on the farm, increasing the paternalistic relationship they have to these people but also becoming increasingly dependent on their choice to stay. Dentzman and Mindes (chapter 6) show that with increasing herbicide resistance, grain farmers are grappling with the increasing shortage of immigrant laborers in the United States, which previously was an issue felt less by the grain industry than its fresh fruit and vegetable counterparts. Minkoff-Zern and Sloat, Schmid, Joassart-Marcelli and Bosco, Khojasteh, and Passidomo and Wood all discuss situations in which employers are themselves immigrants, hiring workers that include other immigrants.

With the recent rise of anti-immigrant sentiment, policies, and enforcement tactics, it is critical to study the relationships between immigrants and employers to assess how the industries that rely on immigrant labor respond to this political climate. Will this be in the ways they historically have—behind closed doors, through lobbying, and nearly always in order to continue to subsidize their profits through immigrant labor and cultural capital, or in ways that speak to newly humanizing alliances, such as the Restaurant Opportunities Center United (ROCU), which advocates for "high road" policies, resulting in employers speaking vocally for the rights of immigrants' presence, fair payment, and protections in the United States and Canada?

Tangible Policy Implications

Our authors point to tangible policy implications that may push for these new types of alliances. Joassart-Marcelli and Bosco, and Khojasteh, tell stories of the intertwined relationship between urban gentrification and capitalized immigrant foodways. Khojasteh expertly points to a gap that this shows in many cities' recent assertions of themselves as "sanctuary cities": simply protecting immigrants from ICE actions through the sanctuary city label, mayors are being reactive rather than proactive in considering what the immigrant experience is or could be in their city. Taking an immigrant-food nexus perspective, we might ask these cities, "So you're a sanctuary city, but are you actually a sanctuary?" The immigrant-food nexus speaks to how generative the area of food could be for mayors to enact progressive policies (such as immigrant use of abandoned land and agricultural spaces, and support for ROCU) that proactively build the belonging of immigrants into the city's political and economic conversations and spaces.

Ostenso, Dring, and Wittman (chapter 11) point to the key to making these types of policies a reality in both US and Canadian contexts: food policy actors must recognize that a racial justice lens is fundamental, not supplemental, to their mission for food system change. Despite the vagueness on food and farming in recent US proposals such as the Green New Deal, we see numerous possibilities for progressive policies here: prioritize the voices and decisions of immigrant representatives in planning decisions on current and new local food centers such as farmers markets or city events (see Joassart-Marcelli and Bosco, and Khojasteh); bolster immigrant farmers' ability to capitalize on the local food economy by explicitly including the concept of translocal food in city and business local food narratives (see Schmid; Huang; Linton; Passidomo and Wood; Alkon and Vang; and the authors of chapter 14); expand available support for immigrant farmers and businesses with the express purpose of reducing barriers to their success embedded within national, state, and local bureaucracy (see Minkoff-Zern and Sloat; Schmid; and Linton); and finally, build state and city support for organizations that are already doing work in this area, such as ROCU, the Coalition of Immokalee Workers, and immigrant-focused law groups. This should include the use of current popular local and sustainable food platforms to assert that immigrants' rights to *belong* and *thrive* are fundamental to what food system reform can *become*: a vital and growing part of a larger movement for environmental, racial, and economic justice.

Volume Summary

Throughout this volume, we have sought to show how a focus on immigration policy and immigrant foodways must be prioritized by the alternative food movement, food scholars, and policymakers. We have likewise shown that national conversations on immigration must become informed on the intersectionalities between this policy arena and all areas of food, from the field, to the restaurant, to the home. Our authors show the importance of humanizing, multiscalar, and comprehensive scholarship that resists the tired binaries that reify so many false narratives and incomplete understandings plaguing current conversations about the "place" of immigrants within our nations. Through ground-up narratives coming out of the immigrant-food nexus, we hope to catalyze more productive and transformative conversations.

Concluding Thoughts

1. These authors have asked to remain anonymous in solidarity with the immigrant participants of this volume, many of whom must remain nameless for their security.

2. Lipsitz's theorization of the white spatial imaginary details the ways in which white Americans understand place based on years of living in illegally segregated spaces, creating a worldview that idealizes homogeneity and remains ignorant of the histories of discrimination and structural racism that have created such segregation (Lipsitz 1998).

Reference

Lipsitz, George. 1998. *The Possessive Investment in Whiteness: How White People Profit from Identity Politics*. Philadelphia: Temple University Press.

Contributors

Julian Agyeman is professor of urban and environmental policy and planning at Tufts University.

Alison Hope Alkon is associate professor of sociology at the University of the Pacific.

Fernando J. Bosco is professor in the Department of Geography at San Diego State University.

Kimberley Curtis is senior lecturer, sustainable communities program, at Northern Arizona University.

Katherine Dentzman is a postdoctoral research associate at the University of Idaho.

Colin Dring is a PhD candidate at the Centre for Sustainable Food Systems, University of British Columbia.

Sydney Giacalone is program associate at The Philanthropic Initiative.

Phoebe Godfrey is in the Department of Sociology at the University of Connecticut.

Sarah D. Huang is a PhD candidate in the Department of Anthropology and Ecological Sciences & Engineering at Purdue University.

Pascale Joassart-Marcelli is professor in the Department of Geography at San Diego State University.

Maryam Khojasteh is a doctoral candidate in the Department of City and Regional Planning at the University of Pennsylvania.

Jillian Linton is a master of arts candidate at the University of Toronto.

Samuel C. H. Mindes is a research associate at the University of Idaho.

Laura-Anne Minkoff-Zern is assistant professor of food studies at Syracuse University.

Christopher Neubert is in the Department of Geography at the University of North Carolina at Chapel Hill.

Fabiola Ortiz Valdez is a PhD candidate in the Department of Anthropology at Syracuse University.

Victoria Ostenso holds a master of science from the Centre for Sustainable Food Systems, the University of British Columbia.

Catarina Passidomo is assistant professor of anthropology and southern studies at the University of Mississippi.

Mary Beth Schmid is affiliated with Western Carolina University.

Sea Sloat is an independent scholar.

Dianisi Torres is a nutrition outreach educator at the Expanded Food and Nutrition Education Program, University of Connecticut, College of Agriculture and Natural Resources.

Kat Vang is an independent scholar.

Hannah Wittman is a professor at the Centre for Sustainable Food Systems, University of British Columbia.

Sara Wood is in the Southern Oral History Program at the University of North Carolina at Chapel Hill.

Food, Health, and the Environment

Series Editor: Robert Gottlieb, Henry R. Luce Professor of Urban and Environmental Policy, Occidental College

Keith Douglass Warner, *Agroecology in Action: Extending Alternative Agriculture through Social Networks*

Christopher M. Bacon, V. Ernesto Méndez, Stephen R. Gliessman, David Goodman, and Jonathan A. Fox, eds., *Confronting the Coffee Crisis: Fair Trade, Sustainable Livelihoods and Ecosystems in Mexico and Central America*

Thomas A. Lyson, G. W. Stevenson, and Rick Welsh, eds., *Food and the Mid-level Farm: Renewing an Agriculture of the Middle*

Jennifer Clapp and Doris Fuchs, eds., *Corporate Power in Global Agrifood Governance*

Robert Gottlieb and Anupama Joshi, *Food Justice*

Jill Lindsey Harrison, *Pesticide Drift and the Pursuit of Environmental Justice*

Alison Hope Alkon and Julian Agyeman, eds., *Cultivating Food Justice: Race, Class, and Sustainability*

Abby Kinchy, *Seeds, Science, and Struggle: The Global Politics of Transgenic Crops*

Vaclav Smil and Kazuhiko Kobayashi, *Japan's Dietary Transition and Its Impacts*

Sally K. Fairfax, Louise Nelson Dyble, Greig Tor Guthey, Lauren Gwin, Monica Moore, and Jennifer Sokolove, *California Cuisine and Just Food*

Brian K. Obach, *Organic Struggle: The Movement for Sustainable Agriculture in the U.S.*

Andrew Fisher, *Big Hunger: The Unholy Alliance between Corporate America and Anti-hunger Groups*

Julian Agyeman, Caitlin Matthews, and Hannah Sobel, eds., *Food Trucks, Cultural Identity, and Social Justice: From Loncheras to Lobsta Love*

Sheldon Krimsky, *GMOs Decoded: A Skeptic's View of Genetically Modified Foods*

Rebecca de Souza, *Feeding the Other: Whiteness, Privilege, and Neoliberal Stigma in Food Pantries*

Bill Winders and Elizabeth Ransom, eds., *Global Meat: The Social and Environmental Consequences of the Expanding Meat Industry*

Laura-Anne Minkoff-Zern, *The New American Farmer: Immigration, Race, and the Struggle for Sustainability*

Julian Agyeman and Sydney Giacalone, eds, *The Immigrant-Food Nexus: Borders, Labor, and Identity in North America*

Index

Abarca, Meredith E., 249, 284
Academic achievement, 35–36, 143, 148
African American farmers, 146, 147
African American labor, 247
Agrarian class mobility, 142–144, 147, 153, 156, 175
Agribusiness: global labor locations/shifts, 46, 161, 165–166; immigrant labor and, 6–7, 10, 41–43, 45–47, 48–55, 143, 144–145, 151, 152, 161; laborer quotations, 152, 161; US policy history, 10, 143, 144–145, 154, 176, 266
Agricultural labor: Canadian immigrants, 5–6, 120–121, 211, 215–217; climate change effects, 22–23; community food systems, 83, 89, 94, 107–110; domestic labor, 10–11, 183–197; family enterprises, Appalachia, 10, 161–177; family farms, 110, 142, 143, 144, 150, 161–177, 211, 214, 220; farmer independence, 148, 149–150, 152, 188; "farmers" and "farmworkers" terminology, 147, 161–162, 290; farm settlement, 212–213, 217, 219; guest workers, 25, 29, 125; immigrant farmers and support programs, 141–156, 176, 217–219, 265–266, 303; immigrant labor, and border security, 21, 22–38, 122–123; immigrant labor, costs, 6, 23, 24, 28, 119, 120, 121–122, 124, 127, 131–133, 150, 184, 218–219, 220; immigrant labor, descriptions, 5, 21, 25–26, 28–29, 30, 109–110, 117, 119–120, 124–125, 130–131, 150, 161, 211–212, 215–219, 220, 293–294; immigrant labor, food scholarship, 1–12, 117–118, 121, 125, 128–135, 143–156, 175, 194, 304; immigrant labor, statistics, 6, 21, 23, 25, 28–29, 119, 122, 161, 184; immigrant labor scholarship, 1–12, 117–118, 121, 125, 128–135, 143–156, 175, 194, 304; immigrant travel, 21, 25–27, 29, 30, 32–33, 127; mechanization and effects, 119–120, 121–122, 131–135; North American dependency on immigrant labor, 1–2, 5–6, 10, 46, 53–54, 83, 119–120, 121–128, 130–131, 184, 217–219, 304; refugee workers, 89, 94, 102, 103, 107–110; state economic change, 41, 46, 104, 246–247, 248–249; surveys, 122, 184; valuation and visibility, 23, 24, 25, 30, 33–34, 53–54, 117, 212–213, 217–218. *See also* Agribusiness; Immigrant labor; Meat processing industry
Agricultural price supports, 145

Agriculture in cities, 65, 213, 215, 216, 230, 290–291
Agriculture programs: Alaska, 102, 103–104, 107–109; needs for more, with greater equity, 175, 265–266; USDA, 10, 141–156
Agriculture sustainability, 102, 104, 146–147, 166
Agriprocessors Inc., 45–47, 47–48
Agyeman, Julian, 1–12, 299–306
Alaska, 10, 102, 103–105
Alkon, Alison Hope, 11, 206, 219, 261–276, 300, 302–303
Allport, Gordon, 45
Alternative food movement, 3, 203, 220n1, 299–300; imaginaries, and white identities, 4, 206, 229, 230, 293, 300; immigrant-led aspects, 117, 145, 148, 165, 174–175, 300, 305; local food movement within, 100, 209
Alt-right movement, 2
Alvarez, Steven, 258n1
American Civil Liberties Union (ACLU), 33
"American dream," 256–257
American identity: American South, 247, 249; populism, nationalism, and xenophobia, 19, 23, 45, 255, 265; rural areas, 44–45; transnational identities, 11, 247, 249–258, 276
American values, 37, 45
Anchorage, Alaska, 10, 99–110
Angolans and Angolan food, 99–100
Antigentrification efforts, 75–76
Anzeluda, Gloria, 284, 285
Appalachian agriculture, 10, 161–177
Arellano, Gustavo, 245, 246
Arendt, Hannah, 34
Arizona, employment law, 32
Arpaio, Joe, 22, 23, 46
Asian cuisines, 262, 263, 269–271, 273–274, 275

Asian immigrants: American South, 11, 248, 254–255; farmers and farming, 215–216, 265–266, 272–273; Hmong people and foodways, 261–276; immigration policy and migration, 25, 90, 228; small businesses, 65
Asparagus, 293–294
Assimilation: American expectations, 91, 210; dietary, 263, 272; Hmong, and academic studies, 264–265, 267, 274–275
Asylum seeking, 36. *See also* Refugee populations
Australia, agriculture, 135
Authenticity: vs. adaptation, immigrant foodways, 251–252, 263, 266–267, 270, 286, 292, 293, 302–303; "authentic" cooking, 251, 285–287, 291, 292, 294, 302; ethnic foodscapes and appropriation, 4, 10, 60, 61, 70–71, 72–73, 74–75; vs. originality, 284, 293, 295–296; unique products, 193, 286

Bakeries, 252–257, 302
Bankston, Carl, 247–248
Barad, K., 285, 289
Barrio Logan neighborhood (San Diego), 62–74, 64f, 67f, 69f
Belonging. *See* Citizenship; Geographic borders; Identity; "Others" and outsiders
Bestor, Theodore, 164
Betancur, Manolo, 253
Bhutanese refugee immigrants, 102, 107–109
Binary configurations: food and health, 100–101, 282; racial, 228, 237, 238, 247, 255
Black/white binary, 228, 247, 255
Body politics, 44

Index

Borderlands: and "beyond the border," 48–53, 249; border security and immigrant labor, 9, 21, 22–38; demographic enclaving, 29–30; and *la frontera* (Anzeluda concept), 284, 288, 292; as sites of struggle in world-making, 24, 34–35, 37–38; spatial segregation, 44, 47; Yuma area interviews, 24, 29, 30–37

Borders and boundaries. *See* Borderlands; Border security; Geographic borders; Walls and wall-building; specific border regions

Border security: narratives and tropes, 47; nativist regimes, 6, 31–34, 37; personal accounts, 31, 195; US history and laws, 2, 6, 22, 25–28, 31, 123. *See also* US Border Patrol; US Customs and Border Protection

Border towns/areas: economic conditions, 28–30, 30t, 85–86, 248; racial demographics, 29, 30

Bosco, Fernando J., 9–10, 59–77, 303, 304

Bourdieu, Pierre, 62

Bracero Program (1942–1964), 5, 26, 121

Brandstad, Terry, 42, 49

Brandt, Deborah, 121

Brown, Wendy, 23

Budech, Keiko, 194

Buffalo, New York, 10, 82, 84–89, 93–94

Bureaucracy, farming support programs, 143, 145, 148, 152

Business associations, 93, 95

Business financing: immigrant farmers, 150, 151–152, 153, 155–156, 172; money lending, 173; payment systems, 171–172; small ethnic businesses, 67, 75

Business incubators, 287

Butz, Earl, 144–145

California: contested ethnic foodscapes, 9–10, 59–60, 62–76; farmers markets, 206; Hmong American foodways, 261–262, 266; immigrant agricultural labor, 148; urban food councils, 226

Canada: agricultural labor/policy, 5–6, 10, 120–121, 125–127, 133–134, 215–219; Canadian identity, 206–207, 209, 210–212, 216; immigration policy, 5–6, 125–128, 207, 209–210, 226, 228; local food policy and attitudes, 11, 205, 206, 207–209, 212–213, 225–240; Ontarian multicultural agrarian narratives, 11, 205–220; populations, 126, 134, 206–207, 209–212, 226, 227–229, 236; Vancouver food systems and inclusion, 11, 225–240

"Canadian dream," 211–212, 215–216, 219

Capacity building: Alaskan agriculture, 104; immigrant communities, 21, 36–37, 254

Capitalist systems: agricultural food production, and race, 43, 54; alternatives, considerations, 165, 168–169; foodscapes and gentrification, 60, 68, 70, 71, 75, 76, 93; immigrant farming management, 2, 162–163, 165, 166, 167–171, 176; immigrant labor value, 82–83, 119

Carr's (store), 106

Catering services, 87, 291–292

Census and population data: American South, 248; Canadian metro areas, 227–229; economic conditions, border towns/areas, 28–30, 30t, 85–86, 248; food business locations, 62, 86, 245; immigrant populations, US, 1, 63–64, 64f, 86, 102, 245, 247, 248, 254, 264; Latino farmers, 142; small town size, 44

Central Intelligence Agency (CIA), 264
Charlotte, North Carolina, 252–255
Chemicals. *See* Herbicides use
Chicago, Illinois, 81
Chicken, 190–191, 269–271
Child care, and domestic work, 187, 188–199
Chinese Exclusion Act (1882), 25
Chinese history, 264
Chomsky, Aviva, 27
Christian values, 36
Chuh, Kandice, 265
Circular migration, 25, 125
Citizenship: American paths, 26, 32, 253; Canadian paths, and belonging, 207, 210, 216, 219; family relations, 31–32; food citizenship, 7, 225; noncitizen human rights, 6–7, 27, 124, 126, 127, 184; US history and race, 145–146, 265. *See also* Permanent residency status
City Heights neighborhood (San Diego), 62–74, 64f, 67f, 69f
Civic worlds and world-making, 9, 24, 34–35, 37
Civil and civic associations: food policy councils, 4, 229–230, 233–234; social ties and community building, 34–35
Civil rights policy, USDA, 142–143, 146–147, 154–156
Class distinctions: changing ethnic foodscapes, 10, 74–75, 107, 275; farm owners and operators, 142–144, 147, 153, 156, 161–162, 175; food and taste judgments, 62, 65, 68, 70–71, 73, 107, 292–293
CLiCK, Inc., 282, 286–292
Climate change: agriculture systems, 102; food transportation and, 209; human migration, 22–23
Coalition identities and power, 169–171
Cobas, Jose, 255

Collective strategy and cooperative practices: barriers, 168; farming enterprises, 10, 162, 163–164, 165, 168–176
Colonialism: Canada, 210–211, 212, 220, 226–227; race: white colonists and visible minorities, 210–212, 227, 237
"Color-blind" and "race-neutral" discourse and policies, 210–212, 227, 230–231, 233, 234–235, 238
Comercio mētis, 162–165, 166–168, 169, 174–175, 176–177
Comfort food, 182, 191, 195–196, 270–271, 294, 301
Commercial kitchens, 11–12, 250, 281–282, 286–292
Commodity crops: agricultural policy, 144–145, 176; grain crop sharing, kin groups, 173–174; labor and herbicide resistance, 10, 119–120, 125, 131, 132, 133–134, 304
Commodity paths, 164
Communication and language. *See* Language and communication
Community and community-building: combatting isolation, 34–35, 183, 289–290, 301; ethnic foodscapes' value, 65–66, 67–68, 73, 91–92, 183, 250, 254–255, 257–258, 262; immigrant communities, 10, 34–36, 66, 67–68, 82–83, 86, 87, 90, 175, 290–293; immigrant farming enterprises, 10, 150–151, 162, 165, 166, 174–177; immigrant food enterprises, and "revitalization," 81–95, 254; seeking commonalities, 36–37, 245, 257–258; small business investments, 75–76, 85
Community gardens, 108, 109–110, 230, 290
Community shared agriculture (CSA), 213
Commute times, labor, 21, 29
Contact hypothesis of prejudice, 45

Contractor systems, labor, 26, 30, 124
Cooking. *See* Home and traditional cooking; Recipes
Cooking from the Heart: The Hmong Kitchen in America (Scripter and Yang), 269, 270, 274
Cooperative practices and collective strategy: barriers, 168; farming enterprises, 10, 162, 163–164, 165, 168–176
Co-op markets, 76
Corn industry and growers, 128, 129*t*, 145
Cosmopolitanism and "foodie" culture, 10, 68, 74–75, 91, 93, 212, 303
Court cases/decisions: immigrant detainees, 27; immigration bans, 2–3
Coutin, Susan, 32
Cravey, A., 249
Crime rates, 28, 51, 52
Criminal economies, border enforcement and effects, 22, 26–27, 31
Criminalization: immigration discourse, 47; immigration policy, 9, 23–24, 26–28, 31, 37
Crisis "manufacturing," 44
Critical race theory, 227, 255, 265
Crop sharing, 173–174
Cultural appropriation. *See under* Whites and whiteness
Cultural assimilation. *See* Assimilation
Cultural foodways. *See* Foodways
Cultural inclusion, food systems, 225–240
Curtis, Kimberley, 9, 21–38, 301, 303

Dairy industry: gender, food, and labor, 10–11, 181–197; labor law, 197n5; working conditions, 181, 184–186
Daniel, P., 146
Deferred Action for Childhood Arrivals (DACA), 2
DeMaster, Kathryn, 273

Demographics: border towns/areas, 29–30, 30*t*; foreign-born populations, 63–64, 64*f*, 86, 102, 245, 247, 248, 254, 264; small town definitions, 44–45
Dentzman, Katherine, 10, 119–136, 304
Deportation: current US immigration policy, 1–2, 6, 28, 31, 32; immigrants' situations and fear, 2, 6, 31, 32, 126, 185, 191, 193; US history, 23, 25–27, 28
DePue, M., 286
Detention and detention centers: immigrant rights, 27; immigrant workplace raids, 46–47, 195
Dietary acculturation, 11, 263, 266–267, 273–274, 292
Direct sales, immigrant farmers, 10, 143, 144, 147, 150–151, 166
Discourse analysis, 208
Displacement, land, 146
Diversity. *See* Multiculturalism and diversity
Documentation, farming, and USDA support, 143, 145, 148, 152
Domestic labor, 10–11, 182–183, 185–197, 287, 288
Domestic terrorism, 2
Downtown development programs, 64, 86
Dring, Colin, 11, 225–240, 305
Drought effects, 22–23
Due process rights, 27
Duke, Elaine, 22
Duncan, Dorothy, 210–211
DuPuis, E. Melanie, 246

Eagle Grove, Iowa, 41–43, 48, 55
The Earth Knows My Name: Food Culture and Sustainability in the Gardens of Ethnic Americans (Klindienst), 290
EBT payment system, 66

Economic growth: government programs, 81, 83, 86, 88–89, 94, 104; immigrant communities' generation, 81–83, 84–86, 92–93, 248, 254; state economies, 41, 46, 104, 246–248, 248–249, 254
Economies of scale, 169–171, 172–173
Edge, John T., 250
Education: English-language, 50–51, 90; immigrant families, 35–36, 143, 269; immigrant farmers, 143, 148, 152, 154
Elections: Canada, 2011, 205; United States, 2016, 43, 48, 123, 181, 299
Empanadillas, 282, 287, 296
Employment and labor law. *See* Labor regulations
Employment Equity Act (Canada; 1995), 240n1
Energy costs, 104
English language: education, 50–51, 90; forms, 153–154; recipes, 282–283; skill and barriers, 90, 109–110, 143, 152–154, 195, 264, 291–292
Entrepreneurs. *See* Small businesses
Environmental issues: agribusiness and public opinion, 42, 49, 51, 53; food policy council work, 229, 230, 236; food production, and environmental justice, 54, 55; herbicide resistance, 119–135, 304; local food benefits, 209
Esquivel, Laura, 285, 288
Ethnic foodscapes: appropriation, gentrification, and resistance, 9–10, 59–60, 68–76, 93; descriptions, 7, 61, 65–71, 73, 86–87, 89–93, 105–106, 302; food locality and familiarity, 10, 99–110; immigrant enterprises, and "revitalization," 10, 81–95, 254–255; San Diego neighborhoods, 9–10, 62–76, 69f; use value, in immigrants' lives, 65–68, 86–87; use value vs. symbolic value, 61–62. *See also* Foodways
"Ethnic markets": constructions and defining, 59, 107; economic conditions, 66–68, 70, 73–74, 83, 90; ethnic diets and, 84, 88, 91–92, 106, 273; food business locations and offerings, 65–66, 67f, 69f, 83, 84–95, 106; wholesale buying, 88, 89
European immigration: Canada, 209, 210–211, 216, 227; United States, 89, 90–93, 247
E-Verify programs, 123, 136n1
Executive orders, 28

Factory farming, 41–42
Faidman, Anne, 265
Fairclough, Norman, 208
Fair Labor Standards Act (1938), 21, 197n2
Familiarity, foodscapes, 99–100, 101, 105–107, 109–110
Family farms: Canadian, 211, 214, 216–217, 220; economic conditions, 150; Hmong growers, 272–273; immigrants, 110, 142, 143, 144, 150, 162, 170, 176–177, 211, 272–273; Mexican American farming families, 161–177; romanticization, 213–214
Family relations: citizenship issues, 31–32; cultural food connections, 182, 196, 270–271, 273, 294; immigrant family structures, 268–269; immigrant stories, 251, 253–254, 256, 261–262, 288; youth education, 35–36
Farm Bill, 145
Farmers. *See* Agricultural labor
Farmers markets: within food industry, 166; immigrants' direct food sales, 10, 143, 144, 147, 150–151, 166; *marketeras* in Appalachia, 162,

Index

165–168; and white imaginaries, 75, 100, 107–109, 206, 212–213
Fascism: body politics, 9, 44; immigration policy and, 25, 26, 27, 37; social preconditions, 26, 34
Feagin, Joe, 255
Fear: immigrant lives, 2, 6, 31–34, 176, 181, 185, 191, 192, 193, 195, 258, 304; political exploitation, 44, 48, 252, 258; white displacement fears, 19, 23, 53–54, 252, 255
Festivals, 30
Financing, business. *See* Business financing
First Nations peoples, 210–211, 212–213, 228, 236–237
Florida, 248
"Food citizenship," 7, 225
"Food deserts," 10, 64–65, 66, 73, 85–86
"Food from home." *See* Home and traditional cooking
"Foodie" culture. *See* Cosmopolitanism and "foodie" culture
Food Justice Working Group (Vancouver), 229, 231–233
Food labeling, 104–105
"Food maps," 183, 190, 193, 195–196
Food policy councils, 4, 11, 225–226, 227–240
Food programs: for farmworkers, 197, 266; healthy food sales, 88–89; immigrant businesses partnerships, 94; local food production, 65, 94, 99, 102; need for more, with greater equity, 176, 305. *See also* Agriculture programs; Food security
Foodscapes. *See* Ethnic foodscapes
Food security: Alaska, 99, 102, 104; Canada, 227, 233–235; immigrant businesses, 84, 266; local food, 99, 209; migrant workers, 7, 126; policy and action programs, 99, 233–235, 236, 266; urban agriculture, 65

Food studies research: food business locations, 62, 65–66, 67f, 83–95, 249–258; food service labor conditions, 82; gaps and shortcomings, 299; geographic borders and, 19; immigrant agricultural labor, 1–12, 117–118, 121, 125, 128–135, 143–156, 175, 194, 304; immigrant food practices, 263, 264–274; journalism reviews, 207–208, 209, 210, 211–220; local food systems and exclusion, 4, 10, 99–110, 205–209, 212–213; public health and, 82, 83–84; use/symbolic value of food, 61–62
Food systems: cultural inclusion, Vancouver, 225–240; food justice and, 3–4; locality, 99–110, 225; planning, 4, 11, 225–226, 227–240; racialization effects, 227; United States, 3–4
"Food voice," 284, 288–289, 290
Foodways, 1, 7; adaptation and blending, 251–252, 263, 266–267, 270, 275, 282, 292, 293, 302–303; immigrant-food nexus, 3, 7–9, 299, 305–306; immigrant lives and importance, 2, 7, 11, 203, 245–246, 269, 301, 302–303; "local food" movement and, 101, 212, 220; of recipes, 281; southern US, 11, 245–246, 249–258; transnational Hmong foodways, 261–276
Forgery, 27, 31
For-profit detention centers, 27
Forth, Ken, 211
Franzen, Lisa, 266
Free trade agreements, 5, 126, 288
Freidberg, Susanne, 164
Fresh fruit and vegetable industry, 150–151, 162, 165–168, 171–172. *See also* Produce
Friedland, William H., 121–122, 131–133, 134, 166

Fruits and vegetables. *See* Fresh fruit and vegetable industry; Produce
Fufu, 99

Gabaccia, Donna, 252
Garcia, Matt, 246
Gardening, 108, 109–110, 230, 270, 290, 294
Gender, food, and labor: dairy industry, 10–11, 181–197; domestic labor, 11, 182–183, 185–197, 287, 288; gender assumptions of labor, 162, 183, 194, 257; gender discrimination, farm programs, 142; sexual power dynamics, 189–190, 194; women and traditional mētis, 164, 174, 288; women-led farming enterprises, 10, 161–177; women-led restaurant enterprises, 249–257, 285–287
Gentrification: defined, 60; ethnic foodscapes, 9–10, 59–60, 68–76, 91, 93, 109–110, 303; resident responses and resistance, 72–73, 75–76, 93
Geographic borders: bodies, and bordering practices, 44; border security and immigrant labor, 21, 22–38, 123; crossing points and crossers, 22, 23–24, 31, 248; food studies research and, 19; legislation, 22; "local food" within, 100, 104–105, 208, 209, 213; maps, *29*; meanings: belonging, 6, 19, 32. *See also* Borderlands; Yuma, Arizona
Geopolitics, 43–44
Gerlach, S. C., 104
Giacalone, Sydney, 1–12, 299–306
"Global crops," 176
Gökariksel, Banu, 44
Goldberg, Anna, 266
Goldman, A., 286
Goodlatte, Bob, 123
Grading, produce, 167–168

Grain crops: crop sharing, kin groups, 173–174; labor and herbicide resistance, 10, 119–120, 125, 131, 132, 133–134, 304
Green cards, 31, 32, 33, 161, 195
Grocery stores. *See* "Ethnic markets"; Supermarkets
Guatemalan immigrants: dairy industry workers, 182, 186–192, 196; meat industry workers, 45–46
Guest worker programs, 5, 25, 29, 123–125, 133–134, 197n5

H-2A and H-2C programs, 6, 123–125, 133, 197n5
Hambly, John, 214
Hanchett, Tom, 254–255
Harrington Seed Destructor, 135
Hauck-Lawson, Annie, 284
Health and nutrition, food: diet acculturation, 11, 263, 266–267, 273–274, 292; ethnicity and diet, 84, 87–88, 91–92, 106, 109–110, 187–188, 230–231, 263, 266–267, 269–270, 271, 273, 275, 292–293; food costs, 84, 88–89, 188, 292, 293; food policy councils, 230–231; government supports, 88–89, 94, 205; guidelines, 84; interventions, 266–267, 275, 292–293; "local food," 4, 100, 205, 209; measurement, 87; organics, 68, 70; store types and locations, 83–84, 85–88, 93–94; taste judgments and class, 68, 70–71, 73, 107, 292–293
Herbicides use: avoidance, 149; herbicide resistance and labor, 10, 119–121, 122, 128–135, 304
Hickman, Jacob, 265
Hmong people: academic literature, 264–268, 274–276; agriculture, 265–266, 272–273; history, 263–264; traditional foods, 262, 263, 269–272,

Index

273–274, 276n1; transnational foodways, 11, 261–276, 302–303
Holidays and celebrations: Hmong New Year, 261, 271; Thanksgiving, 261–262, 267; World Food Day, 209
Home and traditional cooking: cultural and emotional power, 7, 182, 183, 186, 187, 190–191, 195–196, 197, 203, 269–271, 285, 294, 301; familiarity, "food from home," 99, 105, 109–110; food businesses, 188–189, 191–194, 245–246, 249–258, 285–287, 294; methods, 190, 191, 195–196, 250–251, 270, 274; shared-use commercial kitchens, 11–12, 250, 281–296; variety, 187, 192–193, 262, 267, 274, 294; women's domestic labor, 10–11, 182–183, 185–197, 287, 288. *See also* Recipes
Homestead Act (1898), 104
Household incomes, 30t, 67, 268–269
Huang, Sarah D., 10, 99–110
Human migration: circular migration, 25, 125; climate change-related, 22–23; internal displacement, 146. *See also* Immigrant labor; Migrant workers
Human rights: noncitizens, Canada, 126, 127, 135; noncitizens, US, 6–7, 27, 124, 135, 184
Human social needs, 33

I-9 audits, 123, 136n1
ICE. *See* US Immigration and Customs Enforcement (ICE)
Identity: coalition identities, 169; concept breadth, 203; cultural retention and pride, 290–291, 296; farmworker assumptions, 161–162; food's ties, 1, 203, 262, 263, 267–268, 271, 276, 291, 292; hybridization and acculturation, 262, 265, 267; Latinx/Mexican assumptions, 162, 292; national: American, 19, 37, 44–45, 276; national: Canadian, 206–207, 209, 210–212, 216; national: Mexican, 286; translocal and transnational, 11, 203, 249, 262–263, 274–275, 300, 301
Identity documentation: forgery, 27, 31; migrant workers, 31; privacy issues, 235; recipes as, 1, 7, 281; US policy, 26–27, 31, 32
Ignatieff, Michael, 205
Illegal border crossing: legal changes, 27–28; methods, 31; stereotypes and fears, 23–24, 26, 47; U.S. policy and responses, 22, 23–24, 26–28, 31, 123
"Illegality" discourse, and criminalization, 47
"Immigrant-food nexus," 3, 7–9, 299, 305–306
Immigrant labor: agricultural support access/programs, 141–156, 176, 217–219, 265–266, 303; borderlands and immigration policy, 9, 21, 22–24, 29–38; cost-benefit analyses, 82–83, 120, 132–133; family agricultural enterprises, 161–177, 211; federal laws, 5, 25–26, 27, 123, 124; food scholarship, 1–12, 117–118, 121, 125, 128–135, 143–156, 175, 194, 304; immigrants as transformers, 81–95, 169; media coverage/portrayals, 1–2, 4–5, 11, 26, 46, 48, 50–52, 53, 122–123, 215–216; permanent residency programs, 33, 161; public opinion, communication, and stereotypes, 48, 50–51, 52–55, 82–83, 91, 122–123, 147, 161–162, 164, 216, 292; small and ethnic markets, 10, 66–68, 83, 84, 88, 89–91, 94; US, descriptions, 21, 23–24, 25–27, 30, 45–46, 117, 119, 124–125, 150;

Immigrant labor (cont.)
US statistics, 6, 21, 23, 25, 122, 161, 184; workers' legal rights, 6–7, 124, 126, 127, 143, 184, 186, 220. *See also* Agricultural labor; Immigration policy

Immigrant populations: American history and cycles, 5, 83, 85, 89–93, 102, 245, 247–248; anti-immigrant sentiment, 2–3, 50–51, 53–55, 251, 252, 254, 255, 258, 303; Canadian history and present, 209–212, 226, 228, 236; community-building, 10, 34–36, 66, 67–68, 73, 82–83, 86, 87, 90, 175, 290–293; cost-benefit analyses, 82–83; ethnic foodscapes and gentrification, 65–68, 73–74, 75–76, 91, 93, 109–110; food enterprises and community revitalization, 81–95, 169, 248, 254; Hispanic/Latino, 53, 63–64, 245, 247, 248–249, 254, 256, 257, 292; Hmong, 264; local food movements and exclusion, 4, 10, 11, 99–110, 215–216; media portrayals, 1–2, 4–5, 11, 26, 48, 50–52, 53, 102, 122–123, 215–216; personal accounts, 72, 89–93, 251, 252, 253–254, 256–257, 288; push and pull factors, 247–248; recipes for lives, 281–296; sanctuary cities, 304–305; statistics, 1, 63–64, 86, 102, 247, 248; as strangers, 282–283, 285, 293; urban areas, 60–61, 63–76, 84, 85, 89–93, 94–95, 103. *See also* Interviews

Immigration and Nationality Act (1965), 25

Immigration policy: agriculture science and management, 1–2, 7–9, 120–121, 122–128, 133–135, 144, 154; amnesty programs, 253; border security and history, 2, 6, 22, 25–28, 37, 123; Canada, 125–128, 207, 209–210, 226, 228; court cases/decisions, 2–3, 27; criminalization within, 9, 23–24, 25–28, 31, 37; effect, farming families, 164–165, 176–177; federal laws/methods, 5, 25–28, 123, 165, 253, 304; hospitality approaches, 36–37; local business environment, 90–91, 95; public opinion and communication, 36–37, 42, 48, 82–83, 122–123; sanctuary cities, 304–305; Trump administration, 1, 2, 6, 22, 23, 28, 46, 55, 123, 124, 154–155, 258. *See also* Immigrant labor

Immigration Reform and Control Act (1986), 5, 26–27, 123

Implicit bias, 228–229

Indigenous peoples: Canadian foodways, 233; Canadian populations, 210–211, 213, 228, 236–237; Hmong, 263–264, 271; racialization and exclusion, 227

Industrial agrarian model, 144–145, 146, 209. *See also* Agribusiness

Industrial Areas Foundation (IAF), 24

Insiders vs. outsiders. *See* "Others" and outsiders

Integrated weed management, 120

Interviews: agricultural labor and costs, 128–132, 150, 218–219; borderlands and immigration, 24, 29, 31–37, 48, 249; "culinary chats," 249, 281–282, 285–296; dairy industry and domestic labor, 182, 186–197; ethnic foodscapes, 62, 68, 70, 73–74, 91–92, 93, 105–106, 107–109, 187–188, 245–246, 254–255, 268–274, 293; farmer independence, 149–150, 152; food locality, 99–100, 103, 105–106, 107–109; food policy councils, 228, 230–231, 232, 234–235;

Index

immigrant community businesses, 85, 87–88, 90, 93, 245, 249–258, 287–295, 294; immigrant farmers, 144, 149–150, 272–273, 293–294; inland immigrant labor, 48, 187; methodology, 85, 103, 249–250, 268–269, 283, 284; oral history projects, 11, 245–246, 249–258, 284

Iowa: agribusiness and immigrant labor, 9, 41–55; farm labor/herbicide usage, 128, 129–130; political and election activity, 43, 48

Iowa Farm Bureau, 41

Isolation: American South, 247; community building vs., 34–35, 68, 183, 289–290, 301; fascist and totalitarian states, 34; immigrant immobility and invisibility, 31, 32, 33–34, 35, 37, 47, 185, 188, 189, 191, 193–194, 194–197, 218, 304

Italian immigrants and businesses, 89, 90–93

Jewish communities and foodscapes, 61

Joassart-Marcelli, Pascale, 9–10, 59–77, 303, 304

Job creation: community businesses, 75–76, 84; meat processing industry, 41–42, 51

Johns, Laurie, 41

Jordanian immigrants, 105–106

Kancewick, Mary, 103–104
Kentucky, 245–246
Keung, Nicholas, 216–217
Khojasteh, Maryam, 10, 81–95, 300, 302, 304–305
Kin-based collective strategy, 10, 162, 163–164, 165, 168–176
Klindienst, P., 290
Komarnisky, Sara V., 100
Kong, Panyia, 267
Kuhn, Alex, 42, 51

Labor: American history and regions, 246–247; climate and labor patterns, 22–23; contractor systems, 26, 30, 124; and gender, in dairy industry, 10–11, 181–197; and gender, in farm enterprises, 162–164, 167–168; global location shifts, 46, 165–166; immigrant labor regulations, 5, 25–27, 123–127; shortages and solutions, 2, 5–6, 10, 119–120, 121–128, 130–135, 185–186, 193–194, 217, 220; small ethnic businesses, 66–68, 88; state economic change, 41, 46, 104, 246–248, 248–249, 254; women, and traditional métis, 164, 288. *See also* Agricultural labor; Immigrant labor; Job creation; Labor regulations; Migrant workers; Occupational health and safety; Unionized labor

Labor regulations: applications and relevance, 21; corporate compliance, 46–47, 49, 82, 88–89; federal laws, 5, 21, 25–27, 123, 197n2; local laws and processes, 88–89, 197n2

Land grant universities, 266

Language and communication: "Color-blind" and "race-neutral" discourse, 227, 233, 238; English language learning programs, 50–51; English language skill, 90, 108–109, 143, 152–154, 195, 264, 291–292; "farmers" and "farmworkers" terms, 147, 161–162, 290; Latino/a and Latinx, 12n1, 156n1, 177n2; "local food" terminology, 99–101, 104, 206, 208–209; official forms, 153–154; race, and discourse proportionality/power, 48, 50–51, 53, 54–55, 217–218, 228, 232; racist shorthand/dog-whistles, 50, 51, 52, 147, 161–162; recipes, 282–283; vagueness and invisibility, 52–53, 213, 217–218

Laos, 264
Latinos Facing Racism: Discrimination, Resistance, and Endurance (Feagin and Cobas), 255
Latinx farm owners and cultivation practices, 10, 142, 144, 147–151, 161
Law enforcement. *See* Police-immigrant relations; US Immigration and Customs Enforcement (ICE)
Lee, Gary Yia, 263, 272
Lee, Mai Na M., 263
Leitner, Helga, 44–45, 50
Lending. *See* Business financing; Money lending
Lettuce industry, 121–122
Liberal Party of Canada, 205, 210
Like Water for Chocolate (Esquivel), 285, 288
Linton, Jillian, 11, 205–221, 303
Lipsitz, G., 214, 216, 306n2
Literary techniques, 284, 285
Local Food Act (Ontario; 2013), 205
"Local food" and local food movement: definitions and labeling, 104–105, 109, 110, 208–209; exclusion functions, 4, 10, 100–101, 105, 106–110, 205–207, 212–213; fresh market industry, 166; immigrants and refugees within, 4, 99–100, 103, 105, 106, 107–109, 110, 206, 211, 212–216, 219, 282; media coverage and analysis, 207–209, 210, 212–220, 215–216; policy and laws, 205. *See also* Farmers markets
Locality, food systems, 10, 99–110, 225
Lor, Pao, 265
Lorde, A., 289
Loring, Philip A., 104
Los Angeles, California, 226
Lubell, David, 81

Mackey, Eva, 216
Magical realism, 285

Mares, Teresa M., 190
Marketeras, 162, 165–168
Marte, Lidia, 183
Martinez, Zhenia, 250, 252–255
Marx, Karl, 196
Mason City, Iowa, 41–43, 48–53
Massey, Douglas S., 25, 26, 27
McCullen, Christie Grace, 206, 219
Meat processing industry: environmental aspects, 42, 49, 51, 53; local agribusiness economies, 9, 41–43, 45–55; location shifts, 46; nature of work, 47, 53
Mechanization, farm labor, 119–120, 121–122, 131–135
Media coverage: food businesses and food deserts, 64–65, 71; foodways communication, 269; immigrants and immigration, 1–2, 4–5, 11, 26, 46, 48, 50–52, 53, 102, 122–123, 215–216; local economies and industries, 41, 49–52, 53, 55, 71; local food, 207–209, 210, 212–220; public meetings, 49–50, 51, 53
Medieros, 173–174
Melting pot concept: vs. mosaic/salad bowl multiculturalism, 210, 254–255; US descriptions, 92, 93, 210
"Métis," 163–164
Mexican food and food businesses: appropriation and gentrification, 59–60, 65–66, 70–72; authenticity, 285–286, 291, 292–293, 294; home cooking enterprises, 187–188, 190–191, 192–193, 195–196, 287, 288–289, 294; recipes, 281, 285–286; restaurants, 245–246, 250–255, 285–286, 289; small business and entrepreneurship, 85, 89–90, 92–93, 95
Mexico and Mexicans: Appalachian farmers/families, 10, 161–177; border security administration, 23, 123;

cost of living, 28–29; dairy industry workers, 181–182, 186; immigrant farmers/agricultural labor, 6, 10, 21, 23, 25–26, 28–38, 119–125, 126–127, 133, 148, 161–177, 217, 218; immigrant farmers/USDA access, 10, 141–152; immigration policy and immigrants, 25–26, 27, 28, 29–30, 122–123, 125, 248, 288. *See also* Mexican food and food businesses

Mezzadra, Sandro, 24

Miao/Meo people, 264

Microlending, 173

Middle Eastern groceries, 10, 85–86

Migrant workers: Canadian programs and policy, 5–6, 120–121, 125–128, 133–134; climate and labor patterns, 22–23; legal rights, 6–7, 124, 126, 127, 184, 186, 220; US history and legislation, 5, 25–27, 31, 122–125, 133; US regions and statistics, 9, 21, 23, 28–29, 119–120, 122, 130–131, 161, 184; US stereotypes, 162, 164. *See also* Immigrant labor

Migration of foods, 101

Militarization, US border regions, 9, 22, 27–28, 37

Military personnel, 28, 29

Milk consumption, 184

Miller, Todd, 22–23

Mindes, Samuel C. H., 10, 119–136, 304

Minkoff-Zern, Laura-Anne, 141–157, 164, 197, 303, 304

Mitchell, Don, 246

Mobility and travel: checkpoints policy, 32–33; commute times, 21, 29; immigrant labor, 32–33, 127, 165, 185, 188, 191, 193–194, 194–197, 304

Money lending, 173

Mosaic multiculturalism concept, 210, 254–255

Moua, Kao Nou L., 264

Multiculturalism and diversity: Canadian policy and self-identification, 5, 11, 207, 209–212, 215, 216, 219, 227, 228; Canadian populations, 206–207, 209–211, 226, 227, 228, 233, 236; ethnic foodscapes, 74–75, 83–84, 86–87, 89–90, 91–95; food and cultural diversity, 291–292; food system councils and planning, 225–227, 227–228, 230–233, 234–235, 236, 238–239

Municipal governance, 231

Nashville, Tennessee, 250–252

National Agricultural Workers Survey, 184

Nationalism. *See* Populist, nativist, and nationalist movements

Native American and Indigenous farmers, 146, 206

Native cultures and indigenous peoples: Alaska, 103–104; Canada, 210–211, 212–213, 219–220, 227, 228, 236–237; foodways, 212

Nativism. *See* Populist, nativist, and nationalist movements

Neo-Nazis, 2

Neubert, Christopher, 9, 41–55, 303

New Year holidays, 261, 271

New York: dairy workers and industry, 181–182, 184–197; labor laws, 197n2; NYC immigrant residents and cultures, 61

Nielson, Brett, 24

Nixon, Richard, and administration, 144–145

North American Free Trade Agreement (NAFTA), 5, 126, 288

Northern and southern economies, US, 246–249, 254

North Shore Table Matters (Vancouver), 236–237

"Nuevo" American South, 11, 245, 246, 247, 249–258
Nutrition. *See* Health and nutrition, food
Nutritional Environmental Measure Survey, 87

Oakland, California, 226
Obama, Barack, USDA staff/policy, 142–143, 154, 155
Objectification and dehumanization, 217–218
Occupational health and safety: immigrant agricultural workers, 6, 124, 127, 184–185, 186; labor laws and violations, 46–47, 82, 127, 185, 186
Oil prices, 104
Olivos, Edward M., 46
Ontario, Canada: agricultural workers and policy, 127, 215–216, 217–219; local food policy, 205, 213; multicultural agrarian narratives, 11, 205–220
Operation Hold the Line, 27
"Opportunity states," 248
Oral history collections, 11, 245–246, 249–258, 284
Organic produce and goods, 68, 70
Ortega, Argentina, 250, 255–257
Ortiz Valdez, Fabiola, 10–11, 164, 181–197, 301, 302, 303, 304
Ostenso, Victoria, 11, 225–240, 305
"Others" and outsiders: borders and belonging, 6, 19, 32; fear production and exploitation, 44, 48, 252; immigrant/migrant worker assumptions and stereotypes, 5, 161–162, 164, 216, 303; "local food" movement, 11, 100–110, 206, 212–213; narratives' power, 47, 50, 258; national unity vs. indigenous sovereignty, 206–207, 211; objectification, 217–218; racialization, exclusion, and white defaults, 214, 216, 226, 227, 228, 238, 302; sovereignty fears and wall-building, 2, 19, 23, 37, 44, 123; "white" spaces and exclusion, 44–45, 47, 53, 108–109, 230, 302, 306n2. *See also* Binary configurations; Racialization

Park, Yun Joon (Ben), 215
Passidomo, Catarina, 11, 183, 245–259, 302, 304
PATRIOT Act (2001), 27
Payment systems, 171–172
Perdue, Sonny, 154
Permanent residency status: immigrant farmers and experiences, 161; US labor policy, 33
Personal documentation. *See* Identity documentation
Personal lending, 173
Philadelphia, Pennsylvania, 10, 82, 85, 89–93, 94–95
Pittsburgh, Pennsylvania, 81
Police-immigrant relations, 31, 52, 185, 191, 192, 195. *See also* US Immigration and Customs Enforcement (ICE)
Political environment, United States, 8, 43, 48, 181, 186, 191, 192, 252, 258
Political sovereignty: indigenous claims vs. national unity, 206–207, 211; perceived threats, and security measures, 23, 26, 37
Pollan, Michael, 3, 101
Population data. *See* Census and population data
Populist, nativist, and nationalist movements: American political and electoral trends, 8, 43, 44, 48, 252; American regional effects, 23, 31–34; American trends and identity, 6, 19, 23, 265; global, 23

Index

Postcolonial analysis, 263–264
Postville, Iowa, 45–48
Potter, Jonathan, 208
Prejudice theories, 45, 255
Pren, Karen A., 25, 26, 27
Presidential pardons, 23, 46
Prestage, Ron, 49, 53
Prestage Farms, 41–43, 48–55
Price supports, commodity crops, 145
Processed foods, 187–188
Produce: agricultural policy, 145, 151, 152, 176; big-box groceries, 106; buying/payment systems, 167–168, 171–172; direct sales, 144, 150–151; grading, 167–168; immigrant and ethnic markets, 68, 70, 85, 86, 87, 88, 92, 151, 273; immigrant farmers' cultivation and sales, 10, 141, 144, 147, 149, 150–151, 152, 155, 162, 165–172, 175, 215–216, 273; local growers, 107–110, 205, 212, 215–217; organics, 68, 70; serving guidelines, 84; UDSA farming programs, 141, 147, 151–152, 155
Professionalism policies, 231–232, 233, 237
Public health and sanitation: food system planning, 225; immigrants' food environments, 66, 82, 83–84, 263, 266–267, 269, 275; immigrants' housing environments, 184, 185, 186
Public meetings, 49–50, 51, 53
Puerto Ricans, 292–293
Push and pull factors, immigration, 247–248

Quality control, produce, 167–168

Race issues. *See* Binary configurations; "Color-blind" and "race-neutral" discourse and policies; Critical race theory; Racial demographics; Racialization; Racial segregation and enclaving; Racism and discrimination; Whites and whiteness
Racial demographics: border towns, 29, 30; Hispanic identification, census, 53
Racialization: Asiatic, 265; race theory, 227, 228, 255, 265; "visible minorities," 210, 227, 265
Racial justice-based problem solving, 11, 233, 238–239, 305
Racial segregation and enclaving: borderlands, 29–30; farm owners and race, 142, 145–146, 147, 153, 155, 156, 161–162, 206, 219, 227; food system planning, 225–226; spatial segregation, 44, 47, 306n2; "white spaces" and, 45, 47, 53, 55, 108–109, 206, 230, 302, 306n2
Racism and discrimination: accusations, by corporations, 42, 53; discussion challenges, 72–73, 207, 227, 232, 233; formation methods, 47; immigrant and minority businesses, 90–91, 95, 142, 145–146, 147–148, 155, 156; immigration policy and messaging, 23–24, 26, 27, 28, 46, 47, 48, 50, 55, 207, 209–210, 218–219, 227, 252, 258; political exploitation, 44, 48, 252; prejudice theories, 45, 255; racist shorthand/dog-whistles, 50, 51, 52, 147, 161–162; scapegoating, 37, 44, 50; US citizenship policy, 145–146, 265; white racism and white privilege, 44–45, 47, 55, 207, 229, 232, 254, 255, 271–272, 293
Raiteros, 188, 193–194, 195, 286
Ramírez, Laura Patricia, 245–246, 257–258
Reagan, Ronald, 253
Real estate: immigrant residents and economic power, 76, 81, 86, 90, 91, 93; urban industry and neighborhoods, 62–63, 64, 68, 74, 76, 90. *See also* Gentrification

Recipes, 274; cookbooks, 269, 270, 274, 286; food identities and, 7, 203, 246, 249; immigrant lives, 1, 7, 281–296; practices, traditions, and techniques, 190, 196, 250–252, 253, 256, 262, 269–270, 281–282, 289; restaurant offerings, 245, 246, 250–251, 254, 256

Recordkeeping, farming, and USDA support, 143, 145, 148, 152

Redzepi, René, 258

Refugee populations: agricultural producers, 89, 94, 102, 103, 107–110; City Heights, San Diego area, 63; Hmong, 264–265, 267; local business growth, 86; local business ties and welcoming, 66; local food movements and exclusion, 10, 99–110; resettlement programs, 102

Religious organizations, hospitality, 24, 36–37

Restaurant Opportunities Center United, 304, 305

Rice, 231, 274

Richmond, Virginia, 255–257

Richmond Food Security Society (Vancouver), 233–235

Rios, Michael, 262–263

Ross, Hannah, 264–265

Rubashkin, Shalom, 46–47

Ruiz, Karla, 250–252

Rule of law, 28

Rural areas: local food sources, 213–214; objective/subjective defining, 44–45; settlers and settlements, 212–213, 217, 219; as "white spaces," 44–45, 53, 55, 206, 212–213

Sacramento, California, 261, 266

Saldaña, Johnny, 128

Salim, Zia, 61

Salvadorian immigrants, 249, 255–257, 302

Samuelson, Marcus, 267

Sanctuary cities, 304–305

San Diego, California: contested ethnic foodscapes, 9–10, 59–60, 62–76; food business locations, 62, 64–66, 67f, 84; population data, 64f; zoning and industry, 62–63, 64

Sandoval, Gerardo F., 46

Schmid, Mary Elizabeth, 10, 161–177, 183, 300, 304

School lunches, 271–272

Scott, James C., 146, 164

Scripter, Sami, 269, 270

Seasonal Agricultural Worker Program (SAWP; Canada), 5–6, 120–121, 125–127, 134, 217–219

Seasonal workers. *See* Migrant workers

Secure Fence Act (2006), 22

Segregation. *See* Racial segregation and enclaving

Sexual power dynamics and harassment, 189–190, 194

Shared-use kitchens, 11–12, 250, 281–282, 286–292

Simmel, G., 282, 295

Situational Strangers, 281–296

Slaughter industry. *See* Meat processing industry

Sloat, Sea, 141–157, 164, 303, 304

Slow Food movement, 209

Small businesses: community investment and job creation, 75–76, 83, 84, 87, 90, 94; ethnic markets, 66–70, 69f, 73–74, 83, 84, 86–88, 90–93, 93–94, 254–255; farming enterprise tasks, 162–163, 166–168, 172–173; financing, 67, 75, 150, 151–152, 153, 155–156, 173; government promotion/support, 83, 88–89, 94, 305; local labor

regulation, 88, 89; restaurants/food entrepreneurs, 245–246, 249–258, 285–295
Smallholders, agricultural, 145. *See also* Family farms
Small towns: size definitions, 44–45; as "white spaces," 45, 47, 55
Smith, Andrew F., 76
Smith, Barbara Ellen, 248
Smith, Chery, 266
Smith, Eric, 103–104
Smith, Sara, 44
Smuggling and trafficking, 22, 27
Social isolation. *See* Isolation
"Social problems" research approach, 265, 266–267, 275
Southern and northern economies, US, 246–249, 254
Southern Appalachian farming, 10, 161–177
Southern Foodways Alliance, 11, 245–246, 249–258
Southern US foods and cultures, 11, 245–246, 247, 249–258
Soybean industry and growers, 128, 129t, 130, 145
The Spirit Catches You and You Fall Down (Faidman), 265
Stamps, Martha, 250–251
Stang, J., 266
Storm Lake, Iowa, 50–51, 52
Strangers, 282–283, 285, 293, 295
Subsidies, commodity crops, 145
Supermarkets: ethnic food offerings, 106, 107, 191, 193, 195–196, 273, 302; food access and health, 10, 83–84, 85–87, 89, 106; food access and safety, 191, 195–196; product wholesaling, 151, 171
Supreme Court decisions, 2–3
Sustainable agriculture: community communication, 2–3, 230; family and immigrant growers, 166, 273; local food security, 102, 104; USDA programs, 146–147
Symbolic value vs. use value of food, 61–62, 75

Table Matters (Vancouver), 236–237
Tamales, 190
"Taste" judgments: food and authenticity, 285–286, 292, 293; food and class, 62, 65, 68, 70–71, 73, 107, 292–293; food practices: local food, 103, 209, 212
Tax incentive programs, 88–89
Technological innovations: agricultural inventions, 135; foodways and communication, 269, 271; mechanization in farm labor, 119–120, 121–122, 131–135
Terrorism: border security effects, 27, 28; destruction of social spaces/ connection, 34, 37; responses, fear, and racism, 44
Texas, 248
Textual analysis, 208
Thanksgiving traditions and cuisine, 261–262, 267
Time, value, 166
Tomato industry and cultivation, 121, 122, 161, 163, 166, 167
Toronto, Ontario, 11, 207, 210
Toronto Star (newspaper), 207–209, 210, 219–220
Tortillerías, 192–193, 245–246
Traditional cooking. *See* Home and traditional cooking
Translocality, 1, 101–102; cuisines, 7; Hmong translocal placemaking, 11, 262–263, 264
Transnational identities, 203, 249; in alternative food system, 300; American South, 11, 247, 249–258; of immigrants, 7, 203, 249, 300, 301

Transportation services, 188, 193–194, 195, 286
Trudeau, Pierre, 210
Trump, Donald, and administration: border and immigration policy, 1, 2, 6, 22, 23, 28, 46, 55, 123, 124, 154–155, 258; immigrant employees, 125; Muslim immigration ban, 2–3; pardons, 23, 46; political support, 43, 48; rule of law and, 28; UDSA staff and policy, 154–155
Tuck, Eve, 265
Tyson Foods, 50

Underground food business, 288
Undocumented immigrants: agricultural employment and conditions, 1–2, 6–7, 121, 122–124, 126–127, 181–182, 184, 189, 194–196; enforcement programs and resources, 1, 6, 123, 136n1; history and statistics, 1, 5, 6, 26–27, 123; non-farm businesses' employment, 26, 46, 91, 95; workplace raids, 46–47, 136n1, 195. *See also* Illegal border crossing
Unionized labor: dairy industry, 182; hostile environments, 247; organizing, 122, 182, 194; related laws, 197n2; shifts away from, 46
United States Farm Bill, 145
Urban areas: agriculture, 65, 213, 215, 216, 230, 290–291; downtowns' focus and development, 64, 86; "ethnic markets" and food, 59, 60, 62–63, 64–68, 70–76, 84–95, 254–255; food systems and food justice, 100, 225–240; immigrant residents and cultures, 60–61, 63–76, 84, 85, 89–93, 94–95, 103; inner ring regions, 62–63; labor shifts, 46, 62–63, 90, 91
US Border Patrol: court cases, 27; history, 24, 25–27; presence and methodology, 2, 31, 34, 123, 192, 195. *See also* US Customs and Border Protection
US Census Bureau: immigration data by state, 248; small town definitions, 44
US Citizenship and Immigration Services, 23–24
US Citizenship and Naturalization Services, 123
US Customs and Border Protection: international partnering/training, 23; security, and immigrant labor, 22, 23–24, 29–30; Trump border policies, 28. *See also* US Border Patrol
US Department of Agriculture: civil rights policy, 142–143, 146, 147–148, 154–156; Farm Service Agency, 141, 142, 153, 155; financial assistance/loans, 142, 146, 147–148, 151–154, 155–156; food research data, 127–129; immigrant farmers, food programs and resources, 10, 141–156; immigration reports, 123; industrial agriculture, 144–145, 146; lawsuits, 142–143, 147–148, 154, 156n3; local food and labeling, 104; Natural Resources Conservation Service, 141, 151–152; nutrition guidelines, 84; Office of Advocacy and Outreach, 154
US Department of Homeland Security: border protection, 22, 23, 28; immigration enforcement policy, 23–24, 123, 136n1; travel checkpoint documentation, 32–33
US Department of Labor, 123–124
US Immigration and Customs Enforcement (ICE): asylum seekers, 36; detention and deportation, 32, 195; local business/community presence, 95, 191, 193, 195, 304; staffing and funding, 123; workplace raids, 46–48, 195

Index

US National Guard, 28
US Supreme Court, 2–3
Use value vs. symbolic value: ethic foodscapes, for immigrants, 65–68; food, 61–62, 75; urban neighborhoods and buildings, 74

Vancouver, British Columbia, 11, 225–240
Vancouver Food Policy Council, 229–233
Vang, Kat, 11, 261–276, 300, 302–303
Vang, Pa Der, 264
van Leeuwen, Theo, 218
Vegan food, 292, 293, 302
Vegetables. *See* Fresh fruit and vegetable industry; Produce
Vento, Joe, 90–91
Vilsack, Thomas J., 142–143, 154
"Visible minorities": defined, 240n1; racialization, 210, 227, 265
Volunteerism, 290
Vu, Vienne, 61

Walls and wall-building, 2, 19, 23, 37, 44, 123
Wal-Mart, 191, 195, 273, 302
War on Terror, 28
Watkins, Joshua, 262–263
Weeds: herbicide management and resistance, 10, 119–121, 122, 128–135; integrated weed management, 120, 135, 149
Welcoming America Global Network, 81
Wells, M. J., 148
Wetherell, Margaret, 208
Whites and whiteness: alternative food movement and imaginaries, 4, 206, 229, 230, 293, 300; American citizenship policy, 145–146; American culture and xenophobia, 23, 44–45, 53–54, 252, 255; Canadian identity, 209–211, 214, 216, 219, 226–227; "color-blind"-ness harms, 233, 234–235, 238; cultural appropriation and ethnic foodscape gentrification, 9–10, 59–60, 68, 70–73, 74–75, 109–110, 212–213; as defaults, 214, 216, 226, 227, 302; discourse proportionality/power, 48, 50–51, 53, 54–55, 217–218, 228, 232; eating habits and tastes, 187–188, 245, 251, 292–293, 302; farm owners and operators, 142, 145–146, 147, 148, 155, 206, 212–214, 216, 219; food system planning and groups, 226, 229, 231–232, 235–236; "local food" and food systems, 4, 11, 100, 106–110, 205–207, 212–214; racism, 45, 55, 232, 254, 255, 293; spaces and segregation, 45, 47, 53, 55, 108–109, 206, 230, 302, 306n2; white/other binary configurations, 228, 237, 238, 247, 255. *See also* European immigration
White savior complex, 59–60, 213
Wholesaling: food businesses, 88, 89, 151, 171; fruit/vegetable farmers, and outside systems, 150, 166, 171
Wilcox, Hui Niu, 267
Winders, Jamie, 248
Wittman, Hannah, 11, 225–240, 305
Women. *See* Gender, food, and labor
Wood, Sara, 11, 183, 245–259, 302, 304
Work ethic: immigrant business owners, 67–68, 256–257; immigrants and immigration, 211; personal accounts, 256–257, 292
Workplace housing, 182–184, 184, 185, 186, 189, 191–192, 193–195, 198n16, 218
World Food Day, 209

Yang, Sheng, 269, 270
Ying, Chris, 258
Yogurt, 181–182, 184
You and I Eat the Same (Ying and Redzepi), 258
Young, I. M., 227
Yuma, Arizona, *29,* 28–30; agricultural production, 9, 21, 28, 30; asylum seekers, 36; borderlands and immigration policy, 9, 21, 22–24, 29–38; demographic and economic data, 30*t*
Yuma Interfaith, 24, 36–37

Zaragoza, Alex, 71–72